Study Guide with Solutions to Selected Problems

General, Organic, and Biological Chemistry

Organic and Biological Chemistry

Third Edition

H. Stephen Stoker

Danny V. White

Joanne A. White

Houghton Mifflin Company Boston New York

Vice President and Publisher: Charles Hartford
Executive Editor: Richard Stratton
Developmental Editor: Rita Lombard
Editorial Assistant: Rosemary Mack
Manufacturing Manager: Florence Cadran
Project Editor: Merrill Peterson
Senior Marketing Manager: Katherine Greig
Marketing Associate: Alexandra Shaw

Cover photograph © 2002 Stuart Hughs/Getty Images

Printed in the U.S.A.

ISBN: 0-618-26600-3

6789- DBH - 07 06 05

Contents

Preface

If the study of chemistry is new to you, you are about to gain a new perspective on the material world. You will never again look at the objects and substances around you in quite the same way. The invisible structure and organization of matter, the "how and why" of chemical change -- knowledge of these areas will help to demystify many occurrences in the world around you. Chemistry is not an isolated academic study. We use it throughout our lives to appreciate and understand the world and to make responsible choices in that world.

The purpose of this study guide is to help you in your study of the textbook, *General, Organic, and Biological Chemistry*, by providing summaries of the text and additional practice exercises. As you use this Study Guide, we suggest that you follow the steps below.

1. Read the overview for the chapter to get a general idea of the facts and concepts in each chapter.

2. Read the section summaries and work the practice exercises as you come to them. Write out the answers even if you are sure you understand the concepts. This will help you to check your understanding of the material. Refer to the answers at the end of the chapter as soon as you have answered each practice exercise. By checking your answers, you will know whether to review or continue to the next section.

3. When you have finished answering the practice exercises, take the self-test at the end of the chapter. Check your answers with the answer key at the end of the chapter. If there are any questions that you answer incorrectly or do not understand, refer to the chapter section numbers in the answer key and review that material. The Solutions section of this book contains answers to selected exercises and problems from the textbook.

Chemistry is a discipline of patterns and rules. Once your mind has begun to understand and accept these patterns, the time you have spent on repetition and review will be well rewarded by a deeper total picture of the world around you. As teachers, we have enjoyed preparing this study guide and hope that it will assist you in your study of chemistry.

Danny and Joanne White

Chapter Overview

Why is the study of chemistry important to you? Chemistry produces many substances of practical importance to us all: building materials, foods, medicines. For anyone entering one of the life sciences, such as the health sciences, agriculture, or forestry, an understanding of chemistry leads to an understanding of the many life processes.

In this chapter you will be studying some of the fundamental ideas and the language of chemistry. You will characterize three states of matter, differentiate between physical properties and chemical properties, and identify two different types of mixtures. You will describe elements and compounds and practice using symbols and formulas.

Practice Exercises

1.1 **Matter** (Sec. 1.1) exists in three physical states. Complete the following table indicating the properties of each of these states of matter.

State	Definite shape?	Definite volume?
solid (Sec. 1.2)	yes	Yes
liquid (Sec. 1.2)	no	yes
gas (Sec. 1.2)	no	no

1.2 The **physical properties** (Sec. 1.3) of a substance can be observed without changing the identity of the substance. **Chemical properties** (Sec. 1.3) are observed when a substance changes or resists changing to another substance. Complete the following table:

Property	Physical	Chemical	Insufficient information
A liquid boils at 100°C.	yes	no	
A solid forms a gas when heated.	yes		
A metallic solid exposed to air forms a white solid.		yes	

1.3 A **physical change** (Sec. 1.4) is a change in shape or form, but not in composition. A **chemical change** (Sec. 1.4) produces a new substance; that is, the composition is changed.

Classify the following processes as physical or chemical changes by writing the correct word in the second column:

Process	Physical or chemicalchange
An ice cube melts, producing water	Physical
A wood block burns, producing ashes and gases.	Chemical
Salad oil freezes, producing a solid.	Physical
Sugar dissolves in hot tea.	Physical
A wood block is split into smaller pieces.	Physical
Butter becomes rancid.	Chemical

1.4 **Mixtures** (Sec. 1.5) of substances may be either **homogeneous** (Sec. 1.5),one phase, uniform throughout, or **heterogeneous** (Sec. 1.5), visibly different phases (parts). Indicate whether each of the following mixtures is homogeneous or heterogeneous:

Mixture	Homogeneous	Heterogeneous
apple juice (water, sugar, fruit juice)	X	
cornflakes and milk		X
fruit salad (sliced bananas, grapes, oranges)		X
brass (copper and zinc)	X	

1.5 Complete the following diagram organizing some terms from this chapter:

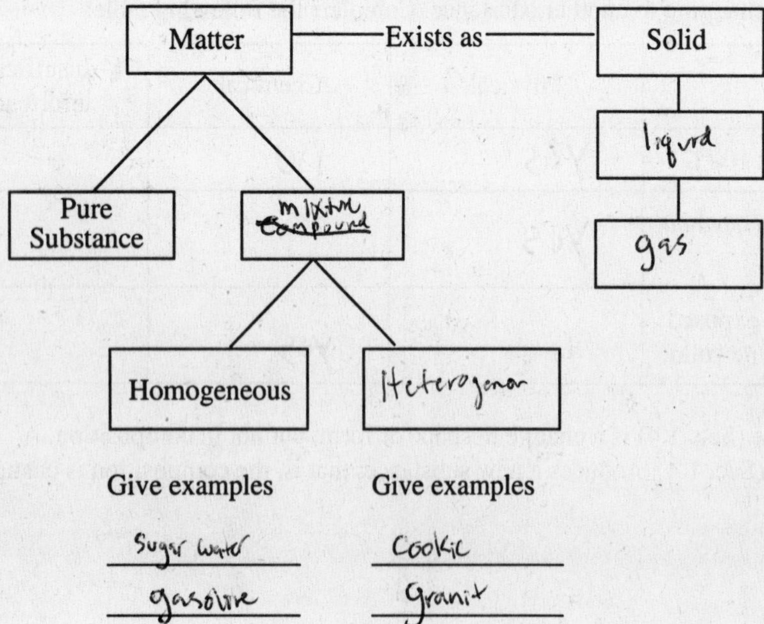

1.6 **Elements** (Sec. 1.6) are **pure substances** (Sec. 1.5) that cannot be broken down into simpler pure substances. **Compounds** (Sec. 1.6) can be broken down into two or more simpler pure substances by chemical means. Complete the following table:

	Substance is an element	Substance is a compound	Insufficient information to make a classification
Substance A reacts violently with water.			
Substance B can be broken into simpler substances by chemical processes.		✓	
Cooling substance C at 350°C turns it from a liquid into a solid.			✓
Substance D cannot decompose into simpler substances by chemical processes.	✓		

1.7 In the following table, write the **chemical symbol** (Sec. 1.8) or name for each element:

Name	Symbol
calcium	Ca
copper	Cu
argon	Ar
nickel	ni
magnesium	mg

Symbol	Name
C	Carbon
Ne	neon
Zr	Zirconium
Pb	lead
Fe	Iron

1.8 An **atom** (Sec. 1.9) is the smallest particle of an element that retains the identity of that element. A **molecule** (Sec. 1.9) is a tightly bound group of two or more atoms that functions as a unit. The **chemical formula** (Sec. 1.10) for a molecule is made up of the symbol for each element in the molecule and a subscript indicating the number of atoms of the element.

In the table below, indicate whether each formula unit is an atom or a molecule, and classify each substance as an element or a compound:

Unit	Atom or molecule?	Element or compound?
N_2	Molecule	~~Element~~ ~~Compound~~ Element
Zn	Atom	Element
HCN	molecule	Compound
Au	Atom	Element
CH_4	molecule	compound

1.9 **Homoatomic molecules** (Sec. 1.9) are made up of atoms of one element; **heteroatomic molecules** (Sec. 1.9) contain atoms of two or more elements.

Indicate whether each of the molecules in the table below is homoatomic or heteroatomic, classify the molecule according to the number of atoms it contains (diatomic, triatomic, etc.), and tell how many atoms of each element are in each molecule.

Molecule	Homoatomic or heteroatomic?	Type of molecule	Number of atoms of each element
NH_3	het	4	1 nitrogen 3 hydrogen
O_2	hom	di	2 oxygen
CO_2	het	tri	1 Carbon 2 oxygen
HNO_3	het	5	1 hydroge 1 nitrogen 3 oxygen
HF	het	2	1 hydrogen 1 flourine

1.10 Write a chemical formula for each of the following compounds based on the information given:

a. A molecule of limonene contains 10 atoms of carbon and 16 atoms of hydrogen.

$C_{10}H_{16}$

b. A molecule of nitric acid is pentatomic and contains the elements hydrogen, nitrogen, and oxygen. Each molecule of nitric acid contains only one atom of hydrogen and one of nitrogen.

HNO_3

Self-Test

True-false: Indicate whether the following statements are true or false. If the statement is false, give the word or phrase that may be substituted for the underlined portion to make the statement true.

1. Matter is anything that has volume and occupies space.
2. Gases have no definite shape or volume.
3. Liquids take the shape of their container and completely occupy the volume of the container.
4. A mixture of oil and water is an example of a homogeneous mixture. hetero
5. The evaporation of water from salt water is an example of a chemical change. physical
6. Elements cannot be broken down into simpler pure substances by chemical means.
7. Sugar dissolving in water is an example of a chemical change.
8. A chemical property describes the ability of a substance to change to form a new substance or to resist such change.
9. A mixture is a chemical combination of two or more pure substances.
10. Synthetic (laboratory-produced) elements are all radioactive.
11. The most common (abundant) element in the universe is oxygen.

12. Two-letter chemical symbols are <u>always</u> the first two letters of the element's name.
13. A compound consists of molecules that are <u>homoatomic</u>.

Multiple choice:

14. One of the three states of matter is the solid state. A solid has:
    a. definite volume but no definite shape
    b. no definite volume and no definite shape
    c. definite volume and shape
    d. no definite volume but definite shape
    e. none of these combinations of characteristics

15. An example of a homogeneous mixture is:
    a. sand and water
    b. salt and water
    c. wood and water
    d. oil and water
    e. none of these

16. An example of a physical change is:
    a. iron rusting
    b. sugar dissolving in coffee
    c. gasoline burning in a car engine
    d. coal burning
    e. none of these

17. An example of a chemical change is:
    a. iron rusting
    b. coal burning
    c. gasoline burning in a car engine
    d. a, b, and c are all correct
    e. none of these

18. The symbol for the element iron is:
    a. FE
    b. Fe
    c. F
    d. Ir
    e. none of these

19. The name for the element Ne is:
    a. neodymium
    b. neon
    c. neptunium
    d. nitrogen
    e. none of these

20. $MgCO_3$ is a compound that is composed of which elements?
    a. magnesium, chlorine, iron
    b. magnesium, carbon, neon
    c. manganese, carbon, oxygen
    d. magnesium, carbon, oxygen
    e. none of these

21. The total number of atoms in one molecule of CH_4O is:
    a. 3
    b. 4
    c. 5
    d. 6
    e. none of these

22. On the basis of its chemical formula, which of the following substances is an element?
    a. NH_3
    b. Cl_2
    c. CO_2
    d. CO
    e. none of these

23. On the basis of its chemical formula, which of the following is a compound?
    a. Fe
    b. Fm
    c. O_2
    d. HI
    e. none of these

Answers to Practice Exercises

1.1

State	Definite shape?	Definite volume?
solid	yes	yes
liquid	no	yes
gas	no	no

1.2

Property	Physical	Chemical	Insufficient information
A liquid boils at 100ºC.	X		
A solid forms a gas when heated.			X
A metallic solid exposed to air forms a white solid.		X	

1.3

Process	Physical or chemicalchange
An ice cube melts, producing water	physical
A wood block burns, producing ashes and gases.	chemical
Salad oil freezes, producing a solid.	physical
Sugar dissolves in hot tea.	physical
A wood block is split into smaller pieces.	physical
Butter becomes rancid.	chemical

1.4

Mixture	Homogeneous	Heterogeneous
apple juice (water, sugar, fruit juice)	X	
cornflakes and milk		X
fruit salad (sliced bananas, grapes, oranges)		X
brass (copper and zinc)	X	

1.5

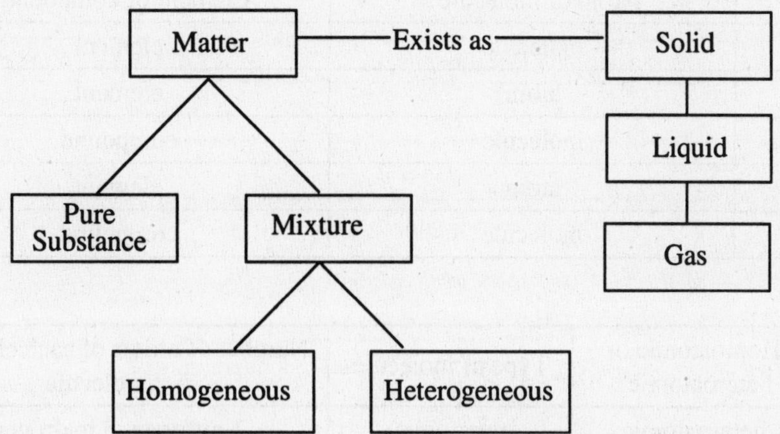

Give examples of homogeneous mixtures:

air

coffee

Give examples of heterogeneous mixtures:

smoke

concrete

(These are a few of many possible examples.)

1.6

	Substance is an element	Substance is a compound	Insufficient information to make a classification
Substance A reacts violently with water.			X
Substance B can be broken into simpler substances by chemical processes.		X	
Cooling substance C at 350°C turns it from a liquid into a solid.			X
Substance D cannot decompose into simpler substances by chemical processes.	X		

1.7

Name	Symbol
calcium	Ca
copper	Cu
argon	Ar
nickel	Ni
magnesium	Mg

Symbol	Name
C	carbon
Ne	neon
Zr	zirconium
Pb	lead
Fe	iron

1.8

Unit	Atom or molecule?	Element or compound?
N_2	molecule	element
Zn	atom	element
HCN	molecule	compound
Au	atom	element
CH_4	molecule	compound

1.9

Molecule	Homoatomic or heteroatomic?	Type of molecule	Number of atoms of each element per molecule
NH_3	heteroatomic	tetratomic	1 nitrogen, 3 hydrogen
O_2	homoatomic	diatomic	2 oxygen
CO_2	heteroatomic	triatomic	1 carbon, 2 oxygen
HNO_3	heteroatomic	pentatomic	1 hydrogen, 1 nitrogen, 3 oxygen
HF	heteroatomic	diatomic	1 hydrogen, 1 fluorine

1.10 a. $C_{10}H_{16}$ b. HNO_3

Answers to Self-Test

The numbers in parentheses refer to sections in your textbook.
1. F; mass (1.1) **2**. T (1.2) **3**. F; gases (1.2) **4**. F; heterogeneous (1.5) **5**. F; physical change (1.4) **6**. T (1.6) **7**. F; physical change (1.4) **8**. T (1.3) **9**. F; physical combination (1.5) **10**. T (1.7) **11**. F; hydrogen (1.7) **12**. F; often (1.8) **13**. F; heteroatomic (1.9) **14**. c (1.2) **15**. b (1.5) **16**. b (1.4) **17**. d (1.4) **18**. b (1.8) **19**. b (1.8) **20**. d (1.8) **21**. d (1.10) **22**. b (1.10) **23**. d (1.10)

Chapter Overview

Measurements are very important in science. In this chapter you will study some of the common units used in measuring length, volume, mass, temperature, and heat in the modern metric (SI) system. You will solve problems involving measurement using the method of dimensional analysis, in which units associated with numbers are used as a guide in setting up the calculations.

You will learn to use scientific notation to express large and small numbers efficiently and will practice using the number of significant figures that corresponds to the accuracy of the measurements being made. You will also use equations to calculate density, to calculate heat loss or gain involved in a temperature change, and to convert temperature from one temperature scale to another.

Practice Exercises

2.1 **Exact numbers** (Sec. 2.3) occur in definitions, in counting, and in simple fractions. **Inexact numbers** (Sec. 2.3) are obtained from measurements. Classify the numbers in the following statements as exact or inexact by marking the correct column.

Number	Exact	Inexact
A bag of sugar weighs 5 pounds.		
The temperature was 104°F in the shade.		
There were 107 people in the airplane.		
An octagon has eight sides.		
One meter equals 100 centimeters.		
1/3 is a simple fraction.		
The swimming pool was 25 meters long.		

2.2 **Significant figures** (Sec. 2.3) are the digits in any **measurement** (Sec. 2.1) that are known with certainty plus one digit that is uncertain. These guidelines will help you in determining the number of significant figures:

1. All nonzero digits are significant. (23.4 m has three significant figures)
2. Zeros in front of nonzero digits are not significant. (0.00025 has two significant figures)
3. Zeros between nonzero digits are significant. (2.005 has four significant figures)
4. Zeros at the end of a number are significant if a decimal point is present (1.60 has three significant figures) but are not significant if there is no decimal point (500 has one significant figure, 500. has three significant figures).

The position of the last significant figure specifies the uncertainty of the measurement. For example, 3.91 has uncertainty in the hundredths place.

In the following table, state the number of significant figures in each number as written and the uncertainty of the measurement.

Number	Number of significant figures	Uncertainty
1578	4	Ones
45932	5	Ones
103.045	6	One thosondth
0.0340	3	10 thousandth
298440.0	7	Tens

2.3 In **rounding off** (Sec. 2.4) a number to a certain number of significant figures:

1. Look at the first digit to be deleted.
2. If that digit is less than 5, drop that digit and all those to the right of it. If that digit is 5 or more, increase the last significant digit by one.

Example: Round 7.3589 to three significant figures. Because 7.35 has the correct number of significant figures and 8 is greater than 5, increase 7.35 to 7.36.

In the following table, round each number to the given number of significant figures.

Number	Rounded to 3 significant figures	Rounded to 2 significant figures
2763	2760	2800
0.003628	.00 363	.0036
65.20	65.2	65
2.989	2.99	~~2.9~~ 3.0

2.4 When multiplying and dividing measurements, the number of significant figures in the answer is the same as the number of significant figures in the measurement that contains the fewest significant figures.

Example: $4.2\ m \times 3.12\ m = 13.104\ m^2 = 13\ m^2$
(Because 4.2 has fewer significant figures, the answer has two significant figures.)

Exact numbers (such as three people or twelve eggs in a dozen) do not limit the number of significant figures.

Carry out each mathematical operation in the following table, and round the answer to the correct number of significant figures (assume there are no exact numbers in any calculation).

Problem	Answer before rounding off	Rounded to correct number of significant figures
160 x 0.32	51.2	51
482 x 0.00358	1.72556	1.7
723 ÷ 4.04		
72 ÷ 1.37		
0.0485 ÷ 88.342		

2.5 When measurements are added and subtracted, the answer can have no more digits to the right of the decimal point than the measurement with the least number of decimal places.

Example: 3.58 m + 7.2 m = 10.78 m = 10.8 m

Carry out the mathematical operations indicated below, and round the answer to the correct number of significant figures.

Problem	Answer before rounding off	Rounded to correct number of significant figures
153 + 4521		
483 + 0.223		
1.097 + 0.34		
744 – 36		
8093 – 0.566		
0.345 – 0.0221		

2.6 **Scientific notation** (Sec. 2.6) is a convenient way of expressing very large or very small numbers in a compact form. To convert a number to scientific notation:

1. Write the original number.
2. Move the decimal point to a position just to the right of the first nonzero digit.
3. Count the number of places the decimal point was moved. This number will be the exponent of 10.
4. If the decimal point is moved to the left, the exponent will be positive; if the decimal point is moved to the right, the exponent will be negative.

Examples:

1. 2300 = 2.3×10^3 (decimal moved 3 places to the left)
2. 43,010,000 = 4.301×10^7 (decimal moved 7 places to the left)
3. 0.0072 = 7.2×10^{-3} (decimal moved 3 places to the right)

Note that only significant figures become a part of the coefficient.

Complete the following tables:

Decimal number	Scientific notation
4378	4.378×10^3
783	7.83×10^2
8400	8.400×10^5
0.00362	3.62×10^{-3}
0.093200	9.3200×10^{-2}

Scientific notation	Decimal number
6.389×10^6	6389000
3.34×10^1	33.4
4.55×10^{-3}	.00455
9.08×10^{-5}	.0000908
2.0200×10^{-2}	.020200

2.7 When numbers in exponential form (as in scientific notation) are added or subtracted, the exponents of 10 must be the same. Use the correct number of significant figures.

Example:

$(1.53 \times 10^{-3}) + (7.2 \times 10^{-4})$ (can be added in this form only with a calculator)

$(1.53 \times 10^{-3}) + (0.72 \times 10^{-3}) = 2.25 \times 10^{-3}$

Carry out the following addition and subtraction problems, and express each answer in scientific notation with the correct number of significant figures.

1. $(4.54 \times 10^{4}) + (1.804 \times 10^{4}) =$ __________
2. $(8.522 \times 10^{-3}) + (1.3 \times 10^{-4}) =$ __________
3. $(5.631 \times 10^{4}) - (1.52 \times 10^{4}) =$ __________
4. $(2.94 \times 10^{-4}) - (5.866 \times 10^{-5}) =$ __________

2.8 When multiplying numbers in exponential form, multiply the coefficients and add the exponents of 10.

Examples:

1. (two positive exponents) $(1.2 \times 10^{3}) \times (4.7 \times 10^{5}) = 5.6 \times 10^{8}$
2. (negative and positive exponents) $(2.3 \times 10^{4}) \times (1.8 \times 10^{-8}) = 4.1 \times 10^{-4}$
3. $(5.1 \times 10^{-2}) \times (7.2 \times 10^{5}) = 37 \times 10^{3} = 3.7 \times 10^{4}$

When dividing numbers in exponential form, divide the coefficients and subtract the exponents of 10.

Examples:

1. $(4.8 \times 10^{8}) \div (3.4 \times 10^{3}) = 1.4 \times 10^{5}$
2. $(6.5 \times 10^{4}) \div (3.7 \times 10^{-3}) = 1.8 \times 10^{7}$

Carry out the mathematical operations below. Express answers to the correct number of significant figures:

1. $(4.155 \times 10^{3}) \times (1.50 \times 10^{6}) =$ __________
2. $(7.36 \times 10^{2}) \times (4.711 \times 10^{3}) =$ __________
3. $(1.7 \times 10^{-3}) \times (3.363 \times 10^{-5}) =$ __________
4. $(4.09 \times 10^{6}) \div (2.9001 \times 10^{3}) =$ __________
5. $(3.1413 \times 10^{4}) \div (7.83 \times 10^{5}) =$ __________
6. $(5.204 \times 10^{-6}) \div (4.1 \times 10^{3}) =$ __________
7. $(6.61 \times 10^{-6}) \div (7.278 \times 10^{-3}) =$ __________

2.9 Use the method of **dimensional analysis** (Sec. 2.6) to solve the problems below. Choose the **conversion factor** (Sec. 2.6) that gives the answer in the correct units. If the wrong conversion factor is used, the units will not cancel.

Example: How many millimeters (mm) are in 2.41 meters (m)?

$$2.41 \text{ m x } \frac{1000 \text{ mm}}{1 \text{ m}} = (2.41 \text{ x } 1000) \left(\frac{\text{m x mm}}{\text{m}}\right) = 2410 \text{ mm}$$

Complete the following table. Use the factors in Table 2.2 of your textbook for conversion between the metric and English systems.

Problem	Relationship with units	Answer with units
3.89 km = ? cm	$3.89 \text{ km x } \frac{1000 \text{ m}}{1 \text{ km}} \text{ x } \frac{100 \text{ cm}}{1 \text{ m}}$	
45.8 mm = ? m	$4.58 \times 10^{1} \text{ mm} \times \left(\frac{10^{-3} \text{ m}}{1 \text{ mm}}\right)$ $4.58 \times 10^{-2} \text{ m}$	$4.58 \times 10^{-2} \text{ m}$
0.987 mm = ? km	$9.87 \times 10^{-1} \text{ mm} \times \left(\frac{10^{-3} \text{ m}}{1 \text{ mm}}\right) \times \frac{10^{-3} \text{ km}}{1 \text{ m}}$	9.87×10
7.89 x10^4 cm = ? m	$7.89 \times 10^{4} \text{ cm} \times \left(\frac{10^{-2} \text{ m}}{1 \text{ cm}}\right)$	$7.89 \times 10^{2} \text{ m}$
4.05 x 10^{-4} cm = ? km	$4.05 \times 10^{-4} \text{ cm} \times \left(\frac{10^{-2} \text{ m}}{1 \text{ cm}}\right)\left(\frac{10^{-3} \text{ k}}{1 \text{ m}}\right)$	$4.05 \times 10^{-9} \text{ K}$
5.5 x 10^5 L = ? mL	$5.5 \times 10^{5} \text{ L} \times \left(\frac{10^{3} \text{ mL}}{1 \text{ L}}\right)$	$5.5 \times 10^{8} \text{ mL}$
3.6 km = ? in.		
8.2 mL = ? qt		
57 fl oz = ? mL		
3.8 oz = ? kg		

2.10 **Density** (Sec. 2.8) is the ratio of the **mass** (Sec. 2.2) of an object to the volume occupied by that object.

Example: An object has a mass of 123 g and a volume of 17 cm^3. Calculate the density of the object.

$$\text{Density} = \frac{\text{mass}}{\text{volume}} = \frac{123 \text{ g}}{17 \text{ cm}^3} = 7.2 \text{ g/cm}^3$$

Example: What mass will a cube of aluminum have if its volume is 32 cm^3? Aluminum has a density of 2.7 g/cm^3.

Using the method of dimensional analysis, we can set up an equation that uses the density as a conversion factor:

$32\ cm^3 = ?\ g$

$$32\ cm^3 \times \frac{2.7\ g}{1.0\ cm^3} = 86\ g$$

Complete the following table:

Problem	Relationship with units	Answer with units
1. What is the density of silver if 38.5 cm^3 has a mass of 404 grams?		
2. Ethanol has a density of 0.789 g/mL. What is the mass of 458 mL of ethanol?		
3. Copper has a density of 8.92 g/cm^3. What is the volume of 8.97 kg of copper?		

2.11 Convert the boiling points of the following compounds to the indicated temperature scales. Use the equations in Section 2.9 of your textbook.

Boiling point	Substituted Equation	Temperature
ethyl acetate (nail polish remover) 77.0°C	°F = 9/5(77.0) + 32	°F
toluene (additive in gasoline) 111°C		°F
isopropyl alcohol (rubbing alcohol) 180°F		°C
naphthalene (moth balls) 424°F		°C
propane (fuel for camping stoves) –42°C		K
methane (natural gas) –260°F		K

2.12 The **specific heat** (Sec. 2.8) of a substance can be used to calculate how much heat is absorbed or given off when the substance changes temperature.

Example:

How many **calories** (Sec. 2.8) of heat would be needed to raise the temperature of 12.4 g of water from 21.0°C to 55.0°C?

1. The specific heat of water is 1.00 cal/g•°C
2. Heat (cal) = specific heat (cal/g•°C) x mass (g) x temperature change (°C)
3. Substitute the numerical values and solve:

Heat (cal) = (1.00 cal/g•°C)(12.4 g)(34.0°C) = 422 cal

Calculate answers to the following problems involving specific heat.

Problem	Substituted equation with units	Answer with units
1. How much heat is required to raise the temperature of 1.00 kg of liquid water 25.0°C?		
2. If the temperature of a 150 g block of aluminum is raised from 22°C to 97°C, how much heat has the aluminum absorbed? (specific heat of Al = 0.21 cal/g•°C)		

2.13 Another unit of heat energy is the joule: 1 calorie (cal) = 4.184 joules (J).

Using this conversion factor, calculate the number of joules absorbed in each of the problems in Practice Exercise 2.12.

1.

2.

Self-Test

True-false: Indicate whether the following statements are true or false. If the statement is false, give the word or phrase that may be substituted for the underlined portion to make the statement true.

1. The basic metric unit of volume is the <u>milliliter</u>.
2. The basic unit of mass in the metric system is the <u>kilogram</u>.
3. In scientific notation, a <u>coefficient</u> between 1 and 10 is multiplied by a power of 10.
4. When numbers in exponential form are <u>multiplied or divided</u>, the exponents of 10 must be the same.
5. In rounding numbers, the number of digits is determined by the number of <u>significant figures</u> in the measurement.
6. The density of an object is the ratio of <u>its weight to its volume</u>.

7. The calorie is a common measure of heat and is defined as the quantity of heat that raises the temperature of 1 gram of water 1°C.
8. In the metric system, the prefix micro- means one-thousandth (0.001, 10^{-3}).
9. The number 0.002010 has four significant figures.
10. On the Kelvin temperature scale, all temperature readings are positive.

Multiple choice:

11. The correct way of expressing 4174 in scientific notation is:

a. 4.174×10^2 b. 4.174×10^3 c. 4.174×10^4
d. 4.2×10^3 e. none of these

12. The number 0.005140 should be written in scientific notation as:

a. 5.140×10^{-2} b. 514×10^{-4} c. 5.14×10^{-3}
d. 5.140×10^{-3} e. none of these

13. The sum of 5472 plus 1946 would be written in scientific notation as:

a. 7418 b. 7.418×10^3 c. 7.4×10^3
d. 7.42×10^3 e. none of these

14. The product of 8311 times 0.01452 would be written in scientific notation as:

a. 120.7 b. 1.21×10^2 c. 1.207×10^2
d. 1.2×10^2 e. none of these

15. When 1.487 is rounded to three significant figures, the correct answer is:

a. 1.480 b. 1.490 c. 1.48 d. 1.49 e. none of these

16. In converting grams to milligrams, the known quantity of grams should be multiplied by which of these conversion factors?

a. 1 g/1000 mg b. 1000 mg/1 g c. 1 g/1000000 µg
d. 1000000 µg/1 g e. none of these

17. How many milliliters are in 25.2 kilograms of a liquid that has a density of 0.833 g/mL?

a. 3.03×10^4 mL b. 21.0 mL c. 2.10×10^4 mL
d. 30.3 mL e. none of these

18. How many calories would be needed to raise the temperature of 420 grams of aluminum from 120°C to 450°C? (See Table 2.4 in your text for specific heat.)

a. 2.6×10^4 cal b. 2.6106×10^4 cal c. 2.9×10^1 cal
d. 1.1×10^4 cal e. none of these

19. Which of the following measurements has three significant figures?

a. 1.050 b. 2301 mL c. 0.0702 g
d. 16.20 m e. none of these

20. A temperature of 22°C would have which of these values on the Fahrenheit scale?

a. –6.6°F b. 54°F c. 97°F d. 72°F e. none of these

Answers to Practice Exercises

2.1

Number	Exact	Inexact
A bag of sugar weighs 5 pounds.		X
The temperature was 104°F in the shade.		X
There were 107 people in the airplane.	X	
An octagon has eight sides.	X	
One meter equals 100 centimeters.	X	
1/3 is a simple fraction.	X	
The swimming pool was 25 meters long.		X

2.2

Number	Number of significant figures	Uncertainty
1578	4	ones
45932	5	ones
103.045	6	thousandths
0.0340	3	ten thousandths
298440.0	7	tenths

2.3

Number	Rounded to 3 significant figures	Rounded to 2 significant figures
2763	2760	2800
0.003628	0.00363	0.0036
65.20	65.2	65
2.989	2.99	3.0

2.4

Problem	Answer before rounding off	Rounded to correct number of significant figures
160 x 0.32	51.2	51
482 x 0.00358	1.72556	1.73
723 ÷ 4.04	178.96039	179
72 ÷ 1.37	52.554744	53
0.0485 ÷ 88.342	0.000549	0.000549

2.5

Problem	Answer before rounding off	Rounded to correct number of significant figures
153 + 4521	4674	4674
483 + 0.223	483.223	483
1.097 + 0.34	1.437	1.44
744 – 36	708	708
8093 – 0.566	8092.434	8092
0.345 – 0.0221	0.3229	0.323

2.6

Decimal number	Scientific notation
4378	4.378×10^3
783	7.83×10^2
8400.	8.400×10^3
0.00362	3.62×10^{-3}
0.093200	9.3200×10^{-2}

Scientific notation	Decimal number
6.389×10^6	6,389,000
3.34×10^1	33.4
4.55×10^{-3}	0.00455
9.08×10^{-5}	0.0000908
2.0200×10^{-2}	0.020200

2.7

1. $(4.54 \times 10^4) + (1.804 \times 10^4) = 6.34 \times 10^4$
2. $(8.522 \times 10^{-3}) + (1.3 \times 10^{-4}) = 8.7 \times 10^{-3}$
3. $(5.631 \times 10^4) - (1.52 \times 10^4) = 4.11 \times 10^4$
4. $(2.94 \times 10^{-4}) - (5.866 \times 10^{-5}) = 2.35 \times 10^{-4}$

2.8

1. $(4.155 \times 10^3) \times (1.50 \times 10^6) = 6.23 \times 10^9$
2. $(7.36 \times 10^2) \times (4.711 \times 10^3) = 34.7 \times 10^5 = 3.47 \times 10^6$
3. $(1.7 \times 10^{-3}) \times (3.363 \times 10^{-5}) = 5.7 \times 10^{-8}$
4. $(4.09 \times 10^6) \div (2.9001 \times 10^3) = 1.41 \times 10^3$
5. $(3.1413 \times 10^4) \div (7.83 \times 10^5) = 0.401 \times 10^{-1} = 4.01 \times 10^{-2}$
6. $(5.204 \times 10^{-6}) \div (4.1 \times 10^3) = 1.3 \times 10^{-9}$
7. $(6.61 \times 10^{-6}) \div (7.278 \times 10^{-3}) = 9.08 \times 10^{-4}$

2.9

Problem	Relationship with units	Answer with units
3.89 km = ? cm	$3.89 \text{ km} \times \frac{1000 \text{ m}}{1 \text{ km}} \times \frac{100 \text{ cm}}{1 \text{ m}}$	3.89×10^5 cm
45.8 mm = ? m	$45.8 \text{ mm} \times \frac{1 \text{ m}}{1000 \text{ mm}}$	4.58×10^{-2} m
0.987 mm = ? km	$0.987 \text{ mm} \times \frac{1 \text{ m}}{1000 \text{ mm}} \times \frac{1 \text{ km}}{1000 \text{ m}}$	9.87×10^{-7} km
7.89×10^4 cm = ? m	$7.89 \times 10^4 \text{ cm} \times \frac{1 \text{ m}}{100 \text{ cm}}$	7.89×10^2 m
4.05×10^{-4} cm = ? km	$4.05 \times 10^{-4} \text{ cm} \times \frac{1 \text{ m}}{100 \text{ cm}} \times \frac{1 \text{ km}}{1000 \text{ m}}$	4.05×10^{-9} km
5.5×10^5 L = ? mL	$5.5 \times 10^5 \text{ L} \times \frac{1000 \text{ mL}}{1 \text{ L}}$	5.5×10^8 mL
3.6 km = ? in.	$3.6 \text{ km} \times \frac{1000 \text{ m}}{1 \text{ km}} \times \frac{39.4 \text{ in.}}{1.00 \text{ m}}$	1.4×10^5 in.
8.2 mL = ? qt	$8.2 \text{ mL} \times \frac{1 \text{ L}}{1000 \text{ mL}} \times \frac{1.00 \text{ qt}}{0.946 \text{ L}}$	8.7×10^{-3} qt
57 fl oz = ? mL	$57 \text{ fl oz} \times \frac{1.00 \text{ mL}}{0.034 \text{ fl oz}}$	1.7×10^3 mL
3.8 oz = ? kg	$3.8 \text{ oz} \times \frac{28.3 \text{ g}}{1.00 \text{ oz}} \times \frac{1 \text{ kg}}{1000 \text{ g}}$	1.1×10^{-1} kg

2.10 1. $\text{density} = \dfrac{\text{mass}}{\text{volume}} = \dfrac{404\ \text{g}}{38.5\ \text{cm}^3} = 10.5\ \text{g/cm}^3$

2. Use density as a conversion factor:.

$$\text{mass} = \frac{0.789\ \text{g}}{1\ \text{mL}} \times 458\ \text{mL} = 361\ \text{g}$$

3. Use two conversion factors to solve this problem.

$$\text{volume} = \frac{1.00\ \text{cm}^3}{8.92\ \text{g}} \times \frac{1000\ \text{g}}{1\ \text{kg}} \times 8.97\ \text{kg} = 1.01 \times 10^3\ \text{cm}^3$$

2.11

Boiling point	Substituted equation	Temperature
ethyl acetate (nail polish remover) 77.0°C	°F = 9/5(77.0) + 32	171°F
toluene (additive in gasoline) 111°C	°F = 9/5(111) + 32	232°F
isopropyl alcohol (rubbing alcohol) 180°F	°C = 5/9(180 – 32)	82°C
naphthalene (moth balls) 424°F	°C = 5/9(424 – 32)	218°C
propane (fuel for camping stoves) –42°C	K = (–42) + 273	231 K
methane (natural gas) –260°F	°C = 5/9(–260–32) = – 162°C K = (–162) + 273	111 K

2.12 1. Heat absorbed (cal) = specific heat (cal/g•°C) x mass (g) x temperature change (°C)

Heat absorbed = 1.00 cal/g•°C x 1000 g x 25.0°C = 2.50×10^4 cal

2. Heat absorbed = 0.21 cal/g•°C x 150 g x 75°C = 2400 cal

2.13 1. $(2.50 \times 10^4\ \text{cal})(4.184\ \text{J/cal}) = 1.05 \times 10^5\ \text{J}$

2. $(2.4 \times 10^3\ \text{cal})(4.184\ \text{J/cal}) = 1.0 \times 10^4\ \text{J}$

Answers to Self-Test

The numbers in parentheses refer to sections in your textbook.
1. F; liter (2.2) **2.** F; gram (2.2) **3.** T (2.6) **4.** F; added or subtracted (2.6) **5.** T (2.4)
6. F; its mass to its volume (2.8) **7.** T (2.9) **8.** F; milli- (2.2) **9.** T (2.4) **10.** T (2.9)
11. b (2.5) **12.** d (2.5) **13.** b (2.5) **14.** c (2.4 and 2.5) **15.** d (2.4) **16.** b (2.7)
17. a (2.8) **18.** e; 2.9×10^4 cal (2.9) **19.** c (2.4) **20.** d (2.9)

Atomic Structure and the Periodic Table — Chapter 3

Chapter Overview

All matter is made of basic building blocks called atoms. As you study the structure of atoms, you can begin to develop an understanding of how atoms bond together to form the many substances that make up our world. By studying the periodic law, you will begin to predict the properties of elements according to their positions in the periodic table.

By the end of this chapter you should be able to describe the three basic particles that make up atoms in terms of mass, charge ,and location and to calculate the number of each of the three types of particles in an atom using the atomic number and the mass number of that atom. You will learn to describe an isotope from its symbol, and you will calculate the average atomic mass of an element. Using the electron configuration of an element and the principle of the distinguishing electron, you will be able to classify the elements into groups with similar properties.

Practice Exercises

3.1 Atoms are made up of even smaller particles called **subatomic particles** (Sec. 3.1). The three types of subatomic particles found in atoms are **electrons, protons,** and **neutrons** (Sec. 3.1). Complete the table below summarizing properties of these particles and their location within atoms.

Type of particle	Relative mass	Charge	Location
electron			
proton			
neutron			

3.2 The **atomic number** (Sec. 3.2) of an **element** (Sec. 3.2) is the number of protons in the **nucleus** (Sec. 3.1) of atoms of that element. Because the number of protons in an atom is equal to the number of electrons, the atomic number also gives the number of electrons in a neutral atom. The **mass number** (Sec. 3.2) is the total number of **nucleons** (protons and neutrons) (Sec. 3.1) in the nucleus. Therefore, the number of neutrons can be found by subtraction:

Mass number – atomic number = number of neutrons

Use the relationships above and the **periodic table** (Sec. 3.4) to complete the following table:

Atomic number	Mass number	Number of protons	Number of neutrons	Number of electrons	Symbol of element
6			6		
	39	19			
			77		Xe
	64			29	
		35	45		

3.3 **Isotopes** (Sec. 3.3) are atoms that have the same number of protons but different numbers of neutrons and, therefore, different mass numbers. Isotopes are usually represented as follows:

$^{14}_{6}C$ The superscript is the mass number, or A.
The subscript is the atomic number, or Z.

Complete the following table:

Isotope	A	Z	Protons	Neutrons	Electrons
$^{40}_{20}Ca$					
$^{40}_{18}Ar$					
$^{23}_{11}Na$					
$^{37}_{17}Cl$					
$^{35}_{17}Cl$					

3.4 The **atomic mass** (Sec. 3.3) of an element is an average mass of the mixture of isotopes that reflects the relative abundance of the isotopes as they occur in nature. The atomic mass can be calculated by multiplying the relative mass of each isotope by its fractional abundance and then totaling the products.

Example:
Magnesium is composed of 78.7% ^{24}Mg, 10.1% ^{25}Mg and 11.2% ^{26}Mg. To find the atomic mass for magnesium, multiply each isotope's mass by the percent abundance and add these products together:

0.787 x 23.99 amu = 18.88 amu
0.101 x 24.99 amu = 2.52 amu
0.112 x 25.98 amu = 2.91 amu
24.31 amu

An element has two common isotopes: 80.4% of the atoms have a mass of 11.01 amu and 19.6% of the atoms have a mass of 10.01 amu. In the space below, set up the equations and calculate the atomic mass for this element. Identify the element.

Atomic mass =_________ Element: ____________

3.5 According to the **periodic law** (Sec. 3.4), when elements are arranged in order of increasing atomic number, elements with similar properties occur at periodic intervals. The periodic table represents this statement graphically: elements with similar properties are found in the same **group** (Sec. 3.4) or vertical column. The horizontal rows are known as **periods** (Sec. 3.4). A steplike line in the periodic table separates the **metals** (Sec. 3.5) on the left from the **nonmetals** (Sec. 3.5) on the right.

Refer to your periodic table for information to complete the table below:

Element	Group	Period	Metal	Nonmetal
Be	IIA	2	X	
Na				
N				
Br				
O				
Sn				
K				

Which two elements in the table above would you expect to have similar chemical properties and why?

3.6 The **electron configuration** (Sec. 3.7) of an atom is a statement of the number of electrons the atom has in each **electron subshell** (Sec. 3.6). A shorthand system is used to show electron configurations. For each subshell occupied by electrons, write the number and the symbol for the subshell with a superscript indicating the number of electrons in that subshell.

Example: The electron configuration for $_6C$ is $1s^22s^22p^2$. This shows that the atom has 2 electrons in the 1*s* subshell, 2 in the 2*s* subshell, and 2 in the 2*p* subshell.

Write the electron configurations for the elements below. Use Figure 3.10 in your textbook to determine the order in which the **electron orbitals** (Sec. 3.6) are filled.

Element	Electron configuration
neon	
chlorine	
iron	

3.7 An **orbital diagram** (Sec. 3.7) is a statement of how many electrons an atom has in each of its orbitals. Each arrow in the diagram indicates an electron. Electron spin is denoted by the direction of the arrow.

	1*s*	2*s*	2*p*		
Example: $_6C$	↓↑	↓↑	↑	↑	

Put in the arrows for the orbital diagram of each element below:

	1s	2s	2p			3s	3p			4s
$_{17}Cl$	☐	☐	☐	☐	☐	☐	☐	☐	☐	☐

	1s	2s	2p			3s	3p			4s
$_{20}Ca$	☐	☐	☐	☐	☐	☐	☐	☐	☐	☐

3.8 We can classify an element by determining the subshell of its **distinguishing electron** (Sec. 3.8), the last electron added in the electron configuration. If the distinguishing electron is added to an *s* or a *p* subshell, the element is a **representative element** (Sec. 3.9), and if the *p* subshell is filled, the element is a **noble gas** (Sec. 3.9). If the distinguishing electron is added to a *d* subshell, the element is a **transition element** (Sec. 3.9); if it is added to an *f* subshell, the element is an **inner transition element** (Sec. 3.9).

Write the level of the distinguishing electron for each element below, and indicate the element's classification.

Element	Distinguishing electron	Noble gas	Representative element	Transition element
Mg	$3s^2$		X	
Ti				
Ar				
S				

3.9 Early experiments on atomic structure indicated that atoms are made up of smaller subatomic particles (Sec. 3.1). Complete the following table summarizing some of these experiments.

Scientist	Experiments performed	What they showed
Goldstein		
Thomson		
Rutherford		

Self-Test

True-false: Indicate whether the following statements are true or false. If the statement is false, give the word or phrase that may be substituted for the underlined portion to make the statement true.

1. The <u>nucleus</u> is the smallest particle of an atom.
2. Most of the mass of an atom is located in the <u>nucleus</u>.
3. The nucleus contains <u>electrons and protons</u>.
4. For a neutral atom, the number of protons <u>equals</u> the number of electrons.
5. The atomic number of an element is the number of <u>protons and neutrons</u>.
6. The mass number is the total number of <u>protons and electrons</u> in the atom.

7. Isotopes of a specific element have different numbers of neutrons in the nuclei of their atoms.
8. Electron orbitals have different shapes: *s*-orbitals are spherical.
9. The periodic law states that when elements are arranged in order of increasing atomic number, elements with similar properties occur at periodic intervals.
10. In the modern periodic table, the horizontal rows are called groups.
11. Metals are substances that have a high luster and are malleable.
12. Metals are on the left side of the periodic table.
13. Nonmetals are good conductors of electricity.
14. The noble gases have the outermost *s* and *p* subshells of electrons filled.
15. The transition elements are characterized by dinstinguishing electrons in the *d*-orbitals.

Multiple choice:

16. The nucleus of an atom contains these basic particles:

    a. electrons and protons b. neutrons and electrons c. protons and neutrons
    d. only neutrons e. none of these

17. Isotopes of a specific element vary in the following manner:

    a. Electron numbers are different. b. Neutron numbers are different.
    c. Proton numbers are different. d. Neutron and proton numbers are different.
    e. none of these

18. The element $^{48}_{22}Ti$ has the following electron configuration:

    a. $1s^22s^22p^63s^23p^63d^{10}4s^2$ b. $1s^22s^22p^63s^23p^63d^4$
    c. $1s^22s^22p^63s^23p^64s^23d^2$ d. $1s^22s^22p^63s^23p^64s^23d^4$
    e. none of these

19. In the isotope $^{81}_{35}Br$ how many neutrons are in the nucleus?

    a. 35 b. 46 c. 81 d. 116 e. none of these

20. How many nucleons are located in an atom of cesium, $^{133}_{55}Cs$?

    a. 55 b. 78 c. 133 d. 188 e. none of these

21. The element that has the electron configuration $1s^22s^22p^63s^23p^64s^23d^{10}4p^6$ is:

    a. $_{10}Ne$ b. $_{54}Xe$ c. $_{18}Ar$ d. $_{36}Kr$ e. none of these

22. In the periodic table, the elements on the far left side are classified as:

    a. metals b. noble gases c. transition metals
    d. nonmetals e. none of these

23. In the periodic table, the elements called noble gases are in:

    a. Group IA b. Group IIA c. Group VIIA
    d. Group VA e. none of these

24. The distinguishing electron for $_{19}K$ would be found in what subshell?

    a. 3*s* b. 2*p* c. 3*d* d. 4*s* e. none of these

Answers to Practice Exercises

3.1

Type of particle	Relative mass	Charge	Location
electron	small	–1	outside the nucleus
proton	large	+1	inside the nucleus
neutron	large	0	inside the nucleus

3.2

Atomic number	Mass number	Number of protons	Number of neutrons	Number of electrons	Symbol of element
6	12	6	6	6	C
19	39	19	20	19	K
54	131	54	77	54	Xe
29	64	29	35	29	Cu
35	80	35	45	35	Br

3.3

Isotope	A	Z	Protons	Neutrons	Electrons
$^{40}_{20}Ca$	40	20	20	20	20
$^{40}_{18}Ar$	40	18	18	22	18
$^{23}_{11}Na$	23	11	11	12	11
$^{37}_{17}Cl$	37	17	17	20	17
$^{35}_{17}Cl$	35	17	17	18	17

3.4 0.804 x 11.01 amu = 8.85 amu
0.196 x 10.01 amu = 1.96 amu
10.81 amu

Atomic mass = 10.81 amu Element: boron

3.5

Element	Group	Period	Metal	Nonmetal
Be	IIA	2	X	
Na	IA	3	X	
N	VA	2		X
Br	VIIA	4		X
O	VIA	2		X
Sn	IVA	5	X	
K	IA	4	X	

We would expect Na and K to have similar chemical properties; they are in the same group of the periodic table.

3.6

Element	Electron configuration
neon	$1s^2 2s^2 2p^6$
chlorine	$1s^2 2s^2 2p^6 3s^2 3p^5$
iron	$1s^2 2s^2 2p^6 3s^2 3p^6 4s^2 3d^6$

3.7

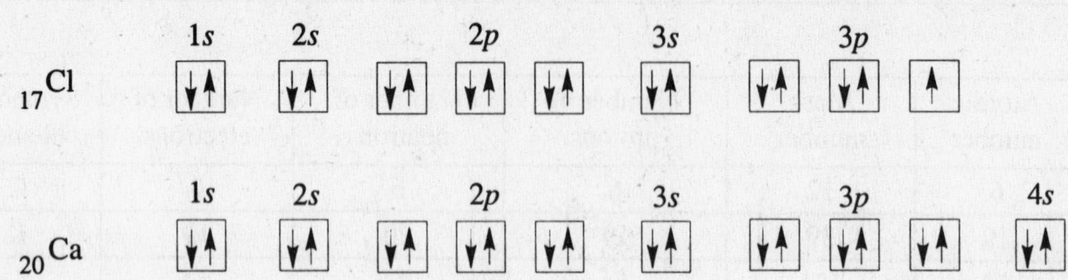

3.8

Element	Distinguishing electron	Noble gas	Representative element	Transition element
Mg	$3s^2$		X	
Ti	$3d^2$			X
Ar	$3p^6$	X		
S	$3p^4$		X	

3.9

Scientist	Experiments performed	What they showed
Goldstein	Discharge tube with a gridlike cathode	Canal rays contained many different types of positive particles.
Thomson	Cathode ray experiments	Negatively charged particles (electrons) present in all matter; electrons embedded in a positively charged mass
Rutherford	Alpha particle experiments with gold foil	New atomic model: small, dense, positively charged nucleus and lots of space containing only tiny, negatively charged electrons

Answers to Self-Test

The numbers in parentheses refer to sections in your textbook.
1. F; electron (3.1) **2**. T (3.1) **3**. F; protons and neutrons (3.1) **4**. T (3.2)
5. F; protons (3.2) **6**. F; protons and neutrons (3.2) **7**. T (3.3) **8**. T (3.6) **9**. T (3.4)
10. F; periods (3.4) **11**. T (3.5) **12**. T (3.5) **13**. F; poor (3.5) **14**. T (3.9) **15**. T (3.9)
16. c (3.1) **17**. b (3.3) **18**. c (3.7) **19**. b (3.3) **20**. c (3.3) **21**. d (3.7)
22. a (3.5) **23**. e; Group VIIIA (3.9) **24**. d (3.8)

Chemical Bonding: The Ionic Bond Model — Chapter 4

Chapter Overview

The electron configuration of the atoms of an element determines the chemical properties of that element. In this chapter you will see how electrons transfer from one atom to another to form an ionic bond.

You will identify the valence electrons of an atom using the electron configurations of the atom and the element's group number in the periodic table. You will draw the Lewis symbols for atoms and use these structures to show electron transfer in ionic bond formation. You will predict the formulas for ionic compounds and learn to name these ionic compounds.

Practice Exercises

4.1 The **valence electrons** (Sec. 4.2) of an atom are the electrons in the outermost electron shell. Write the electron configuration and give the number of valence electrons and the Group number in the periodic table for atoms of each of the following elements:

Element	Electron configuration	Number of valence electrons	Group number
lithium			
beryllium			
boron			
phosphorus			
sulfur			

4.2 **Lewis symbols** (Sec. 4.2) are atomic symbols with one dot for each valence electron placed around the element's symbol. Give the number of valence electrons and write the Lewis symbol for each of the elements or groups below. (Use the symbol X as a group symbol.)

Group	Valence electrons	Lewis symbol
Group IA		X
Group IVA		X
Group VIIA		X

Element	Valence electrons	Lewis symbol
sulfur		S
bromine		
magnesium		

4.3 According to the **octet rule** (Sec. 4.3), in compound formation, atoms of elements lose, gain, or share electrons in such a way that their electron configurations become identical to that of the noble gas nearest them in the periodic table. Atoms on the left side of the periodic table tend to lose electrons and become positively charged **monatomic ions** (Sec. 4.4 and 4.10), and atoms on the right side of the periodic table gain electrons to become negatively charged monatomic ions.

Complete the table below showing ion formation, and identify the noble gas that is **isoelectronic** (Sec. 4.5) with the ion formed.

Element	Group number	Electrons lost/gained	Ion formed	Noble gas
Na				
Br				
S				
Ca				
N				

4.4 Some metals have a variable ionic charge; they can form more than one type of ion. Complete the following tables for these metals with variable ionic charge:

Element	Electrons lost	Symbol for ion
tin	2	Sn^{2+}
tin	4	Sn^{4+}
cobalt	2	Co^{2+}

Element	Electrons lost	Symbol for ion
cobalt	3	Co^{3+}
iron	2	Fe^{2+}
iron	3	Fe^{3+}

4.5 **Ionic bonds** (Sec. 4.1) form when electrons are transferred from metal atoms to nonmetal atoms. Formation of ionic compounds requires a charge balance: the same number of electrons must be lost as are gained. Show the valence electrons for the individual atoms of these ionic compounds and then the formation of the ionic compounds by electron transfer.

Chemical formula	Lewis symbols for atoms		Formation of ionic compound
KBr	K	Br	
CaI_2	Ca	I	
SrS	Sr	S	

4.6 A **binary ionic compound** (Sec. 4.9) is made up of a metal and a nonmetal. Because ionic compounds consist of an alternating array of positive and negative charges, the term **formula unit** (Sec. 4.8) is used to refer to the smallest unit of the ionic compound.

Practice balancing charges by writing a formula unit for the ionic compound formed from the elements given in the table below. A binary ionic compound is named by naming the metal first, followed by the stem of the nonmetal with the ending *-ide*.

Elements	Ions formed	Formula unit	Name of ionic compound
potassium and chlorine			
beryllium and iodine			
sodium and sulfur			
strontium and oxygen			
aluminum and fluorine		AlF_3	
			cesium bromide
			calcium oxide
			aluminum sulfide

4.7 A **polyatomic ion** (Sec. 4.10) is a **covalently bonded** (Sec. 4.1) group of atoms having a charge. In writing the formula for an ionic compound, treat a polyatomic ion as a unit. If more than one of these ions is required for charge balance, enclose the ion in parentheses and put the number of ions outside the parentheses. Give the formulas for the ionic compounds prepared by combining the following ions:

	Bromide	Nitrate	Carbonate	Phosphate
Sodium	NaBr			
Calcium		$Ca(NO_3)_2$		
Ammonium				
Aluminum				

4.8 In naming a compound containing a metal with a variable ionic charge, use a Roman numeral after the metal name to indicate the charge on the metal ion. Give the formulas and the names of the ionic compounds prepared by combining the following ions:

	F^-	N^{3-}	SO_4^{2-}	ClO_2^-
K^+		K_3N potassium nitride		
Pb^{2+}				
Fe^{3+}				
Sn^{4+}				

Self-Test

True-false: Indicate whether the following statements are true or false. If the statement is false, give the word or phrase that may be substituted for the underlined portion to make the statement true.

1. Lewis symbols show the number of <u>inner electrons</u> of an atom.
2. Valence electrons determine the <u>chemical properties</u> of an element.
3. A negative ion is formed when an element <u>loses</u> an electron.
4. Metals tend to <u>gain</u> electrons to attain the configuration of a noble gas.
5. Bromine would accept an electron to attain the configuration of the noble gas <u>krypton</u>.
6. <u>An ionic bond</u> results from the sharing of one or more pairs of electrons between atoms.
7. The maximum number of valence electrons for any element is <u>four</u>.
8. The most stable electron configuration is that of <u>the noble gases</u>.
9. <u>An ionic compound</u> is formed from a metal that can donate electrons and a nonmetal that can accept electrons.
10. In naming binary ionic compounds, the full name of the metallic element is given <u>first</u>.
11. In binary ionic compounds, the fixed-charge metals are generally found in <u>Groups VIIA and VIIIA</u>.
12. A polyatomic ion is a group of atoms that is held together by <u>ionic bonds</u> and has acquired a charge.

Multiple choice:

13. The binary ionic compound RbI would be called:

    a. rubidium(I) iodide b. rubidium iodate c. rubidium iodine
    d. rubidium iodide e. none of these

14. The formula for the binary ionic compound silver sulfide is:

    a. SiS b. AgS c. Ag_2S d. AgS_2 e. none of these

15. In the electron configuration for sulfur, $1s^22s^22p^63s^23p^4$, what electron shell number determines the valence electrons?

    a. 1 b. 2 c. 3 d. 2 and 3 e. none of these

16. Which of these representative elements would be isoelectronic with calcium?

    a. potassium b. barium c. scandium
    d. aluminum e. none of these

17. The electron configuration of a noble gas is:

    a. $1s^22s^2$ b. $1s^22s^22p^4$ c. $1s^22s^22p^63s^23p^2$
    d. $1s^22s^22p^63s^23p^6$ e. none of these

18. The electron configuration of the ion S^{2-} is:

    a. $1s^22s^22p^6$ b. $1s^22s^22p^63s^23p^4$ c. $1s^22s^22p^63s^23p^6$
    d. $1s^22s^22p^63s^23p^64s^2$ e. none of these

19. In the formula Na_3N, the total number of electrons accepted by the nitrogen is:

a. 1 b. 2 c. 3 d. 4 e. none of these

20. In the ionic compound calcium phosphate, how many polyatomic ions (phosphate ions) are in 1 formula unit?

a. 1 b. 2 c. 3 d. 4 e. none of these

21. The Lewis symbol for a Group VA element would have dots representing the following number of valence electrons:

a. 2 b. 3 c. 4 d. 5 e. none of these

22. Which of the following pairs of elements would form a binary ionic compound?

a. sulfur and oxygen
b. bromine and chlorine
c. magnesium and bromine
d. oxygen and hydrogen
e. none of these

23. At room temperature, an ionic compound would be in which physical state?

a. gas
b. liquid
c. solid
d. gas and liquid
e. none of these

Answers to Practice Exercises

4.1

Element	Electron configuration	Number of valence electrons	Group number
lithium	$1s^22s^1$	1	IA
beryllium	$1s^22s^2$	2	IIA
boron	$1s^22s^22p^1$	3	IIIA
phosphorus	$1s^22s^22p^63s^23p^3$	5	VA
sulfur	$1s^22s^22p^63s^23p^4$	6	VIA

4.2

Group	Valence electrons	Lewis symbol
Group IA	1	$\underset{\cdot}{X}$
Group IVA	4	$\cdot\underset{\cdot}{\dot{X}}\cdot$
Group VIIA	7	$:\underset{\cdot\cdot}{\dot{X}}:$

Element	Valence electrons	Lewis symbol
sulfur	6	$:\underset{\cdot}{\dot{S}}:$
bromine	7	$\cdot\underset{\cdot\cdot}{\overset{\cdot\cdot}{Br}}:$
magnesium	2	$\cdot Mg\cdot$

4.3

Element	Group number	Electrons lost/gained	Ion formed	Noble gas
Na	IA	1 lost	Na^{+}	Ne
Br	VIIA	1 gained	Br^{-}	Kr
S	VIA	2 gained	S^{2-}	Ar
Ca	IIA	2 lost	Ca^{2+}	Ar
N	VA	3 gained	N^{3-}	Ne

4.4

Element	Electrons lost	Symbol for ion
tin	2	Sn^{2+}
tin	4	Sn^{4+}
cobalt	2	Co^{2+}

Element	Electrons lost	Symbol for ion
cobalt	3	Co^{3+}
iron	2	Fe^{2+}
iron	3	Fe^{3+}

4.5

Lewis symbols for atoms		Formation of ionic compound (Lewis structures)
K·	·Br:	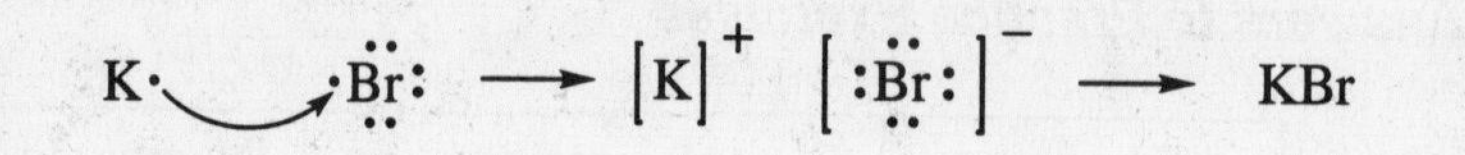 K· ·Br: → $[K]^{+}$ $[:Br:]^{-}$ → KBr
·Ca·	·I:	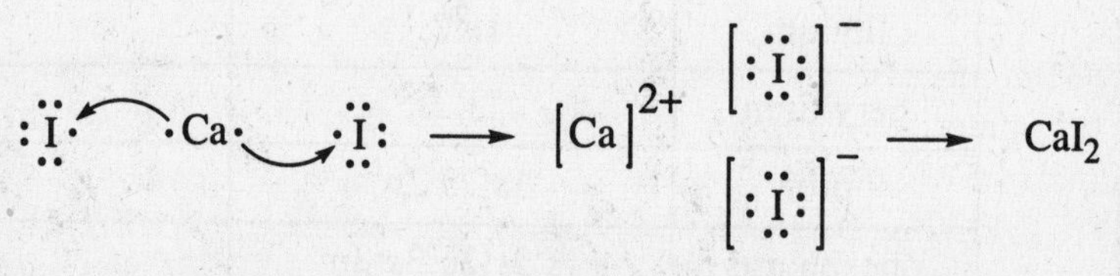 :I· ·Ca· ·I: → $[Ca]^{2+}$ $[:I:]^{-}$ $[:I:]^{-}$ → CaI_2
Sr·	·S:	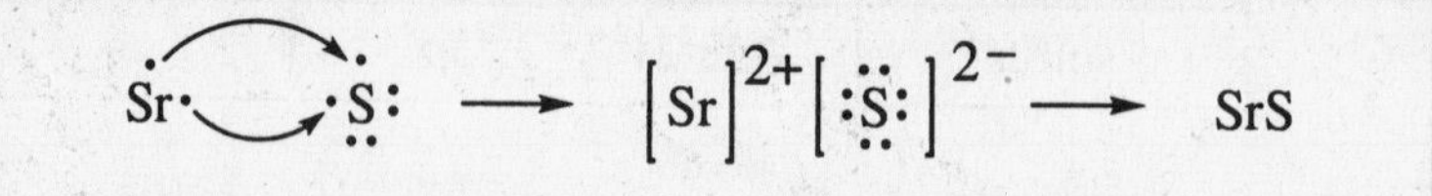 Sr· ·S: → $[Sr]^{2+}$ $[:S:]^{2-}$ → SrS

4.6

Elements	Ions formed	Formula unit	Name of ionic compound
potassium and chlorine	K^{+}, Cl^{-}	KCl	potassium chloride
beryllium and iodine	Be^{2+}. I^{-}	BeI_2	beryllium iodide
sodium and sulfur	Na^{+}. S^{2-}	Na_2S	sodium sulfide
strontium and oxygen	Sr^{+2}, O^{2-}	SrO	strontium oxide
aluminum and fluorine	Al^{3+}, F^{-}	AlF_3	aluminum fluoride
cesium and bromine	Cs^{+}, Br^{-}	CsBr	cesium bromide
calcium and oxygen	Ca^{2+}, O^{2-}	CaO	calcium oxide
aluminum and sulfur	Al^{3+}, S^{2-}	Al_2S_3	aluminum sulfide

4.7

	Bromide	Nitrate	Carbonate	Phosphate
Sodium	$NaBr$	$NaNO_3$	Na_2CO_3	Na_3PO_4
Calcium	$CaBr_2$	$Ca(NO_3)_2$	$CaCO_3$	$Ca_3(PO_4)_2$
Ammonium	NH_4Br	NH_4NO_3	$(NH_4)_2CO_3$	$(NH_4)_3PO_4$
Aluminum	$AlBr_3$	$Al(NO_3)_3$	$Al_2(CO_3)_3$	$AlPO_4$

4.8

	F^-	N^{3-}	SO_4^{2-}	ClO_2^-
K^+	KF potassium fluoride	K_3N potassium nitride	K_2SO_4 potassium sulfate	$KClO_2$ potassium chlorite
Pb^{2+}	PbF_2 lead(II) fluoride	Pb_3N_2 lead(II) nitride	$PbSO_4$ lead(II) sulfate	$Pb(ClO_2)_2$ lead(II) chlorite
Fe^{3+}	FeF_3 iron(III) fluoride	FeN iron(III) nitride	$Fe_2(SO_4)_3$ iron(III) sulfate	$Fe(ClO_2)_3$ iron(III) chlorite
Sn^{4+}	SnF_4 tin(IV) fluoride	Sn_3N_4 tin(IV) nitride	$Sn(SO_4)_2$ tin(IV) sulfate	$Sn(ClO_2)_4$ tin(IV) chlorite

Answers to Self-Test

The numbers in parentheses refer to sections in your textbook.
1. F; outermost or valence electrons (4.2) **2.** T (4.2) **3.** F; gains (4.4) **4.** F; lose (4.5) **5.** T (4.5) **6.** F; covalent (4.1) **7.** F; eight (4.5) **8.** T (4.3) **9.** T (4.6) **10.** T (4.9) **11.** F; Groups IA and IIA (4.9) **12.** F; covalent bonds (4.10) **13.** d (4.9) **14.** c (4.9) **15.** c (4.2) **16.** b (4.2) **17.** d (4.3) **18.** c (4.5) **19.** c (4.7) **20.** b (4.11) **21.** d (4.2) **22.** c (4.5, 4.9) **23.** c (4.1)

Chemical Bonding: The Covalent Bond Model

Chapter 5

Chapter Overview

Molecular compounds are the result of the sharing of electrons between atoms in molecules. Covalent bonds join nonmetallic atoms together to form molecules.

In this chapter you will use Lewis structures to indicate the various types of covalent bonds in molecules. You will study the concept of electronegativity differences between atoms and how this determines whether a bond is ionic or covalent, polar or nonpolar. You will use VSEPR theory to predict the three-dimensional shape of molecules and determine molecular polarity.

Practice Exercises

5.1 Remember that the valence electrons of an atom, those in the atom's outermost shell, are the electrons involved in bond formation. Draw Lewis symbols showing valence electrons for the following elements:

·Mg·				
Mg	C	P	Br	Ar

5.2 **Bonding electrons** (Sec. 5.2) are pairs of valence electrons that are shared between atoms in a covalent bond. **Nonbonding electrons** (Sec. 5.2) are pairs of electrons that are not involved in sharing. Molecules tend to be stable when each atom in the molecule shares in an octet of electrons.

Draw the Lewis structures for one molecule of each of these molecular compounds. Circle the bonding electrons in each **single covalent bond** (Sec. 5.3) in the molecule:

H⊙Br:				
HBr	F_2	BrI	H_2O	CH_4

5.3 In a **double covalent bond** (Sec. 5.3), two atoms share two pairs of electrons; in a **triple covalent bond** (Sec. 5.3), two atoms share three pairs of electrons.

Draw Lewis structures showing the single, double, and triple covalent bonds in these molecules, as well as the nonbonding electrons. For help in determining which atom should be the central atom of a Lewis structure, see Section 5.6 of your textbook.

H:C::C:H H H				
C_2H_4	CS_2	HCN	H_2CO	C_2H_2

5.4 Because a covalent bond consists of a pair of shared electrons, we would expect an atom of nitrogen, having three unpaired electrons, to share in three single bonds, or in one single and one double bond, or in one triple bond. According to this concept, how many bonds would you expect each of the following atoms to form?

Oxygen ____

Carbon ____

5.5 In order that all atoms in a molecule share in an octet of electrons, it is possible for one atom to supply both electrons of the shared pair of a bond. This is called a **coordinate covalent bond** (Sec. 5.5). The electron pair for this bond comes from one of the nonbonding electron pairs of one of the atoms. Draw Lewis structures for the following molecules and circle the coordinate covalent bonds:

H:C̈l⊙Ö: (with nonbonding pairs shown as dots; coordinate bond circled)	
HClO	$HBrO_2$

5.6 In Lewis structures for molecules, the shared electron pairs may be represented with dashes. Rewrite the structures in Exercise 5.3 by replacing the bonding electron pairs with a dash to show the covalent bond between atoms. Include the nonbonding electron pairs as dots:

H—C=C—H
    | |
    H H				

C_2H_4	CS_2	HCN	H_2CO	C_2H_2

5.7 Polyatomic ions consist of covalently bonded atoms acting as a unit with a charge on it. The Lewis structure for a polyatomic ion is drawn in the same way as it is for a molecule, except that the total number of electrons is increased or decreased according to the charge on the ion. Draw Lewis structures for the following polyatomic ions:

$[:\ddot{O}:\ddot{N}:\ddot{O}:]^-$		
NO_2^-	NO_3^-	ClO_4^-

5.8 According to **VSEPR theory** (Sec. 5.8), the geometry of a molecule is determined by the number of electron groups (including nonbonding pairs) around the central atom of a molecule. Double and triple bonds each count as a single electron group. The **molecular geometry** (Sec. 5.8) is the one in which the electron groups are farthest from one another.

The shapes that will give the greatest distance between electron groups are: for two electron pairs, 180° bond angle (linear); for three electron pairs, 120° bond angle (trigonal planar or angular); for four electron pairs, 109° bond angle (tetrahedral, trigonal pyramidal, or angular).

Use these guidelines to predict the shape of the following molecules (See the Chemistry at a Glance in your textbook). For each molecule, draw the Lewis structure and count the electron groups around the central atom. Use this information to predict the geometry of the molecule.

Molecular formula	Lewis structure	Number of electron groups around central atom	Molecular geometry
CBr_4			
CH_2O			
CS_2			
H_2S			
NCl_3			

5.9 **Electronegativity** (Sec. 5.9) is a measure of the relative attraction that an atom has for the shared electrons in a bond. The electronegativity difference between the two bonded atoms is a measure of the polarity of the bond. If the difference is 2.0 or greater, the bond is ionic. If the difference is less than 2.0, the bond is **polar covalent** (Sec. 5.10), and if the difference is zero, it is **nonpolar covalent** (Sec. 5.10).

Calculate the electronegativity difference for each element pair, and indicate whether the bond formed between them will be ionic, nonpolar covalent,or polar covalent. (Electronegativities are given in Figure 5.11 of your textbook.)

Pair of elements	Electronegativity difference	Ionic bond	Nonpolar covalent bond	Polar covalent bond
sodium and fluorine				
bromine and bromine				
sulfur and oxygen				
phosphorus and bromine				

5.10 **Molecular polarity** (Sec. 5.11) is a measure of the total electron distribution over a molecule, rather than over just one bond. A molecule whose bonds are polar may be a **polar molecule** or a **nonpolar molecule** (Sec. 5.11), depending on its molecular geometry. Individual bond polarities may cancel one another in a highly symmetrical molecule, resulting in a nonpolar molecule.

Classify the molecules in Practice Exercise 5.8 as polar molecules or nonpolar molecules.

Molecular formula	Molecular geometry	Number of polar bonds	Polar or nonpolar molecule?

5.11 In naming binary molecular compounds, name the element of lower electronegativity first, followed by the stem of the more electronegative nonmetal and the suffix *-ide*. Include prefixes to indicate the number of atoms of each nonmetal. Name the following molecular compounds:

a. CCl_4 ____________________

b. CS_2 ____________________

c. NCl_3 ____________________

d. N_4S_4 ____________________

Self-Test

True-false: Indicate whether the following statements are true or false. If the statement is false, give the word or phrase that may be substituted for the underlined portion to make the statement true.

1. Carbon dioxide is a polar molecule.
2. Covalent bond formation between nonmetal atoms involves electron transfer.
3. Nonbonding electrons are pairs of valence electrons that are not shared between atoms having a covalent bond.
4. A nitrogen molecule, N_2, would have a double covalent bond between the two nitrogen atoms.
5. Carbon can form multiple covalent bonds with other nonmetallic elements.
6. A coordinate covalent bond is a covalent bond formed when both electrons of a shared pair are donated by one atom.
7. According to VSEPR, the electron groups in the valence shell arrange themselves to maximize the repulsion between the electron groups.
8. According to VSEPR, a water molecule, H_2O, would have a linear arrangement of the valence electron groups.

9. According to VSEPR theory convention, single and triple bonds are equal.
10. Electronegativity is a measure of the relative repulsion that an atom has for the shared electrons in a bond.
11. Electronegativity values increase from left to right across periods in the periodic table.
12. The bond between fluorine and bromine would be a nonpolar covalent bond.
13. Nonbonding electron pairs are important in determining the shape of a molecule.

Multiple choice:

14. Which of the following pairs of elements would form a covalent bond?

    a. sulfur and oxygen b. potassium and iodine c. magnesium and bromine
    d. calcium and fluorine e. none of these

15. Which of the following pairs of elements would form a nonpolar covalent bond?

    a. nitrogen and oxygen b. fluorine and fluorine c. calcium and iodine
    d. potassium and bromine e. none of these

16. Which of the following pairs of elements would form a polar covalent bond?

    a. carbon and carbon b. bromine and bromine c. potassium and fluorine
    d. sodium and oxygen e. none of these

17. How many valence electrons are found in a triple covalent bond?

    a. 2 b. 3 c. 4 d. 6 e. none of these

18. An element that can form a triple covalent bond may be found in:

    a. Group VA b. Group IIA c. Group VIIA
    d. Group VIIIA e. none of these

19. What types of electron pairs are used in VSEPR calculations?

    a. core electrons b. bonding electron pairs c. nonbonding electron pairs
    d. both b and c e. none of these

20. How many VSEPR electron groups are found in ammonia, NH_3?

    a. 1 b. 2 c. 3 d. 4 e. none of these

21. Which element would be more electronegative than chlorine?

    a. sulfur b. lithium c. bromine d. carbon e. none of these

22. The correct name for the binary molecular compound SO_3 is:

    a. sulfur oxide b. sulfur trioxygen c. sulfur trioxide
    d. trioxygen sulfide e. none of these

Answers to Practice Exercises

5.1

·Mg·	·Ċ· (dots on all four sides)	·Ṗ: (one dot top, left, bottom; pair right)	:Br: (one dot top; pairs left, right, bottom)	:Ar: (pairs on all four sides)

5.2

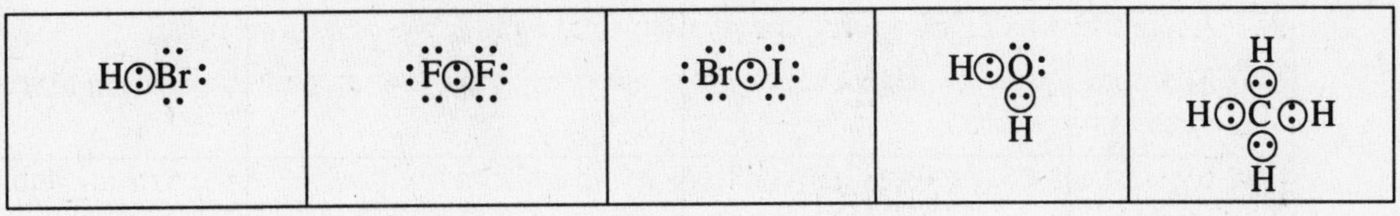

H⊙Br:	:F⊙F:	:Br⊙I:	H⊙O: ⊙ H	H ⊙ H⊙C⊙H ⊙ H

5.3

H:C::C:H H H	:S::C::S:	H:C:::N:	H:C::O H	H:C:::C:H

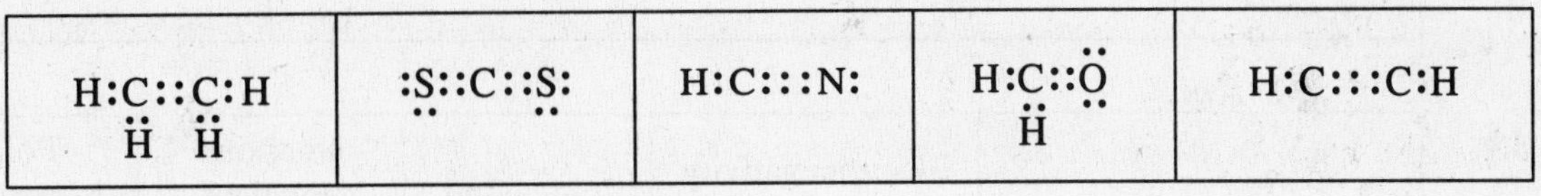

5.4 Oxygen – 2 bonds; two single or one double
Carbon – 4 bonds; four single, or two double, or two single and one double, or one single and one triple.

5.5

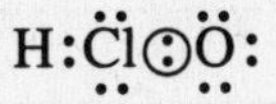

H:Cl⊙O:	:O: ⊙ H:Br⊙O:
HClO	$HBrO_2$

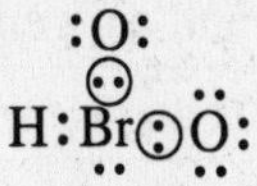

5.6

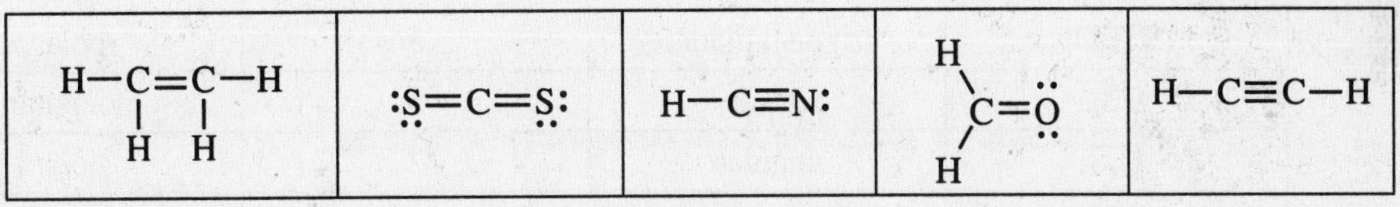

H—C=C—H, H H	:S=C=S:	H—C≡N:	H, H, C=O	H—C≡C—H

5.7

[:O:N:O:]⁻	[:O:N:O: :O:]⁻	[:O: :O:Cl:O: :O:]⁻
NO_2^-	NO_3^-	ClO_4^-

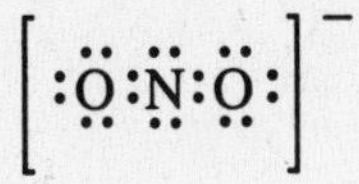

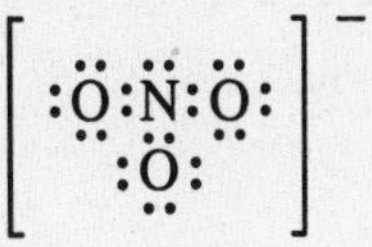

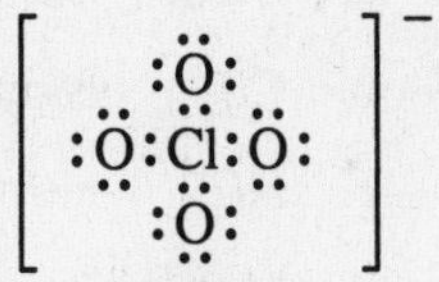

5.8

Molecular formula	Lewis structure	Number of electron groups around central atom	Molecular geometry
CBr_4	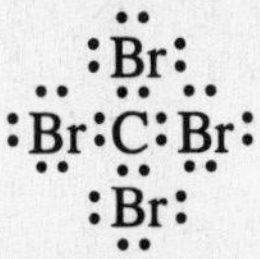 :Br: :Br:C:Br: :Br:	4	tetrahedral
CH_2O	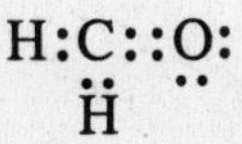 H:C::O: H	3	trigonal planar
CS_2	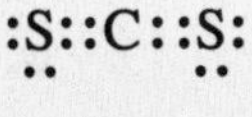 :S::C::S:	2	linear

H_2S	H:S: H	4	angular
NCl_3	:Cl:N:Cl: :Cl:	4	trigonal pyramid

5.9

Pair of elements	Electronegativity difference	Ionic bond	Nonpolar covalent bond	Polar covalent bond
sodium and fluorine	3.1	X		
bromine and bromine	0.0		X	
sulfur and oxygen	1.0			X
phosphorus and bromine	0.7			X

5.10

Molecular formula	Molecular geometry	Number of polar bonds	Polar or nonpolar molecule?
CBr_4	tetrahedral	4	nonpolar
CH_2O	trigonal planar	3	polar
CS_2	linear	2	nonpolar
H_2S	angular	2	polar
NCl_3	trigonal pyramid	3	polar

5.11 a. CCl_4 carbon tetrachloride

b. CS_2 carbon disulfide

c. NCl_3 nitrogen trichloride

d. N_4S_4 tetranitrogen tetrasulfide

Answers to Self-Test

The numbers in parentheses refer to sections in your textbook.
1. F; nonpolar (5.11) **2**. F; sharing (5.1) **3**. T (5.2) **4**. F; triple (5.3) **5**. T (5.4) **6**. T (5.5) **7**. F; minimize (5.8) **8**. F; an angular (5.8) **9**. T (5.8) **10**. F; attraction (5.9) **11**. T (5.9) **12**. F; a polar covalent (5.10) **13**. T (5.8) **14**. a (5.10) **15**. b (5.10) **16**. e (5.10) **17**. d (5.3) **18**. a (5.4) **19**. d (5.8) **20**. d (5.8) **21**. e (5.9) **22**. c (5.12)

Chemical Calculations: Formula Masses, Moles, and Chemical Equations — Chapter 6

Chapter Overview

Calculation of the ratios and masses of the substances involved in chemical reactions is very important in many chemical processes. Central to these calculations is the concept of the mole, a convenient counting unit for atoms and molecules.

In this chapter you will learn to determine the formula mass of substances and the number of moles of substances. You will practice writing and balancing chemical equations, and you will learn to use these equations in determining amounts of substances that react and are produced in chemical reactions.

Practice Exercises

6.1 The **formula mass** (Sec. 6.1) of a compound is the sum of the atomic masses of the atoms in the chemical formula of the substance. Calculate the formula mass for each of these compounds:

Formula	Calculations with atomic masses	Formula mass
KBr		
$CaCl_2$		
Na_2CO_3		
$(NH_4)_3PO_4$		

6.2 The **mole** (Sec. 6.2 and 6.3) is a useful unit for counting numbers of atoms and molecules. The number of particles in a mole is 6.02×10^{23}, which is known as **Avogadro's number** (Sec. 6.2). Set up and complete the calculations below:

Moles	Relationship with units	Number of atoms
1.00 mole of helium atoms		
2.60 moles of sodium atoms		
0.316 mole of argon atoms		

6.3 The mass of 1 mole, called the **molar mass** (Sec. 6.3), of any substance is its formula mass expressed in grams.

Using dimensional analysis and the correct conversion factor (either moles/gram or grams/mole), find either the mass or the number of moles for the substances below:

Given quantity	Relationship with units	Calculated
1.00 mole H_2O		g H_2O
2.53 moles H_2O		g H_2O
0.519 mole H_2O		g H_2O
1.00 g NaBr		mole NaBr
417 g NaBr		mole NaBr
0.322 g NaBr		mole NaBr

6.4 The subscripts in a chemical formula show the number of atoms of each element per formula unit. They also show the number of moles of atoms of each element in one mole of the substance (atoms/molecule = moles of atoms/moles of molecules). Determine the moles of carbon atoms in the following problems:

Moles of compound	Carbon atoms/molecule	Moles of carbon atoms
1.00 mole of $C_{10}H_{16}$ (limonene)		
13.5 moles of C_2H_6O (ethanol)		
0.705 mole of C_5H_{12} (pentane)		

6.5 Remember that the number of grams of one substance cannot be compared directly to the number of grams of another substance; however, moles can be related to moles quite easily by looking at subscripts in chemical formulas. Figure 6.7 in your textbook gives you a map of the steps to take in solving problems involving grams and moles.

a. How many molecules of KCl would be found in 0.125 g of KCl?

b. Calculate the number of fluorine atoms in 1.77 g of AlF_3.

c. Calculate the number of grams of Cl in 12.5 g of KCl.

d. Calculate the number of grams of F in 1.77 g of AlF_3.

6.6 The description of a chemical reaction can be expressed efficiently with the formulas and symbols of a **chemical equation** (Sec. 6.6). The substances that react (the reactants) are placed on the left, and those that are produced (products) are on the right. The arrow in the chemical equation is read as "to produce." Plus signs on the left side mean "reacts with," and plus signs on the right are read as "and."

Write chemical equations for the following chemical reactions:

a. Hydrogen chloride reacts with sodium hydroxide to produce sodium chloride and water.

b. Silver nitrate and potassium bromide react with one another to produce silver bromide and potassium nitrate.

6.7 To be most useful, chemical equations must be **balanced** (Sec. 6.6); that is, the number of atoms of each element must be the same on both sides of the equation. A suggested method for balancing equations is in Section 6.6 of your textbook. Remember: use the **coefficients** (Sec. 6.6) to balance equations, but do not change the subscripts within the formulas.

Balance the following equations:

a. $H_2 + Cl_2 \rightarrow HCl$

b. $AgNO_3 + H_2S \rightarrow Ag_2S + HNO_3$

c. $P + O_2 \rightarrow P_2O_3$

d. $HCl + Ba(OH)_2 \rightarrow BaCl_2 + H_2O$

6.8 Balanced chemical equations are useful in telling us what amounts of products we can expect from a given amount of reactant, or how much reactant to use for a specific amount of product. The chemical equation tells us the ratio of the numbers of atoms and molecules involved in the reaction, but it also tells us the ratio of the numbers of moles of the substances involved.

The mole diagram in Figure 6.9 of your textbook will help you to determine the sequence of steps to use in solving the following types of problems.

Balanced equation: $4Na + O_2 \rightarrow 2Na_2O$

a. How many moles of Na_2O could be produced from 0.300 mole of Na?

b. How many moles of Na_2O could be produced from 2.18 g of sodium?

c. How many grams of Na_2O could be produced from 5.15 g of Na?

6.9 Balance the equation: $Al + Cl_2 \rightarrow AlCl_3$

a. How many moles of chlorine gas will react with 0.160 mole of aluminum?

b. How many grams of aluminum chloride could be produced from 5.27 moles of aluminum?

c What is the maximum number of grams of aluminum chloride that could be produced from 14.0 g of chlorine gas?

d. How many grams of chlorine gas would be needed to react with 0.746 g of aluminum?

Self-Test

True-false: Indicate whether the following statements are true or false. If the statement is false, give the word or phrase that may be substituted for the underlined portion to make the statement true.

1. The mass of 1 mole of helium atoms would be <u>the same as</u> the mass of 1 mole of gold atoms.
2. In a balanced equation, the total number of atoms on the reactants side is <u>equal to</u> the total number of atoms on the products side.
3. Formula masses are calculated on the $^{16}_{8}O$ <u>relative-mass</u> scale.
4. In balancing a chemical equation, do not change the <u>coefficients</u> within the formulas.
5. Atomic mass and formula mass are both expressed in <u>amu</u>.
6. In a chemical equation, the <u>reactants</u> are the materials to the right of the arrow.
7. The number of <u>atoms</u> in a mole of H_2O is equal to 6.02×10^{23}.
8. One mole of glucose ($C_6H_{12}O_6$) contains <u>6 moles</u> of carbon atoms.
9. In a chemical equation, the products are the materials that are <u>consumed</u>.
10. A mole of nitrogen gas contains 6.02×10^{23} nitrogen <u>atoms</u>.

Multiple choice:

11. How many atoms are contained in 6.8 moles of calcium?

a. 4.1×10^{24} b. 1.6×10^{26} c. 6.2×10^{22} d. 3.3×10^{23} e. none of these

12. What is the formula mass for iron(III) carbonate?

a. 115.86 g/mole b. 287.57 g/mole c. 291.73 g/mole
d. 171.71 g/mole e. none of these

13. How many grams are contained in 4.72 moles of $NaHCO_3$?

a. 84.1 b. 283 c. 264 d. 396 e. none of these

14. A mole of butane contains 4 moles of carbon atoms and 10 moles of hydrogen atoms. Its formula is:

a. C_6H_6 b. H_6C_{10} c. C_4H_{10} d. H_4C_6 e. none of these

15. What is the total number of moles of all atoms in 6.55 moles of $(NH_4)_2CO_3$?

a. 52.4 b. 91.7 c. 111 d. 157 e. none of these

16. The conversion factor used in changing grams of O_2 to moles of O_2 is:

a. 16.00 g/1 mole b. 1 mole/16.00 g c. 32.00 mole/1 g
d. 1 mole/32.00 g e. none of these

17. When oxygen gas and hydrogen gas combine to form water, which of the following is true? (Hint: Write the balanced equation.)

a. 2 moles of O_2 produce 1 mole of H_2O
b. 2 moles of H_2 react with 1 mole of H_2O
c. 1 mole of O_2 produces 1 mole of H_2O
d. 2 moles of H_2 produce 2 moles of H_2O
e. none of these

18. Using the equation you wrote in Question 17, find the mass of water in grams that would be produced by the complete reaction of 4.00 g of oxygen gas.

a. 4.50 g b. 36.0 g c. 7.32 g d. 14.7 g e. none of these

19. The mass of 0.560 mole of methanol, CH_4O, would equal:

a. 32.0 amu b. 32.0 g c. 17.9 amu d. 17.9 g e. none of these

20. 42.0 g of ethanol, C_2H_6O, would equal:

a. 1.10 moles b. 0.911 mole c. 1.10 amu
d. 0.911 amu e. none of these

21. After balancing the equation below, calculate the sum of all the coefficients in the balanced equation:

$$C_2H_6 + O_2 \rightarrow CO_2 + H_2O$$

a. 4 b. 9 c. 15 d. 19 e. none of these

Use the following balanced equation to answer Questions 22 through 24.

$$CH_4 + 2O_2 \rightarrow CO_2 + 2H_2O$$

22. How many moles of water would be produced from 0.420 mole of methane (CH_4)?

a. 2.00 moles b. 0.210 mole c. 0.420 mole
d. 0.840 mole e. none of these

23. How many grams of carbon dioxide could be produced by the reaction of 6.90 g of oxygen gas?

a. 4.74 g b. 6.92 g c. 9.55 g d. 19.0 g e. none of these

24. How many grams of methane would be used to produce 10.0 g of water?

a. 2.22 g b. 4.44 g c. 8.88 g d. 10.0 g e. none of these

Answers to Practice Exercises

6.1

Formula	Calculations with atomic masses	Formula mass
KBr	39.10 + 79.90	119.00 amu
$CaCl_2$	40.08 + 2(35.45)	110.98 amu
Na_2CO_3	2(22.99) + 12.01 + 3(16.00)	105.99 amu
$(NH_4)_3PO_4$	3[14.01 + 4(1.01)] + 30.97 + 4(16.00)	149.12 amu

6.2

Moles	Relationship with units	Number of atoms
1.00 mole of helium atoms	$1.00 \text{ mole x } \dfrac{6.02 \text{ x } 10^{23} \text{ atoms}}{1 \text{ mole}} =$	$6.02 \text{ x } 10^{23}$
2.60 moles of sodium atoms	$2.60 \text{ moles x } \dfrac{6.02 \text{ x } 10^{23} \text{ atoms}}{1 \text{ mole}} =$	$1.57 \text{ x } 10^{24}$
0.316 mole of argon atoms	$0.316 \text{ mole x } \dfrac{6.02 \text{ x } 10^{23} \text{ atoms}}{1 \text{ mole}} =$	$1.90 \text{ x } 10^{23}$

6.3

Given quantity	Relationship with units	Calculated
1.00 mole H_2O	$1.00 \text{ mole } H_2O \text{ x } \dfrac{18.02 \text{ g } H_2O}{1.00 \text{ mole } H_2O} =$	18.0 g H_2O
2.53 moles H_2O	$2.53 \text{ mole } H_2O \text{ x } \dfrac{18.02 \text{ g } H_2O}{1.00 \text{ mole } H_2O} =$	45.6 g H_2O
0.519 mole H_2O	$0.519 \text{ mole } H_2O \text{ x } \dfrac{18.02 \text{ g } H_2O}{1.00 \text{ mole } H_2O} =$	9.35 g H_2O
1.00 g NaBr	$1.00 \text{ g NaBr x } \dfrac{1.00 \text{ mole NaBr}}{102.89 \text{ g NaBr}} =$	0.00972 mole NaBr
417 g NaBr	$417 \text{ g NaBr x } \dfrac{1.00 \text{ mole NaBr}}{102.89 \text{ g NaBr}}$	4.05 mole NaBr
0.322 g NaBr	$0.322 \text{ g NaBr x } \dfrac{1.00 \text{ mole NaBr}}{102.89 \text{ g NaBr}} =$	0.00313 mole NaBr

6.4

Moles of compound	Carbon atoms/molecule	Moles of carbon atoms
1.00 mole of $C_{10}H_{16}$ (limonene)	10	10(1.00) = 10.0 moles
13.5 moles of C_2H_6O (ethanol)	2	2(13.5) = 27.0 moles
0.705 mole of C_5H_{12} (pentane)	5	5(0.705) = 3.53 moles

6.5 a. $0.125 \text{ g KCl} \times \frac{1.00 \text{ mole KCl}}{74.55 \text{ g KCl}} \times \frac{6.02 \times 10^{23} \text{ KCl molecules}}{1.00 \text{ mole KCl}} = 1.01 \times 10^{21} \text{ KCl molecules}$

b. $1.77 \text{ g AlF}_3 \times \frac{1 \text{ mole AlF}_3}{83.98 \text{ g AlF}_3} \times \frac{6.02 \times 10^{23} \text{ molecules AlF}_3}{1 \text{ mole AlF}_3} \times \frac{3 \text{ atoms F}}{1 \text{ molecule AlF}_3}$

$= 3.81 \times 10^{22} \text{ atoms F}$

c. $12.5 \text{ g KCl} \times \frac{1 \text{ mole KCl}}{74.55 \text{ g KCl}} \times \frac{1 \text{ mole Cl}}{1 \text{ mole KCl}} \times \frac{35.45 \text{ g Cl}}{1 \text{ mole Cl}} = 5.94 \text{ g Cl}$

d. $1.77 \text{ g AlF}_3 \times \frac{1 \text{ mole AlF}_3}{83.98 \text{ g AlF}_3} \times \frac{3 \text{ moles F}}{1 \text{ mole AlF}_3} \times \frac{19.00 \text{ g F}}{1 \text{ mole F}} = 1.20 \text{ g F}$

6.6 a. $HCl + NaOH \rightarrow NaCl + H_2O$

b. $AgNO_3 + KBr \rightarrow AgBr + KNO_3$

6.7 a. $H_2 + Cl_2 \rightarrow 2HCl$

b. $2AgNO_3 + H_2S \rightarrow Ag_2S + 2HNO_3$

c. $4P + 3O_2 \rightarrow 2P_2O_3$

d. $2HCl + Ba(OH)_2 \rightarrow BaCl_2 + 2H_2O$

6.8 Balanced equation: $4Na + O_2 \rightarrow 2Na_2O$

a. $0.300 \text{ mole Na} \times \frac{2 \text{ moles Na}_2\text{O}}{4 \text{ moles Na}} = 0.150 \text{ mole Na}_2\text{O}$

b. $2.18 \text{ g Na} \times \frac{1 \text{ mole Na}}{22.99 \text{ g Na}} \times \frac{2 \text{ moles Na}_2\text{O}}{4 \text{ moles Na}} = 0.0474 \text{ mole Na}_2\text{O}$

c. $5.15 \text{ g Na} \times \frac{1 \text{ mole Na}}{22.99 \text{ g Na}} \times \frac{2 \text{ moles Na}_2\text{O}}{4 \text{ moles Na}} \times \frac{61.98 \text{ g Na}_2\text{O}}{1 \text{ mole Na}_2\text{O}} = 6.94 \text{ g Na}_2\text{O}$

6.9 $2Al + 3Cl_2 \rightarrow 2AlCl_3$ (molar mass of $AlCl_3$: 133.33 g/mole)

a. $0.160 \text{ mole Al} \times \frac{3 \text{ moles Cl}_2}{2 \text{ moles Al}} = 0.240 \text{ mole Cl}_2$

b. $5.27 \text{ moles Al} \times \frac{2 \text{ moles AlCl}_3}{2 \text{ moles Al}} \times \frac{133.33 \text{ g AlCl}_3}{1 \text{ mole AlCl}_3} = 703 \text{ g AlCl}_3$

c. $14.0 \text{ g Cl}_2 \times \frac{1 \text{ mole Cl}_2}{70.90 \text{ g}} \times \frac{2 \text{ moles AlCl}_3}{3 \text{ moles Cl}_2} \times \frac{133.33 \text{ g AlCl}_3}{1 \text{ mole AlCl}_3} = 17.6 \text{ g AlCl}_3$

d. $0.746 \text{ g Al} \times \frac{1 \text{ mole Al}}{26.98 \text{ g Al}} \times \frac{3 \text{ moles Cl}_2}{2 \text{ moles Al}} \times \frac{70.90 \text{ g Cl}_2}{1 \text{ mole Cl}_2} = 2.94 \text{ g Cl}_2$

Answer to Self-Test

1. F; different from (6.3) **2.** T (6.6) **3.** F; $^{12}_{6}C$ relative mass (6.1) **4.** F; subscripts (6.6) **5.** T (6.1) **6.** F; products (6.6) **7.** F; molecules (6.4) **8.** T (6.4) **9.** F; produced (6.6) **10.** F; molecules (6.4) **11.** a (6.4) **12.** c (6.1) **13.** d (6.4) **14.** c (6.4) **15.** b (6.4) **16.** d (6.5) **17.** d (6.6) **18.** a (6.8) **19.** d (6.3) **20.** b (6.3) **21.** d (6.6) **22.** d (6.8) **23.** a (6.8) **24.** b (6.8)

Gases, Liquids, and Solids — Chapter 7

Chapter Overview

The physical states of matter and the behavior of matter in these states are determined by the behavior of the particles (atoms, molecules, ions) of which matter is made. The movements and interactions of these particles are described by the kinetic molecular theory of matter.

In this chapter you will study the five statements of the **kinetic molecular theory** (Sec. 7.1) and the ways in which these statements explain the physical behavior of matter. You will use the gas laws to describe quantitatively various changes in the conditions of pressure, temperature, and volume of matter in the gaseous state. You will study three types of **intermolecular forces** (Sec. 7.13) that affect liquids and solids and their changes of state.

Practice Exercises

7.1 According to the **kinetic molecular theory of matter** (Sec. 7.1), the differing physical properties of the **solid, liquid, and gaseous states** of matter (Sec. 7.2) are determined by the **potential energy** (cohesive forces) and the **kinetic energy** (disruptive forces) of that state.

In the table below, indicate which form of energy is dominant in a given state and write in a word or two to describe how this dominance affects the physical properties listed:

State	Cohesive forces	Disruptive forces	Density	Compressibility	Thermal expansion
Gas					
Liquid					
Solid					

7.2 Gases can be described by quantitative relationships called **gas laws** (Sec. 7.3). According to **Boyle's law** (Sec. 7.4), the volume of a fixed amount of gas is inversely proportional to the **pressure** (Sec. 7.3) of the gas if the temperature is constant: $P_1 \times V_1 = P_2 \times V_2$

Complete the following problems using Boyle's law. Rearrange the equation to solve for the needed variable.

a. The pressure on 2.45 L of helium is changed from 2340 mm Hg to 3580 mm Hg at 50.5°C. What is the new volume?

b. The pressure on 12.5 L of nitrogen gas is doubled from 1.00 atm to 2.00 atm, and the temperature is held constant. What is the new volume of the nitrogen gas?

c. The volume of 8.24 L of gas at 3630 mm Hg is increased to 16.4 L. If the temperature is held constant, what is the new pressure?

7.3 According to **Charles's law** (Sec. 7.5), the volume of a fixed amount of gas at constant pressure is proportional to the Kelvin temperature of the gas:

$$V_1/T_1 = V_2/T_2$$

Complete the following problems using Charles's law. Rearrange the equation to solve for the needed variable.

a. The temperature of 4.71 L of gas is reduced from 278°C to 122°C. If the pressure remains constant, what is the new volume of the gas?

b. The volume of 14.5 L of a gas at 345 K is increased to 20.5 L, and pressure is held constant. What is the new temperature of the gas?

7.4 The gas laws can be combined into a single equation called the **combined gas law** (Sec. 7.6):

$$\frac{P_1 \times V_1}{T_1} = \frac{P_2 \times V_2}{T_2}$$

Rearrange this equation to solve for the correct variable in completing the following combined gas law problems:

a. The volume of a fixed amount of gas is 5.72 L at 30°C and 1.25 atm. If the gas is heated to 50°C and compressed to a volume of 4.50 L, what will be the new pressure?

b. A fixed amount of gas at 514 K and 338 mm Hg is heated to 311°C and 507 mm Hg. What will be the final volume of the gas, if the initial volume is 14.2 L?

7.5 Combining the three gas laws gives an equation that describes the state of a gas at a single set of conditions. This equation, $PV = nRT$, is called the **ideal gas equation** (Sec. 7.7). T is measured on the Kelvin scale, and the value of R (the ideal gas constant) is 0.0821 atm·L/mole·K.

Rearrange the ideal gas equation to solve for the correct variable in completing the following problems.

a. What is the volume of 1.49 moles of helium with a pressure of 1.21 atm at 224°C?

b. What is the temperature of neon gas, when 0.339 mole of neon gas is in a 5.72-liter tank and the pressure gauge reads 2.53 atm?

7.6 **Dalton's law of partial pressures** (Sec. 7.8) states that the total pressure exerted by a mixture of gases is the sum of the **partial pressures** (Sec. 7.8) of the individual gases:

$$P_T = P_1 + P_2 + P_3 + \cdots$$

Using Dalton's law of partial pressures, complete the following problems:

a. What is the total pressure exerted by a mixture of helium and argon? The partial pressures of helium and argon are $P_{He} = 270$ mm Hg and $P_{Ar} = 400$ mm Hg.

b. What is the partial pressure of oxygen gas in a mixture of O_2, CO_2 ($P_{CO_2} = 341$ mm Hg), and CO ($P_{CO} = 114$ mm Hg) if the total pressure of the mixture is 744 mm Hg?

7.7 Use the following table summarizing the gas laws to review your knowledge of this chapter:

Law	Quantities held constant	Variables	Equation
Boyle's law		Pressure Volume	$P_1V_1 = P_2V_2$
Charles		Volume Temperature	$\frac{V_1}{T_1} = \frac{V_2}{T_2}$
Combined gas law		Pressure volume Temperature	$\frac{P_1V_1}{T_1} = \frac{P_2V_2}{T_2}$
Ideal gas law	$R = 0.0821$ L·atm/mole·K		$PV = nRT$
Dalton's law of partial pressures			

7.8 A **change of state** (Sec. 7.9) is a process in which matter changes from one state to another. If heat is absorbed, the change is **endothermic** (Sec. 7.9); if heat is released during the process, the change is **exothermic** (Sec. 7.9).

a. Complete the following table with the correct term for the physical change involved.

	To solid	To liquid	To gas
From solid	XXXXXXX		
From liquid		XXXXXXX	
From gas			XXXXXXX

b. In the table above, indicate for each physical change whether it is endothermic (A) or exothermic (B).

7.9 **Hydrogen bonds** (Sec. 7.13) are strong **dipole-dipole interactions** (Sec. 7.13) that occur when hydrogen is bonded to fluorine, oxygen, or nitrogen. The hydrogen atom in this case is almost a "bare" nucleus, and it is very strongly attracted to a pair of electrons on an electronegative atom of another molecule.

Complete the following table by indicating for each substance whether hydrogen bonding can occur between individual molecules or with a water molecule.

Molecule	Hydrogen bonding between individual molecules	Hydrogen bonding with water molecules
HF		
HI		
NH_3		
CH_4		
CO		
CH_3CH_2OH		
$CH_3–O–CH_3$		

Self-Test

True-false: Indicate whether the following statements are true or false. If the statement is false, give the word or phrase that may be substituted for the underlined portion to make the statement true.

1. Boyle's law states that for a given mass of gas at constant temperature, the volume of the gas varies directly with pressure.
2. Charles's law states that for a given mass of gas at constant pressure, the volume of the gas varies directly with temperature.
3. Gases cool when they are compressed.
4. Assuming that the temperature and number of moles of gas remain constant, doubling the volume of a gas will double the pressure of the gas.

5. <u>One atmosphere</u> is the pressure required to support 760 mm of Hg.
6. The total pressure exerted by a mixture of gases is <u>equal to</u> the sum of the partial pressures.
7. The vapor pressure of a liquid <u>decreases</u> as temperature increases.
8. As temperature increases, the velocity of molecules in a liquid <u>decreases</u>.
9. The energy resulting from the attractions and repulsions between particles in matter is a part of that matter's <u>potential energy</u>.
10. <u>Liquids</u> are very compressible because there is a lot of empty space between particles.
11. Foods cook faster in a pressure cooker because the boiling point of water is <u>lower</u> than it is at normal atmospheric pressure.
12. A volatile liquid is one that has a <u>high</u> vapor pressure.
13. A state of equilibrium may exist between a liquid and a gas in a <u>closed</u> container.

Multiple choice:

14. A fixed amount of oxygen gas with a volume of 5.00 L at 1 atm and 273 K was heated to 402 K and the pressure was doubled. What was the new volume of the oxygen gas?

    a. 3.68 L b. 5.00 L c. 1.71 L d. 14.7 L e. none of these

15. Two gases, nitrogen and oxygen, in the same container, have a total pressure of 600 mm Hg. If the partial pressure of oxygen equals the partial pressure of nitrogen, what is the partial pressure of oxygen?

    a. 600 mm Hg b. 400 mm Hg c. 300 mm Hg
    d. 200 mm Hg e. none of these

16. How many moles of helium are in a 28.4 L balloon at 45°C and 1.03 atm?

    a. 0.271 mole b. 0.0453 mole c. 6.98 moles
    d. 1.12 moles e. none of these

17. The strongest intermolecular forces between water molecules are:

    a. ionic bonds b. covalent bonds c. hydrogen bonds
    d. London forces e. none of these

18. Hydrogen bonding would *not* occur between two molecules of which of these compounds?

    a. HF b. CH_4 c. CH_3NH_2
    d. H_2O e. CH_3OH

19. London forces would be the strongest attractive forces between two molecules of which of these substances?

    a. HF b. BrCl c. F_2
    d. H_2O e. none of these

20. Compared to liquids that have no hydrogen bonding,, liquids that have significant hydrogen bonding, have a:

    a. higher vapor pressure b. higher boiling point
    c. lower condensation temperature d. greater tendency to evaporate
    e. none of these

21. The pressure on 526 mL of gas is increased from 755 mm Hg to 974 mm Hg. If the temperature remains constant, what is the new volume?

a. 408 mL b. 633 mL c. 215 mL
d. 387 mL e. none of these

22. Which of the following changes is endothermic?

a. condensation b. freezing c. sublimation
d. deposition e. none of these

23. What will increase the pressure of a gas in a closed container?

a. decreasing the temperature of the gas
b. adding more gas to the container
c. increasing the volume of the container
d. replacing the gas with the same number of moles of a different gas
e. both b and c

Answers to Practice Exercises

7.1

State	Cohesive forces	Disruptive forces	Density	Compressibility	Thermal expansion
Gas		X	low	large	moderate
Liquid	X	X	high	small	small
Solid	X		high	small	very small

7.2 a. $V_2 = \frac{P_1 \times V_1}{P_2} = \frac{2340 \text{ mm Hg} \times 2.45 \text{ L}}{3580 \text{ mm Hg}} = 1.60 \text{ L}$

b. $V_2 = \frac{P_1 \times V_1}{P_2} = \frac{1.00 \text{ atm} \times 12.5 \text{ L}}{2.00 \text{ atm}} = 6.25 \text{ L}$

c. $P_2 = \frac{P_1 \times V_1}{V_2} = \frac{3630 \text{ mm Hg} \times 8.24 \text{ L}}{16.4 \text{ L}} = 1820 \text{ mm Hg}$

7.3 a. $V_2 = \frac{V_1 \times T_2}{T_1} = \frac{4.71 \text{ L} \times 395 \text{ K}}{551 \text{ K}} = 3.38 \text{ L}$

b. $T_2 = \frac{V_2 \times T_1}{V_1} = \frac{20.5 \text{ L} \times 345 \text{ K}}{14.5 \text{ L}} = 488 \text{ K}$

7.4 a. $P_2 = \frac{P_1 \times V_1 \times T_2}{T_1 \times V_2} = \frac{1.25 \text{ atm} \times 5.72 \text{ L} \times 323 \text{ K}}{4.50 \text{ L} \times 303 \text{ K}} = 1.69 \text{ atm}$

b. $V_2 = \frac{P_1 \times V_1 \times T_2}{P_2 \times T_1} = \frac{338 \text{ mm Hg} \times 14.2 \text{ L} \times 584 \text{ K}}{507 \text{ mm Hg} \times 514 \text{ K}} = 10.8 \text{ L}$

7.5 a. $V = \frac{n \times R \times T}{P} = \frac{1.49 \text{ moles} \times 0.0821 \text{ atm L/mole K} \times 497 \text{ K}}{1.21 \text{ atm}} = 50.2 \text{ L}$

b. $T = \frac{P \times V}{n \times R} = \frac{2.53 \text{ atm} \times 5.72 \text{ L}}{0.339 \text{ mole} \times 0.0821 \text{ atm L / mole K}} = 5.20 \times 10^2 \text{ K}$

7.6 a. $P_{total} = P_{He} + P_{Ar} = 270 \text{ mm Hg} + 400 \text{ mm Hg} = 670 \text{ mm Hg}$

b. $P_{total} = P_{O_2} + P_{CO_2} + P_{CO}$

$P_{O_2} = P_{total} - P_{CO_2} - P_{CO} = 744 \text{ mm Hg} - 114 \text{ mm Hg} - 341 \text{ mm Hg} = 289 \text{ mm Hg}$

7.7

Law	Quantities held constant	Variables	Equation
Boyle's law	temperature, number of moles of gas	pressure, volume	$P_1V_1 = P_2V_2$
Charles's law	pressure, number of moles of gas	volume, temperature	$\frac{V_1}{T_1} = \frac{V_2}{T_2}$
Combined gas law	number of moles of gas	pressure, volume, temperature	$\frac{P_1 \times V_1}{T_1} = \frac{P_2 \times V_2}{T_2}$
Ideal gas law	$R = 0.0821$ L·atm/mole·K	pressure, volume, temperature, number of moles of gas	$PV = nRT$
Dalton's law of partial pressures	volume, temperature, number of moles of gas	partial pressures, total pressure	$P_T = P_1 + P_2 + P_3 + \cdots$

7.8

	To solid	To liquid	To gas
From solid	XXXXX	melting, A	sublimation, A
From liquid	freezing, B	XXXXX	evaporation, A
From gas	deposition, B	condensation, B	XXXXX

7.9

Molecule	Hydrogen bonding between individual molecules	Hydrogen bonding with water molecules
HF	yes	yes
HI	no	no
NH_3	yes	yes
CH_4	no	no
CO	no	yes
CH_3CH_2OH	yes	yes
CH_3–O–CH_3	no	yes

Answers to Self-Test

The numbers in parentheses refer to sections in your textbook.
1. F; varies inversely (7.4) **2**. T (7.5) **3**. F; become hotter (7.5) **4**. F; halve (7.4)
5. T (7.3) **6**. T (7.8) **7**. F; increases (7.11) **8**. F; increases (7.1) **9**. T (7.1)
10. F; gases (7.2) **11**. F; higher (7.12) **12**. T (7.11) **13**. T (7.11) **14**. a (7.6) **15**. c (7.8)
16. d (7.7) **17**. c (7.13) **18**. b (7.13) **19**. c (7.13) **20**. b (7.13) **21**. a (7.4) **22**. c (7.9)
23. b (7.8)

Solutions — Chapter 8

Chapter Overview

Many chemical reactions take place in **solutions** (Sec. 8.1), particularly in water solutions. The properties of water make it a vital part of all living systems.

In this chapter you will define terms associated with solutions, study how solutions form, and calculate the concentrations of solutions using various units. You will study osmotic pressure and the factors that control the important process of osmosis.

Practice Exercises

8.1 A **solution** (Sec. 8.1) is a homogeneous mixture consisting of a **solvent** (Sec. 8.1) and one or more **solutes** (Sec. 8.1). The solvent is the substance present in the greatest amount.

In the table below, identify the solute and the solvent in each of the solutions:

Solution	Solute	Solvent
10.0 g of potassium chloride in 70.0 g of water		
80.0 g of ethyl alcohol in 50.0 g of water		
40.0 g of potassium iodide in 55.0 g of water		
30.0 mL of ethyl alcohol in 40.0 mL of methyl alcohol		

8.2 The **solubility** (Sec. 8.2) of a substance depends on many factors: the temperature of the solution, the pressure on the solution, the nature of the solvent. The rate at which a substance dissolves to form a solution depends on how fast the particles come in contact with the solvent.

In the table below, indicate whether each of the changes in conditions would affect the solubility or the rate of solution of a given solid substance.

Change of conditions	Solubility in water	Rate of solution in water
raising the temperature		
crushing or grinding the solid		
adding more of the solid		
agitating the solid/solvent mixture		

8.3 The solubility of a substance can be predicted to some extent by the generalization that substances of like polarity tend to be more soluble in each other than substances that differ in polarity: "like dissolves like." However, the solubility of ionic compounds is more complex. Table 8.2 in your textbook gives water solubilities of ionic compounds.

Indicate the solubility of each of the substances below in the two **solvents** (Sec. 8.1) water and benzene.

Substance	Water (polar)	Benzene (nonpolar)
$CaCO_3$ (ionic solid)		
$NaNO_3$ (ionic solid)		
K_2SO_4 (ionic solid)		
petroleum jelly (nonpolar solid)		
butane (nonpolar liquid)		
acetone (polar liquid)		

8.4 The **concentration** (Sec. 8.5) of a solution is the amount of **solute** (Sec. 8.1) present in a specified amount of solution. One way of expressing concentration is **percent by mass** (Sec. 8.5), the mass of solute divided by the mass of solution multiplied by 100.

$$\%(m/m) = \frac{\text{mass of solute}}{\text{mass of solution}} \times 100$$

a. What is the percent by mass, %(m/m), concentration of NaCl in a solution prepared by dissolving 14.8 g of NaCl in 122 g of water?

b. How many grams of NaCl were added to 225 g of water to prepare a 7.52%(m/m) NaCl solution?

8.5 **Percent by volume** (Sec. 8.5) is a percentage unit used when the solute and the solvent are both liquids or both gases.

$$\text{Percent by volume} = \%(v/v) = \frac{\text{volume of solute}}{\text{volume of solution}} \times 100$$

a. Calculate the percent by volume for the following solution:
25.0 mL of ethyl alcohol is added to enough water to make 155 mL of solution.

$$\frac{25.0}{155} \times 100 = 3875$$

b. If 165 mL of ethyl alcohol is added to enough water to give a volume of 425 mL, what is the percent by volume of the resulting solution?

$$\frac{165\text{mL}}{425\text{mL}} \times 100 = 38.82$$

8.6 Another commonly used concentration unit is **mass-volume percent** (Sec. 8.5), the mass of solute divided by the volume of solution:

$$\text{Mass-volume percent} = \%(\text{m/v}) = \frac{\text{mass of solute (g)}}{\text{volume of solution (mL)}} \times 100$$

a. How many grams of KCl must be added to 250.0 mL of water to prepare a 9.82%(m/v) solution?

b. Calculate the mass-volume percent of the following solution:
10.5 g of sugar added to enough water to make a solution having a volume of 164 mL.

8.7 The **molarity** (Sec. 8.5) of a solution is a ratio giving the number of moles of solute per liter of solution:

$$\text{Molarity (M)} = \frac{\text{moles of solute}}{\text{liters of solution}}$$

a. What is the molarity of a solution that contains 1.44 moles of NaCl in 2.50 L of solution?

b. A solution with a volume of 425 mL is prepared by dissolving 2.64 moles of $CaCl_2$ in water. What is the molarity?

c. If 7.21 g of KCl is dissolved in enough water to prepare 0.333 L of solution, what is the molarity?

8.8 The molarity of a solution can be used as a conversion factor to relate liters of solution to moles of solute. Use dimensional analysis to solve for the correct variable in the following problems.

a. How many moles of $CaCl_2$ were dissolved in 3.55 L of a 1.47 M $CaCl_2$ solution?

b. How many grams of $CaCl_2$ were used to prepare 0.250 L of a 0.143 M $CaCl_2$ solution?

c. How many liters of solution would be needed to produce 21.5 g of $CaCl_2$ from a 0.842 M $CaCl_2$ solution?

8.9 **Dilution** (Sec. 8.5) is the process in which more solvent is added to a solution in order to lower its concentration. The simple relationship used for dilution is as follows: the concentration of the stock solution times the volume of the stock solution is equal to the concentration of the **diluted solution** (Sec. 8.2) times the volume of the diluted solution.

$$C_s \times V_s = C_d \times V_d$$

a. If 255 mL of water is added to 325 mL of a 0.477 M solution, what is the molarity of the new solution?

b. How many milliliters of water would have to be added to 250.0 mL of a 2.33 M solution to prepare a 0.551 M solution?

8.10 **Colligative properties** (Sec. 8.7) of solutions are those physical properties affected by the concentration of a solute. In the following table, tell whether a given change in a solution's condition will cause the value of each colligative property to increase or decrease.

Change in conditions	Boiling point	Freezing point	Vapor pressure
adding NaCl to an aqueous solution			
adding water to an aqueous sugar solution			
putting antifreeze in the radiator of a car			

8.11 **Osmosis** (Sec. 8.8) is the movement of water across a **semipermeable membrane** (Sec. 8.8) from a more dilute solution to a more **concentrated solution** (Sec. 8.2). **Osmotic pressure** (Sec. 8.8), the amount of pressure necessary to stop the net flow of water, depends on the number of particles of solute in the solutions. **Osmolarity** (Sec. 8.8) is the product of the molarity of the solution and the number of particles produced when the solute dissociates:

$$\text{Osmolarity} = \text{molarity} \times i$$

Complete the following table on osmolarity.

Molarity of solutions	Osmolarity
3 M KCl	
2 M KCl + 1 M glucose	
3 M $CaBr_2$ + 2 M glucose	
2 M $CaBr_2$ + 1 M KBr	

8.12 Water flows across the semipermeable membrane of a cell from a solution of higher solute concentration to one of lower solute concentration. A solution outside the cell is classified with reference to the solution within the cell: a **hypotonic** solution has a lower concentration than the concentration of the solution within the cell, a **hypertonic** solution has a higher concentration of solute, and an **isotonic** solution has the same solute concentration (Sec. 8.8).

Indicate which way water will flow across the cell membranes of red blood cells under each of these conditions:

Condition	Water flows into cells	Water flows out of cells
cells immersed in concentrated NaCl solution		
solution around cells is hypotonic		
cells immersed in an isotonic solution		
hypertonic solution surrounds cells		
cells immersed in pure water		
cells immersed in physiological saline solution 0.9%(m/v)		

Self-Test

True-false: Indicate whether the following statements are true or false. If the statement is false, give the word or phrase that may be substituted for the underlined portion to make the statement true.

1. A saturated solution contains the maximum amount of <u>solute</u> that will dissolve in the solution.
2. Colligative properties are properties that depend on the <u>amount of solute</u> dissolved in a given mass of solution.
3. Osmotic semipermeable membranes permit only <u>dissolved salt</u> to flow through.
4. Hypertonic solutions contain a <u>smaller</u> number of solute molecules than the intracellular fluid.
5. In a salt-water solution, the <u>solute</u> is water.
6. If the solubility of a substance is 120 g/100 mL of water, a solution containing 65 g of the substance dissolved in 50 mL of water would be <u>unsaturated</u>.
7. Undissolved solute is in equilibrium with dissolved solute in a <u>saturated solution</u>.
8. An unsaturated solution is <u>always</u> a dilute solution.
9. Carbon dioxide is <u>more</u> soluble in water when pressure increases.
10. A polar gas, such as NO_2, is <u>insoluble</u> in water.
11. Percent by mass (%m/m) is mass of solute divided by mass of <u>solvent</u>, x 100.
12. Red blood cells in a hypotonic solution may undergo <u>hemolysis</u>.
13. <u>The same number of moles</u> of NaCl are in 225 mL of a 1.55 M NaCl solution as are in 450 mL of a 1.55 M NaCl solution.

14. Dialysis can be used to remove dissolved ions from solutions containing large molecules.

Multiple choice:

15. Which of the following compounds would *not* dissolve in water?

a. NaCl b. CCl_4 c. $CaCl_2$
d. HCl e. all would dissolve

16. A solution containing 10.0 g of NaCl in 0.500 L of solution would have what molarity?

a. 0.342 M b. 0.200 M c. 0.500 M
d. 0.174 M e. none of these

17. How many milliliters of 4.57 M potassium bromide solution would be needed to prepare 1.00 L of 2.08 M potassium bromide solution?

a. 155 mL b. 325 mL c. 455 mL
d. 695 mL e. none of these

18. How much solute is present in 215 mL of a 0.500 M solution of HCl in water?

a. 1.12 moles b. 0.0566 mole c. 0.752 mole
d. 0.108 mole e. none of these

19. If 53.0 mL of a 3.00 M NaCl solution is diluted to give a solution whose molarity is 0.150 M, what is the volume of the new solution?

a. 0.520 L b. 1.06 L c. 835 mL
d. 626 mL e. none of these

20. The osmolarity of a solution that is 2 M $CaCl_2$ and 2 M glucose is:

a. 4 M b. 6 M c. 8 M
d. 10 M e. none of these

21. Which of the following solutions would be isotonic with 0.1 M NaCl?

a. 0.5 M $CaCl_2$ b. 0.2 M glucose c. 0.1 M sucrose
d. 0.05 M $Ca(NO_3)_2$ e. none of these

22. Water flows out of red blood cells placed in which of the following solutions?

a. hypotonic b. isotonic c. hypertonic
d. both a and c e. none of these

23. What mass-volume percent %(m/v) would result from dissolving 5.00 g of NaCl in enough water to form 50.0 mL of saline solution?

a. 5.00%(m/v) b. 9.09%(m/v) c. 10.0%(m/v)
d. 20.0%(m/v) e. none of these

24. Dissolving 7.5 g of NaCl in 50.3 g of water would yield a solution that is what percent by mass, %(m/m)?

a. 7.50%(m/m) b. 14.9%(m/m) c. 13.0%(m/m)
d. 74.6%(m/m) e. none of these

Answers to Practice Exercises

8.1

Solution	Solute	Solvent
10.0 g of potassium chloride in 70.0 g of water	10.0 g potassium chloride	70.0 g of water
80.0 g of ethyl alcohol in 50.0 g of water	50.0 g of water	80.0 g of ethyl alcohol
40.0 g of potassium iodide in 55.0 g of water	40.0 g of potassium iodide	55.0 g of water
30.0 mL of ethyl alcohol in 40.0 mL of methyl alcohol	30.0 mL of ethyl alcohol	40.0 mL of methyl alcohol

8.2

Change of conditions	Solubility in water	Rate of solution in water
raising the temperature	depends on solid (may increase or decrease)	increases
crushing or grinding the solid	no effect	increases
adding more of the solid	no effect	increases
agitating the solid/solvent mixture	no effect	increases

8.3

Substance	Water (polar)	Benzene (nonpolar)
$CaCO_3$ (ionic solid)	insoluble	insoluble
$NaNO_3$ (ionic solid)	soluble	insoluble
K_2SO_4 (ionic solid)	soluble	insoluble
petroleum jelly (nonpolar solid)	insoluble	soluble
butane (nonpolar liquid)	insoluble	soluble
acetone (polar liquid)	soluble	soluble

8.4 a. $$\%(m/m) = \frac{\text{mass of solute}}{\text{mass of solution}} \times 100 = \frac{14.8 \text{ g NaCl}}{14.8 \text{ g NaCl} + 122 \text{ g } H_2O} \times 100 = 10.8\%(m/m)$$

b. 100 g solution – 7.52 g NaCl = 92.48 g H_2O

$$225 \text{ g } H_2O \times \frac{7.52 \text{ g NaCl}}{92.48 \text{ g } H_2O} = 18.3 \text{ g NaCl}$$

8.5 a. $$\%(v/v) = \frac{\text{volume of solute}}{\text{volume of solution}} \times 100 = \frac{25.0 \text{ mL of solute}}{155 \text{ mL of solution}} \times 100 = 16.1\%$$

b. $$\%(v/v) = \frac{\text{volume of solute}}{\text{volume of solution}} \times 100 = \frac{165 \text{ mL of solute}}{425 \text{ mL of solution}} \times 100 = 38.8\%$$

8.6 a. $\%(m/v) = 9.82\% = \dfrac{9.82 \text{ g KCl}}{100 \text{ mL solution}} \times 100$

mass of KCl = %(m/v) x volume of solution

$$\text{mass of KCl} = \frac{9.82 \text{ g KCl}}{100 \text{ mL solution}} \times 250 \text{ mL of solution} = 24.6 \text{ g KCl}$$

b. $\%(m/v) = \dfrac{\text{mass of solute (g)}}{\text{volume of solution (mL)}} \times 100 = \dfrac{10.5 \text{ g solute}}{164 \text{ mL solution}} \times 100 = 6.40\%$

8.7 a. $\text{Molarity (M)} = \dfrac{\text{moles of solute}}{\text{liters of solution}} = \dfrac{1.44 \text{ moles NaCl}}{2.50 \text{ L}} = 0.576 \text{ M}$

b. $425 \text{ mL} \times \dfrac{1 \text{ L}}{1000 \text{ mL}} = 0.425 \text{ L}$

$$\text{Molarity (M)} = \frac{\text{moles of solute}}{\text{liters of solution}} = \frac{2.64 \text{ moles } CaCl_2}{0.425 \text{ L}} = 6.21 \text{ M}$$

c. $7.21 \text{ g} \times \dfrac{1.00 \text{ mole}}{74.6 \text{ g}} = 9.66 \times 10^{-2} \text{ moles}$

$$\text{M} = \frac{\text{moles of solute}}{\text{liters of solution}} = \frac{9.66 \times 10^{-2} \text{ moles}}{0.333 \text{ L}} = 0.290 \text{ M}$$

8.8 a. $\dfrac{1.47 \text{ moles}}{1 \text{ L}} \times 3.55 \text{ L} = 5.22 \text{ moles } CaCl_2$

b. $\dfrac{0.143 \text{ mole}}{1 \text{ L}} \times \dfrac{111 \text{ g}}{1 \text{ mole}} \times 0.250 \text{ L} = 3.97 \text{ g } CaCl_2$

c. $\text{liters of solution} = \text{moles} \times \dfrac{1}{\text{M}} = \text{moles} \times \dfrac{\text{liters of solution}}{\text{moles}}$

$$\text{liters} = 21.5 \text{ g } CaCl_2 \times \frac{1 \text{ mole}}{111 \text{ g } CaCl_2} \times \frac{1 \text{ L}}{0.842 \text{ mole}} = 0.230 \text{ L}$$

8.9 a. $(C_s \times V_s = C_d \times V_d)$

$$C_d = \frac{C_s \times V_s}{V_d} = \frac{0.477 \text{ M} \times 325 \text{ mL}}{325 \text{ mL} + 255 \text{ mL}} = 0.267 \text{ M}$$

b. $V_d = \dfrac{C_s \times V_s}{C_d} = \dfrac{2.33 \text{ M} \times 250.0 \text{ mL}}{0.551 \text{ M}} = 1057 \text{ mL}$

$$V_d - V_s = 1057 \text{ mL} - 250.0 \text{ mL} = 807 \text{ mL water added}$$

8.10

Change in conditions	Boiling point	Freezing point	Vapor pressure
adding NaCl to an aqueous solution	increases	decreases	decreases
adding water to an aqueous sugar solution	decreases	increases	increases
putting antifreeze in the radiator of a car	increases	decreases	decreases

8.11

Molarity of solutions	Osmolarity
3 M KCl	6
2 M KCl + 1 M glucose	5
3 M $CaBr_2$ + 2 M glucose	11
2 M $CaBr_2$ + 1 M KBr	8

8.12

Condition	Water flows into cells	Water flows out of cells
cells immersed in concentrated NaCl solution		X
solution around cells is hypotonic	X	
cells immersed in an isotonic solution	no flow	in or out
hypertonic solution surrounds cells		X
cells immersed in pure water	X	
cells immersed in physiological saline solution (0.9 % m/v)	no flow	in or out

Answers to Self-Test

The numbers in parentheses refer to sections in your textbook.
1. T (8.2) **2**. F; number of particles (8.7) **3**. F; ions and small molecules (8.8)
4. F; larger (8.8) **5**. F; solvent (8.1) **6**. F; supersaturated (8.2) **7**. T (8.2)
8. F; sometimes (8.2) **9**. T (8.2) **10**. F; soluble (8.4) **11**. F; solution (8.5) **12**. T (8.8)
13. F; fewer moles (8.5) **14**. T (8.9) **15**. b (8.4) **16**. a (8.5) **17**. c (8.5) **18**. d (8.5)
19. b (8.5) **20**. c (8.8) **21**. b (8.8) **22**. c (8.8) **23**. c (8.5) **24**. c (8.5)

Chemical Reactions — Chapter 9

Chapter Overview

Chemical reactions are the means by which new substances are formed. The concepts of collision theory explain how and under what conditions reactions take place.

In this chapter you will learn to recognize five basic types of chemical reactions. You will identify oxidizing agents and reducing agents in redox reactions. You will study factors that affect the rate of a chemical reaction. Not all chemical reactions go to completion; you will learn to calculate the concentrations of reactants and products in an equilibrium state.

Practice Exercises

9.1 In a **chemical reaction** (Sec. 9.1) at least one new substance is produced as the result of chemical change. Most chemical reactions can be classified in five categories. Classify the following reactions as **combination, decomposition, single-replacement, double-replacement,** or **combustion reactions** (Sec. 9.1):

Reaction	Classification
a. $2NaNO_3 \rightarrow 2NaNO_2 + O_2$	
b. $H_2 + Cl_2 \rightarrow 2HCl$	
c. $2C_2H_6 + 7O_2 \rightarrow 4CO_2 + 6H_2O$	
d. $AgNO_3 + KBr \rightarrow AgBr + KNO_3$	
e. $Cu + 2AgNO_3 \rightarrow 2Ag + Cu(NO_3)_2$	

9.2 The **oxidation number** (Sec. 9.2) of an atom represents the charge that the atom would have if all the electrons in each of its bonds were transferred to the more electronegative atom of the two atoms in the bond.

Assign oxidation numbers for all the atoms in the following substances, using the rules for determining oxidation numbers found in Section 9.2 of your textbook.

Substance	Oxidation number
Fe	
Ne	
Br_2	
KBr	
MgO	

Substance	Oxidation number
NO_2	
NO_2^-	
PO_4^{3-}	
SO_4^{2-}	
NH_4^+	

9.3 In **oxidation-reduction (redox) reactions** (Sec. 9.2) electrons are transferred from one reactant to another reactant. Electrons are lost by the substance being **oxidized** (Sec. 9.3), so its oxidation number is increased. A substance being **reduced** (Sec. 9.3) gains electrons, and its oxidation number decreases.

In the following equations, assign oxidation numbers to each atom in the reactants and products. Looking at oxidation number changes, classify the reaction as a redox or a nonredox reaction.

Reaction	Redox or nonredox
a. $KOH + HBr \rightarrow KBr + HOH$	
b. $2NaNO_3 \rightarrow 2NaNO_2 + O_2$	
c. $Cu + 2AgNO_3 \rightarrow 2Ag + Cu(NO_3)_2$	

9.4 In redox reactions an **oxidizing agent** (Sec. 9.3) accepts electrons and is reduced. A **reducing agent** (Sec. 9.3) loses electrons and is oxidized. For the equations below, first assign the oxidation numbers, and then identify the oxidizing and reducing agents and the substances oxidized and reduced.

Equation	Substance oxidized	Substance reduced	Oxidizing agent	Reducing agent
a. $4Na(s) + O_2(g) \rightarrow 2Na_2O(s)$				
b. $Ca(s) + S(s) \rightarrow CaS(s)$				
c. $Mg(ClO_3)_2 \rightarrow MgCl_2 + 3O_2$				

9.5 According to **collision theory** (Sec. 9.4), a chemical reaction takes place when two reactant particles collide with a certain minimum amount of energy, called **activation energy** (Sec. 9.4), and the proper orientation. In an energy diagram, the activation energy is the energy difference between the energy of the reactants and the top of the energy "hill." Some of this energy is regained during the reaction; in an **exothermic reaction** (Sec. 9.5) energy is given off in product formation, but in an **endothermic reaction** (Sec. 9.5) energy is absorbed, so that the products are at a higher energy level than the reactants.

Sketch two energy diagrams below and label these parts on each diagram: a. average energy of reactants, b. average energy of products, c. energy absorbed or given off during the reaction, and d. activation energy.

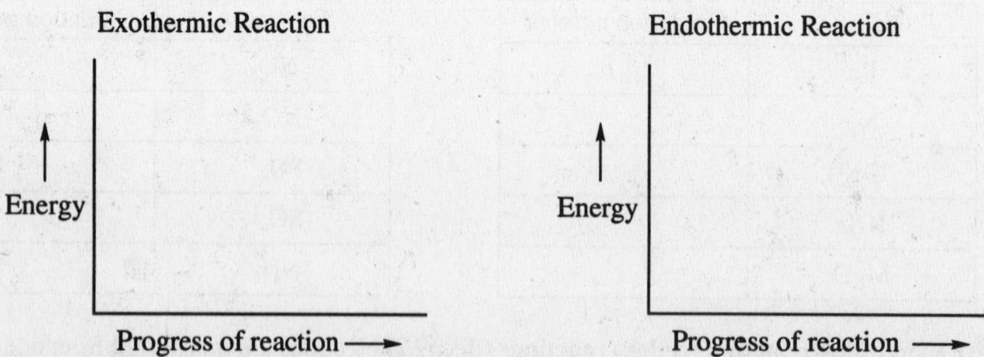

9.6 The **rate of a chemical reaction** (Sec. 9.6) is the rate at which reactants are consumed or products are formed in a given time period. Various factors affect the rate of a reaction: the physical nature of reactants, reactant concentrations, reaction temperature, the presence of a **catalyst** (Sec. 9.6).

Indicate whether the listed conditions would increase or decrease the rate of the following reaction:

A(solid) + B → C + D (heat)

Change in condition	Change in rate	Explanation
Decreasing the concentration of reactants		
Decreasing the temperature of the reaction		
Introduction of an effective catalyst for this reaction		
Increasing the surface area of the solid reactant by dividing the solid into smaller particles		

9.7 A **reversible reaction** (Sec. 9.7) is a chemical reaction in which two reactions (the forward reaction and the reverse reaction) occur simultaneously. When these two opposing chemical reactions occur at the same rate, the system is said to be at **chemical equilibrium** (Sec. 9.7).

$$wA + xB \rightleftharpoons yC + zD$$

Because the rates of the forward and reverse reactions are the same, the concentrations of the reactants and the products remain constant. An **equilibrium constant** (Sec. 9.8) that describes numerically the extent of the reaction can be obtained by writing an equilibrium constant expression and evaluating it numerically.

$$K_{eq} = \frac{[C]^y[D]^z}{[A]^w[B]^x}$$

Write the equilibrium constant expression for each of the equations below. Rules for writing these expressions are found in Section 9.8 of your textbook.

a. $2P + 3I_2 \rightleftharpoons 2PI_3$

b. $CH_4(g) + 2O_2(g) \rightleftharpoons CO_2(g) + 2H_2O(g)$

9.8 If the concentrations of reactants and products are known for a given reaction at equilibrium, the equilibrium constant can be evaluated. Write the equilibrium constant expression for the following equation and substitute the given molarities to calculate a numerical value for K_{eq}.

$2HI(g) \rightleftharpoons H_2(g) + I_2(g)$ In a 1.0 L container, there are 2.3 moles HI, 0.45 mole H_2 and 0.24 mole I_2

9.9 **Le Châtelier's principle** (Sec. 9.8) considers the effects of outside forces on systems at chemical equilibrium. According to this principle, a stress applied to the system can favor the reaction that will reduce the stress -- either the forward reaction, in which case more product is formed, or the reverse reaction, in which case more reactants form. Some of the stresses that can cause this readjustment of the equilibrium are concentration changes, temperature changes, and pressure changes.

a. Indicate what effect each of the conditions below would have on the following exothermic reaction at equilibrium:

$CH_4(g) + 2O_2(g) \rightleftharpoons CO_2(g) + 2H_2O(g) + \text{heat}$

Condition	Change in equilibrium	Explanation
increasing the concentration of O_2		
increasing the temperature of the reaction		
introduction of an effective catalyst for this reaction		
increasing the pressure exerted on the reaction		

b. Indicate what effect each of the conditions below would have on the following endothermic reaction:

$N_2(g) + 2O_2(g) + \text{heat} \rightleftharpoons 2NO_2(g)$

Condition	Change in equilibrium	Explanation
increasing the concentration of O_2		
increasing the temperature of the reaction		
introduction of an effective catalyst for this reaction		
increasing the pressure exerted on the reaction		

Self-Test

True-false: Indicate whether the following statements are true or false. If the statement is false, give the word or phrase that may be substituted for the underlined portion to make the statement true.

1. The reaction $2CuO \rightarrow 2Cu + O_2$ is an example of a <u>single-replacement</u> reaction.
2. The oxidation number of a metal in its elemental state is always <u>positive</u>.
3. The oxidation number of oxygen in most compounds is <u>–2</u>.
4. A substance that is <u>oxidized</u> loses electrons.
5. A reducing agent <u>gains</u> electrons.
6. Adding heat to an <u>endothermic</u> reaction helps the reaction to go toward the products side.
7. The addition of a catalyst <u>will not change</u> the equilibrium position of a reaction.
8. The rate of a reaction is <u>not affected</u> by the addition of a catalyst.
9. Increasing the concentration of products in an equilibrium reaction shifts the equilibrium toward the <u>product side</u> of the reaction.
10. In an equilibrium constant expression, the concentrations of the reactants are found in the <u>numerator</u>.
11. A large equilibrium constant indicates that the equilibrium position is to the <u>right</u> side of the equation.
12. In writing equilibrium constants, we consider that concentrations of <u>pure solids and pure liquids</u> remain constant.

Multiple choice:

13. The equation $X + YZ \rightarrow Y + XZ$ is a general equation for which type of reaction?

    a. combination b. single-displacement c. double-displacement
    d. decomposition e. combustion

14. The oxidation number of chromium in the compound $K_2Cr_2O_7$ is:

    a. +6 b. –7 c. +5 d. –3 e. none of these

15. In the reaction $Zn + Cu(NO_3)_2 \rightarrow Zn(NO_3)_2 + Cu$, the oxidizing agent is:

    a. Zn b. $Cu(NO_3)_2$ c. $Zn(NO_3)_2$ d. Cu e. none of these

16. Which of these factors does *not* affect the rate of a reaction?

    a. the frequency of the collisions b. the energy of the collisions
    c. the orientation of the collisions d. the product of the collisions
    e. all of these affect rate

17. For a reaction at equilibrium, the concentrations of the products:

    a. increase rapidly b. increase slowly c. remain the same
    d. decrease slowly e. none of these

Answer Questions 18 through 21 using the general equilibrium equation:

$$A + B \rightleftharpoons C + D + \text{heat}$$

18. What is the equilibrium constant for this reaction, if the following concentrations are measured at equilibrium: [A] = 0.20 M; [B] = 1.5 M; [C] = 5.2 M; [D] = 3.7 M?
    a. 64 b. 0.016 c. 5.7 d. 0.25 e. none of these

19. The rate of the forward reaction could be increased by:
    a. decreasing the concentration of B
    b. increasing the concentration of A
    c. increasing the concentration of C
    d. both b and c
    e. none of these

20. The value of the equilibrium constant could be increased by:
    a. increasing the temperature of the reaction mixture
    b. increasing the concentration of A
    c. decreasing the concentration of B
    d. decreasing the temperature of the reaction mixture
    e. none of these

21. If more A is added to the reaction mixture at equilibrium:
    a. the amount of C will increase
    b. the amount of B will increase
    c. the amount of B will decrease
    d. both a and c
    e. none of these

22. In the reaction $2Mg + O_2 \rightarrow 2MgO$, the magnesium is:
    a. reduced and is the oxidizing agent
    b. reduced and is the reducing agent
    c. oxidized and is the oxidizing agent
    d. oxidized and is the reducing agent
    e. none of these

Answers to Practice Exercises

9.1

Reaction	Classification
a. $2NaNO_3 \rightarrow 2NaNO_2 + O_2$	decomposition
b. $H_2 + Cl_2 \rightarrow 2HCl$	combination
c. $2C_2H_6 + 7O_2 \rightarrow 4CO_2 + 6H_2O$	combustion
d. $AgNO_3 + KBr \rightarrow AgBr + KNO_3$	double-replacement
e. $Cu + 2AgNO_3 \rightarrow 2Ag + Cu(NO_3)_2$	single-replacement

9.2

Substance	Oxidation number
Fe	0
Ne	0
Br_2	0
KBr	K (+1), Br (–1)
MgO	Mg (+2), O (–2)

Substance	Oxidation number
NO_2	N (+4), O (–2)
NO_2^-	N (+3), O (–2)
PO_4^{3-}	P (+5), 0 (–2)
SO_4^{2-}	S (+6), O (–2)
NH_4^+	N (–3), H (+1)

9.3

	Oxidation numbers for all atoms	Redox or nonredox
a.	$KOH + HBr \rightarrow KBr + H_2O$ +1,–2,+1 +1,–1 +1,–1 +1,–2	nonredox, no oxidation numbers change
b.	$2NaNO_3 \rightarrow 2NaNO_2 + O_2$ +1,+5,–2 +1,+3,–2 0	redox, N (+5 → +3) O (–2 → 0)
c.	$Cu + 2AgNO_3 \rightarrow 2Ag + Cu(NO_3)_2$ 0 +1,+5,–2 0 +2,+5,–2	redox, Cu (0 → +2) Ag (+1 → 0)

9.4

Equation	Substance oxidized	Substance reduced	Oxidizing agent	Reducing agent
a. $4Na(s) + O_2(g) \rightarrow 2Na_2O(s)$ 0 0 +1,–2	sodium	oxygen	oxygen	sodium
b. $Ca(s) + S(s) \rightarrow CaS(s)$ 0 0 +2,–2	calcium	sulfur	sulfur	calcium
c. $Mg(ClO_3)_2 \rightarrow MgCl_2 + 3O_2$ +2,+5,–2 +2,–1 0	oxygen	chlorine	chlorine	oxygen

9.5 a. average energy of reactants, b. average energy of products, c. energy absorbed or given off during the reaction, and d. activation energy.

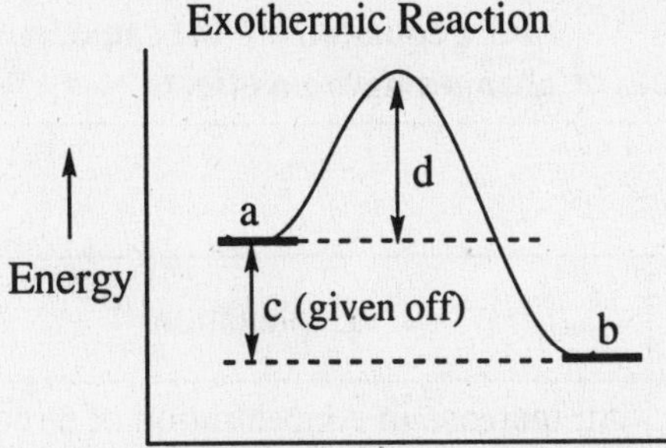

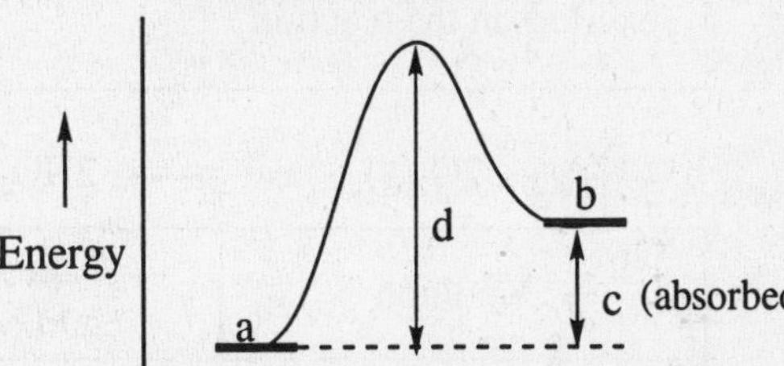

9.6

Change in conditions	Change in rate; explanation
Decreasing the concentration of reactants	Decreases reaction rate; fewer molecules collide, fewer molecules react.
Decreasing the temperature of the reaction	Decreases reaction rate; lower kinetic energy, lower collision energy, so fewer collisions are effective.
Introduction of an effective catalyst for this reaction	Increases reaction rate; catalysts provide alternative reaction pathways that have lower energies of activation.
Increasing the surface area of the solid reactant by dividing the solid into smaller particles	Increases reaction rate; larger surface area of solid provides more chances for collision.

9.7 a. $K_{eq} = \dfrac{\text{Products}}{\text{Reactants}} = \dfrac{[PI_3]^2}{[P]^2[I_2]^3}$

b. $K_{eq} = \frac{\text{Products}}{\text{Reactants}} = \frac{[CO_2][H_2O]^2}{[CH_4][O_2]^2}$

9.8 $K_{eq} = \frac{[H_2][I_2]}{[HI]^2} = \frac{(0.45) \times (0.24)}{(2.3)^2} = 0.020$

9.9 a. $CH_4(g) + 2O_2(g) \rightleftharpoons CO_2(g) + 2H_2O(g) + \text{heat}$

Condition	Change in equilibrium	Explanation
increasing the concentration of O_2	shift to right	An increase in concentration of a reactant produces more products.
increasing the temperature of the reaction	shift to left	The equilibrium shifts to decrease the amount of heat produced.
introduction of an effective catalyst for this reaction	no change	A catalyst cannot change the position of the equilibrium because it only lowers the energy of activation.
increasing the pressure exerted on the reaction	no change	Because the moles of gas on each side of the equation are the same, pressure changes have no effect.

b. $N_2(g) + 2O_2(g) + \text{heat} \rightleftharpoons 2NO_2\ (g)$

Condition	Change in equilibrium	Explanation
increasing the concentration of O_2	shift to right	An increase in concentration of a reactant produces more products.
increasing the temperature of the reaction	shift to right	For an endothermic reaction the equilibrium shifts to the product side, increasing the amount of heat consumed.
introduction of an effective catalyst for this reaction	no change	A catalyst cannot change the position of the equilibrium because it only lowers the energy of activation.
increasing the pressure exerted on the reaction	shift to right	An increase in the forward reaction would relieve pressure because there are fewer moles of gas on the right.

Answers to Self-Test

The numbers in parentheses refer to sections in your textbook.
1. F; decomposition (9.1) **2.** F; zero (9.2) **3.** T (9.2) **4.** T (9.3) **5.** F; loses (9.3) **6.** T (9.5) **7.** T (9.5) **8.** F; increased (9.6) **9.** F; reactants (9.9) **10.** F; denominator (9.8) **11.** T (9.8) **12.** T (9.8) **13.** b (9.1) **14.** a (9.2) **15.** b (9.3) **16.** d (9.4) **17.** c (9.7) **18.** a (9.8) **19.** b (9.6) **20.** d (9.8) **21.** d (9.9) **22.** d (9.3)

Acids, Bases, and Salts — Chapter 10

Chapter Overview

Acids, bases, and salts play a central role in much of the chemistry that affects our daily lives. Learning the terms and concepts associated with these compounds will give you a greater understanding of the chemistry of the human body, and of the ways in which chemicals are manufactured.

In this chapter you will learn to identify acids and bases according to the Arrhenius and Brønsted-Lowry definitions, write equations for acid and base dissociations in water, and calculate pH, a measure of acidity. You will write equations for the hydrolysis of the salt of a weak acid or a weak base, and you will study the actions of buffers.

Practice Exercises

10.1 According to the Arrhenius acid-base theory, the **dissociation** (Sec. 10.1) of an **Arrhenius acid** (Sec.10.1) in water produces hydrogen ions (H^+), and the dissociation of an **Arrhenius base** in water produces hydroxide ions (OH^-). Arrhenius acids and bases have certain properties that help us identify them. In the table below, specify whether each property is that of an acid or of a base.

Property	Acid	Base
has a sour taste	X	
turns blue litmus red	X	
has a slippery feel		X
turns red litmus blue		X
has a bitter taste		X

10.2 According to the **Brønsted-Lowry** (Sec. 10.2) definitions, an acid is a proton donor and a base is a proton acceptor. The **conjugate base** (Sec. 10.2) of an acid is the species that remains when an acid loses a proton. The **conjugate acid** (Sec. 10.2) of a base is the species formed when a base accepts a proton.

$$\underset{\text{Acid}}{HA} + \underset{\text{Base}}{B} \rightleftharpoons \underset{\text{Conjugate acid}}{HB^+} + \underset{\text{Conjugate base}}{A^-}$$

Give the formula of the conjugate acid or base for the following substances:

Base	Conjugate acid
NH_3	NH_4^+
BrO_3^-	$HBrO_3$
HCO_3^-	H_2CO_3

Acid	Conjugate base
HCO_3^-	CO_3^{-2}
$HClO_2$	ClO_2^-
HNO_3	NO_3^-

10.3 Identify the acid and base in the reactants in the following equations. Identify the conjugate acid and conjugate base in the products.

a. $HClO_3 + H_2O \rightarrow ClO_3^- + H_3O^+$

b. $HNO_2 + OH^- \rightarrow NO_2^- + H_2O$

10.4 A **monoprotic acid** (Sec. 10.3) transfers one H^+ ion per molecule during an acid-base reaction, but a **diprotic acid** can transfer two H^+ ions (protons) per molecule, and a **triprotic acid** can transfer three H^+ ions per molecule in acid-base reactions. Complete the following reactions involving a diprotic acid, and label the acids, bases, conjugate acids, and conjugate bases. Use one mole of OH^- per mole of acid in each equation.

a. $H_2CO_3(aq) + OH^-(aq) \rightleftharpoons$

b. $HCO_3^-(aq) + OH^-(aq) \rightleftharpoons$

10.5 An **acid ionization constant** (Sec. 10.5), K_a, is the equilibrium constant corresponding to the **ionization** (Sec. 10.1) of an acid. It gives a measure of the **strength** (Sec. 10.4) of the acid. The **base ionization constant** (Sec. 10.5), K_b, gives a measure of the strength of a base.

a. Write the ionization equation and the ionization constant expression for the ionization of nitrous acid, HNO_2, in water.

b. Write the ionization equation and the ionization constant expression (K_b) for the ionization of ethylamine, $C_2H_5NH_2$, in water. (The nitrogen atom accepts a proton.)

10.6 The acid ionization constant can be calculated for an acid if its concentration and percent ionization are known.

A 0.0150 M solution of an acid, HA, is 22% ionized at equilibrium. Determine the individual ion concentrations, and then calculate the K_a for this acid.

10.7 **Salts** (Sec. 10.6) are compounds made up of positive metal or polyatomic ions, and negative nonmetal or polyatomic (except hydroxide) ions. Identify each of the following compounds as an acid, a base, or a salt.

Compound	Acid	Base	Salt
HCl			
NaCl			
H_2SO_4			
NaOH			
$CaBr_2$			
$Ba(OH)_2$			

10.8 Soluble salts dissolved in water are completely dissociated into ions in solution. Write a balanced equation for the **dissociation** (Sec. 10.1) of the following soluble ionic compounds in water.

a. KI K I

b. Na_3PO_4 3na 4PO

c. CaI_2 Ca 2I

d. Na_2CO_3 2na 3CO

10.9 **Neutralization** (Sec. 10.7) is the reaction between an acid and a hydroxide base to form a salt and water. Complete the following neutralization equations by adding the missing products or reactants. Balance the equations, keeping in mind that H^+ and OH^- ions react in a one-to-one ratio to form water. Under each reactant molecule, write acid or base, and under each product molecule, write salt or water.

a. $HCl + NaOH \rightarrow$

b. $\rightarrow CaCl_2 + H_2O$

c. $\rightarrow Sr_3(PO_4)_2 + H_2O$

10.10 In pure water an extremely small number of molecules transfer protons to form the ions H_3O^+ and OH^-.

$$H_2O + H_2O \rightleftharpoons H_3O^+ + OH^-$$

Equal concentrations of H_3O^+ and OH^- are produced by this self-ionization, and each is equal to 1×10^{-7} M. This value can be used to calculate the **ion product constant** (Sec. 10.7).

$$\text{Ion product constant for water} = [H_3O^+] \times [OH^-] = (1 \times 10^{-7})(1 \times 10^{-7}) = 1 \times 10^{-14}$$

The ion product constant relationship is true for water solutions as well as pure water, so it can be used to calculate the concentration of either H_3O^+ or OH^- if the concentration of the other ion is known.

Using the ion product constant for water, determine the following concentrations:

Given concentration	Substituted equation	Answer
$[H_3O^+] = 2.4 \times 10^{-6}$ M		$[OH^-] =$
$[OH^-] = 3.2 \times 10^{-8}$ M		$[H_3O^+] =$

10.11 Because the hydronium ion concentrations (measure of solution acidity) in aqueous solutions have a very large range of values, a more practical way to represent acidity is by using the **pH scale** (Sec. 10.9):

$$pH = -\log[H_3O^+]$$

Using your calculator, complete the following pH relationships:

$[H_3O^+]$	pH	Acidic, basic, or neutral
1.0×10^{-4}		
1.0×10^{-9}		
2.8×10^{-3}		
7.9×10^{-8}		

10.12 Using the definition of pH and the ion product constant, calculate the following concentrations of ions:

pH	$[H_3O^+]$	$[OH^-]$
5.00		
2.00		
3.80		
10.40		

10.13 Acid strength can also be expressed in terms of pK_a:

$$pK_a = -\log K_a$$

a. Determine the pK_a of an acid whose ionization constant is $K_a = 1.02 \times 10^{-7}$.

b. If Acid A has a pK_a of 8.69 and Acid B has a pK_a of 11.62, which is the stronger acid?

10.14 **Hydrolysis** (Sec. 10.10) is the reaction of a substance with water to produce hydronium ion or hydroxide ion or both. Salts formed from weak acids or weak bases hydrolyze in water to form their "parent" weak acids or weak bases.

Write the aqueous hydrolysis equations for each of the following ions:

a. HCO_3^- (proton acceptor)

b. NH_4^+ (proton donor)

c. NO_2^- (proton acceptor)

10.15 A **buffer** (Sec. 10.12) solution is a solution that resists a change in pH when small amounts of acid or base are added to it. Buffers (the solutes in buffer solutions) consist of one of the following combinations: a weak acid and the salt of its conjugate base or a weak base and the salt of its conjugate acid. These are known as **conjugate acid-base pairs** (Sec. 10.2).

Predict whether each of the following pairs of substances could function as a buffer in an aqueous solution. Explain your answer.

Pair of substances	Explanation
KOH, KCl	no, strong ~~acid~~ Base, strong ~~Base~~ acid - strong base salt
HI, NaI	no strong acid, strong base - strong base salt
NH_4I, NH_3	Yes, weak base and strong acid - weak base salt
NaH_2PO_4, H_3PO_4	Yes, week ~~base~~ acid and weak acid - stron base salt

10.16 Buffers contain a substance that reacts with and removes added base and a substance that reacts with and removes added acid. Write an equation to show the buffering action in each of the following aqueous solutions.

a. NH_4I/NH_3 in a basic solution, OH^-

b. NaH_2PO_4/H_3PO_4 in an acidic solution, H_3O^+

10.17 The pH of a buffered solution may be determined by using the Henderson-Hasselbalch equation:

$$pH = pK_a + \log\frac{[A^-]}{[HA]}$$

Calculate the pH of each of these buffer solutions:

Given values	Substituted equation	Answer
[HA] = 0.34 M $[A^-]$ = 0.51 M pK_a = 5.48		pH =
[HA] = 0.27 M $[A^-]$ = 0.55 M $K_a = 8.4 \times 10^{-5}$		pH =
[HA] = 0.25 M $[A^-]$ = 0.37 M $K_a = 6.2 \times 10^{-7}$		pH =

10.18 An **electrolyte** (Sec. 10.14) is a substance that forms a solution in water that conducts electricity. A **strong electrolyte** (Sec. 10.14) is a substance that completely dissociates into ions in aqueous solution. Salts and strong acids and strong bases are strong electrolytes. A **weak electrolyte** (Sec. 10.14) is a substance that is only partially ionized in aqueous solution. Weak acids and weak bases are weak electrolytes.

Classify each of the substances below, dissolved in aqueous solution, as a weak electrolyte or a strong electrolyte.

Formula	Weak electrolyte	Strong electrolyte
H_2SO_4		
NH_3		
NH_4Cl		
MgI_2		

10.19 **Acid-base titration** (Sec. 10.15) is a procedure used to determine the concentration of an acid or base solution. A measured volume of an acid (or a base) of known concentration is exactly reacted with a measured volume of a base (or an acid) of unknown concentration. The unknown concentration can be calculated using dimensional analysis.

a. Determine the molarity of an unknown HCl solution if 21.9 mL of 0.338 M NaOH was needed to neutralize 41.6 mL of the HCl solution. Hint: Write the balanced neutralization reaction equation.

b. What is the molarity of an unknown sulfuric acid solution if 33.2 mL of 0.225 M NaOH is needed to neutralize 13.8 mL of the H_2SO_4 solution? Hint: Write the balanced neutralization reaction equation.

Self-Test

True-false: Indicate whether the following statements are true or false. If the statement is false, give the word or phrase that may be substituted for the underlined portion to make the statement true.

1. The value of the ion product constant of water is always <u>1 x 10^{-10}</u>.
2. Arrhenius defined a base as a substance that, in water, produces <u>hydroxide ions</u>.
3. According the Brønsted-Lowry theory, NH_3 is <u>an acid</u>.
4. Aqueous solutions of acids have a hydronium ion concentration <u>less than</u> 1 x 10^{-7} moles per liter.
5. A diprotic acid can transfer <u>two protons</u> per molecule during an acid-base reaction.
6. The pH of an acid is the negative logarithm of the <u>hydronium ion concentration</u>.
7. A neutralization reaction produces a salt and <u>a base</u>.
8. Hydrolysis of the salt of a weak acid and a strong base produces a solution that is <u>basic</u>.
9. A solution whose hydronium ion concentration is 1.0 x 10^{-4} has a pH of <u>1.4</u>.
10. A buffer is a weak acid plus the salt of its conjugate <u>base</u>.
11. An amphoteric substance can function as an acid or a <u>salt</u>.

Multiple choice:

12. The pH of a solution in which $[H_3O^+]$ = 1.0 x 10^{-5} is:

a. –5.00 b. –1.50 c. 5.00 d. 1.50 e. none of these

13. The pH of a solution in which $[OH^-]$ = 1.0 x 10^{-8} is:

a. 6.00 b. 8.00 c. –8.00 d. –1.80 e. none of these

14. If the pH of a solution is 4.80, the hydronium ion concentration is:

a. 4.0 x 10^{-8} M b. 8.3 x 10^{-4} M c. 3.2 x 10^{-12} M
d. 1.6 x 10^{-5} M e. none of these

15. If the pH of a solution is 9.20, the hydroxide ion concentration is:

a. 1.6 x 10^{-5} M b. 2.1 x 10^{-9} M c. 9.3 x 10^{-2} M
d. 7.5 x 10^{-10} e. none of these

16. How many milliliters of 0.512 M HCl would be required to neutralize 35.8 mL of 1.50 M KOH?

a. 11.3 mL b. 35.8 mL c. 105 mL
d. 183 mL e. none of these

17. The conjugate base of the acid HNO_2 is:

a. H_2O b. H_3O^+ c. NO_2^- d. OH^- e. none of these

18. A 0.400 M solution of a monoprotic acid is 9.0% ionized. The value of K_a for this acid is:

a. 3.6×10^{-3} b. 9.8×10^{-2} c. 2.1×10^{-4}
d. 8.9×10^{-4} e. none of these

19. The salt K_2CO_3 is produced by the reaction of:

a. a weak acid with a weak base b. a weak acid with a strong base
c. a strong acid with a weak base d. a strong acid with a strong bas
e. none of these

20. Which of the following is a weak electrolyte?

a. K_3PO_4 b. HNO_3 c. $Ba(OH)_2$ d. H_2CO_3 e. none of these

21. If a small amount of hydroxide ion is added to the buffer H_2CO_3/HCO_3^-, the reaction will produce more:

a. H_3O^+ b. HCO_3^- c. H_2CO_3 d. H_2 e. none of these

Answers to Practice Exercises

10.1

Property	Acid	Base
has a sour taste	X	
turns blue litmus red	X	
has a slippery feel		X
turns red litmus blue		X
has a bitter taste		X

10.2

Base	Conjugate acid
NH_3	NH_4^+
BrO_3^-	$HBrO_3$
HCO_3^-	H_2CO_3

Acid	Conjugate base
HCO_3^-	CO_3^{2-}
$HClO_2$	ClO_2^-
HNO_3	NO_3^-

10.3 a. $HClO_3$ + H_2O → ClO_3^- + H_3O^+
acid base conjugate base conjugate acid

b. HNO_2 + OH^- → NO_2^- + H_2O
acid base conjugate base conjugate acid

10.4

a. $H_2CO_3(aq) + OH^-(aq) \rightleftharpoons HCO_3^-(aq) + H_2O(l)$

Acid, Base, Conjugate acid, Conjugate base

b. $HCO_3^-(aq) + OH^-(aq) \rightleftharpoons CO_3^{2-}(aq) + H_2O(l)$

Acid, Base, Conjugate acid, Conjugate base

10.5

a. $HNO_2 + H_2O \rightleftharpoons NO_2^- + H_3O^+$ $\quad K_a = \frac{[H_3O^+][NO_2^-]}{[HNO_2]}$

b. $C_2H_5NH_2 + H_2O \rightleftharpoons C_2H_5NH_3^+ + OH^-$ $\quad K_b = \frac{[C_2H_5NH_3^+][OH^-]}{[C_2H_5NH_2]}$

10.6 $HA + H_2O \rightleftharpoons H_3O^+ + A^-$

$$K_a = \frac{[H_3O^+][A^-]}{[HA]} = \frac{[0.0150 \text{ x } 0.22][0.0150 \text{ x } 0.22]}{[(0.0150)-(0.0150 \text{ x } 0.22)]} = \frac{[0.0033][0.0033]}{[0.0117]} = 9.3 \text{ x } 10^{-4}$$

10.7

Compound	Acid	Base	Salt
HCl	X		
NaCl			X
H_2SO_4	X		
NaOH		X	
$CaBr_2$			X
$Ba(OH)_2$		X	

10.8

a. $KI \rightarrow K^+(aq) + I^-(aq)$

b. $Na_3PO_4 \rightarrow 3Na^+(aq) + PO_4^{3-}(aq)$

c. $CaI_2 \rightarrow Ca^{2+}(aq) + 2I^-(aq)$

d. $Na_2CO_3 \rightarrow 2Na^+(aq) + CO_3^{2-}(aq)$

10.9

a. $HCl + NaOH \rightarrow NaCl + H_2O$

acid, base, salt, water

b. $2HCl + Ca(OH)_2 \rightarrow CaCl_2 + 2H_2O$

acid, base, salt, water

c. $2H_3PO_4 + 3Sr(OH)_2 \rightarrow Sr_3(PO_4)_2 + 3H_2O$

acid, base, salt, water

10.10

Given concentration	Substituted equation	Answer
$[H_3O^+] = 2.4 \times 10^{-6}$	$[OH^-] = \frac{1.00 \times 10^{-14}}{[2.4 \times 10^{-6}]} =$	$[OH^-] = 4.2 \times 10^{-9}$
$[OH^-] = 3.2 \times 10^{-8}$	$[H_3O^+] = \frac{1.00 \times 10^{-14}}{[3.2 \times 10^{-8}]} =$	$[H_3O^+] = 3.1 \times 10^{-7}$

10.11

$[H_3O^+]$	pH	Acidic, basic, or neutral
1.0×10^{-4}	4.00	acidic
1.0×10^{-9}	9.00	basic
2.8×10^{-3}	2.55	acidic
7.9×10^{-8}	7.10	basic

10.12

pH	$[H_3O^+]$	$[OH^-]$
5.00	1.0×10^{-5}	1.0×10^{-9}
2.00	1.0×10^{-2}	1.0×10^{-12}
3.80	1.6×10^{-4}	6.3×10^{-11}
10.40	4.0×10^{-11}	2.5×10^{-4}

10.13 a. $pK_a = 6.91$

b. Acid A ($pK_a = 8.69$) is the stronger acid.

10.14 a. $HCO_3^- + H_2O \rightarrow H_2CO_3 + OH^-$

b. $NH_4^+ + H_2O \rightarrow H_3O^+ + NH_3$

c. $NO_2^- + H_2O \rightarrow HNO_2 + OH^-$

10.15

Pair of substances	Explanation
KOH, KCl	No. Strong base and strong acid-strong base salt
HI, NaI	No. Strong acid and strong acid-strong base salt
NH_4I, NH_3	Yes. Weak base and strong acid-weak base salt
NaH_2PO_4, H_3PO_4	Yes. Weak acid and weak acid-strong base salt

10.16 a. NH_4I/NH_3 in a basic solution, OH^-

$$NH_4^+ + OH^- \rightarrow NH_3 + H_2O$$

b. NaH_2PO_4/H_3PO_4 in an acidic solution, H_3O^+

$$H_2PO_4^- + H_3O^+ \rightarrow H_3PO_4 + H_2O$$

10.17

Given values	Substituted equation	Answer
[HA] = 0.34 M $[A^-]$ = 0.51 M pK_a = 5.48	$\text{pH} = 5.48 + \log\left(\frac{0.51}{0.34}\right) = 5.66$	pH = 5.66
[HA] = 0.27 M $[A^-]$ = 0.55 M $K_a = 8.4 \times 10^{-5}$	$\text{pH} = 4.08 + \log\left(\frac{0.55}{0.27}\right) = 4.39$	pH = 439
[HA] = 0.25 M $[A^-]$ = 0.37 M $K_a = 6.2 \times 10^{-7}$	$\text{pH} = 6.21 + \log\left(\frac{0.37}{0.25}\right) = 6.38$	pH = 6.38

10.18

Formula	Weak electrolyte	Strong electrolyte
H_2SO_4		X
NH_3	X	
NH_4Cl		X
MgI_2		X

10.19 a. $HCl + NaOH \rightarrow NaCl + H_2O$

$$\text{M HCl} = 21.9 \text{ mL NaOH} \times \frac{0.338 \text{ mole NaOH}}{1000 \text{ mL NaOH}} \times \frac{1 \text{ mole HCl}}{1 \text{ mole NaOH}} \times \frac{1000 \text{ mL}}{0.0416 \text{ L HCl}} = 0.178 \text{ M HCl}$$

b. $H_2SO_4 + 2NaOH \rightarrow Na_2SO_4 + 2H_2O$

$$\text{M } H_2SO_4 = 33.2 \text{ mL NaOH} \times \frac{0.225 \text{ mole NaOH}}{1000 \text{ mL NaOH}} \times \frac{1 \text{ mole } H_2SO_4}{2 \text{ moles NaOH}} \times \frac{1000 \text{ mL}}{0.0138 \text{ L } H_2SO_4} = 0.271 \text{ M } H_2SO_4$$

Answers to Self-Test

The numbers in parentheses refer to sections in your textbook.
1. F; 1.00×10^{-14} (10.8) **2.** T (10.1) **3.** F; base (10.2) **4.** F; more than (10.8) **5.** T (10.3) **6.** T (10.9) **7.** F; water (10.7) **8.** T (10.11) **9.** F; 4 (10.9) **10.** T (10.12) **11.** F; base (10.2) **12.** c (10.9) **13.** a (10.9) **14.** d (10.9) **15.** a (10.9) **16.** c (10.15) **17.** c (10.2) **18.** a (10.5) **19.** b (10.7) **20.** d (10.14) **21.** b (10.12)

Nuclear Chemistry — Chapter 11

Chapter Overview

Chemical reactions involve the exchange or sharing of the electrons of atoms as chemical bonds are broken or formed. In **nuclear reactions** (Sec. 11.1) an atom's nucleus changes, absorbing or emitting particles or rays.

In this chapter you will compare three kinds of nuclear radiation and write equations for radioactive decay. You will study the rate of radioactive decay, defining the concept of the half-life of a radionuclide and using it in calculations. You will compare nuclear fission and nuclear fusion, study the effects of ionizing radiation on the human body, and learn about some of the ways in which radiation and radionuclides are used in medicine.

Practice Exercises

11.1 The **radioactive decay** (Sec. 11.3) of naturally radioactive substances results in the emission of three types of radiation: **alpha particles, beta particles,** and **gamma rays** (Sec. 11.2). They differ in mass and charge. Complete the following table summarizing these three types of radiation:

Type of radiation	Mass number	Charge	Symbol
alpha particles	4	+2	${}^{4}_{2}\alpha$
beta particles	0	-1	${}^{0}_{-1}\beta$
gamma rays	0	0	${}^{0}_{0}\gamma$

11.2 The emission of an alpha particle from a nucleus results in the formation of a **nuclide** (Sec. 11.1) of a different element. The **daughter nuclide** (Sec. 11.3) has an atomic number that is two less and a mass number that is four less than the **parent nuclide** (Sec. 11.3).

Write the **balanced nuclear equations** (Sec. 11.3) for the alpha particle decay of the following nuclides. Give the complete symbol of the new nuclide produced in each process.

a. ${}^{149}_{65}\text{Tb} \rightarrow$

b. ${}^{231}_{91}\text{Pa} \rightarrow$

11.3 Beta particle decay results in the formation of a nuclide of a different element. The daughter nuclide has a mass number that is the same and an atomic number that is one greater than the parent nuclide.

Write the balanced nuclear equations for the beta particle decay of the following nuclides. Give the complete symbol for the new nuclide produced in each process.

a. ${}^{31}_{14}\text{Si} \rightarrow$

b. ${}^{59}_{26}\text{Fe} \rightarrow$

11.4 Not all **radioactive nuclides** (Sec. 11.1) decay at the same rate; the more unstable the nucleus, the faster it decays. The **half-life** (Sec. 11.4) of a substance – the amount of time for half of a given quantity of nuclide to decay – is a measure of the nuclide's stability.

The amount of radioactive material remaining after radioactive decay can be calculated from the following formula, where n = number of half-lives.

$$\begin{pmatrix}\text{Amount of radionuclide} \\ \text{undecayed after n half - lives}\end{pmatrix} = \begin{pmatrix}\text{original amount} \\ \text{of radionuclide}\end{pmatrix} \times \left(\frac{1}{2^n}\right)$$

In the following problems, an original sample of 0.43 mg of plutonium-239 is used (half-life = 24,400 years).

a. How much plutonium-239 would remain after three half-lives?

b. How much plutonium-239 would remain after 122,000 years?

11.5 The number of half-lives that have elapsed since the original measurement can be determined by the fraction of the original nuclide that remains.

A sample of iron-59 after 135 days has 1/8 of the iron-59 of the original sample. What is the half-life of iron-59?

11.6 **Transmutation** (Sec. 11.5) of some nuclides into nuclides of other elements can be attained by bombardment of the nuclei with small particles traveling at very high speeds.

Complete these **bombardment reaction** (Sec. 11.5) equations by supplying the missing parts:

a. $^{10}_{5}B + \_\_\_ \rightarrow {}^{7}_{3}Li + {}^{4}_{2}\alpha$

b. $^{7}_{3}Li + {}^{1}_{1}p \rightarrow \_\_\_ + {}^{4}_{2}\alpha$

c. $^{10}_{5}B + {}^{4}_{2}\alpha \rightarrow {}^{13}_{7}N + \_\_\_$

d. $\_\_\_ + {}^{14}_{7}N \rightarrow {}^{247}_{99}Es + 5\,{}^{1}_{0}n$

11.7 Radionuclides with high atomic numbers decay through a series of steps to reach a stable nuclide of lower atomic number. A few of these steps in a **radioactive decay series** (Sec. 11.6) are shown below. Complete the equations by filling in the missing parts:

step 1: $^{210}_{82}Pb \rightarrow \underline{{}^{210}_{83}} + {}^{0}_{-1}\beta$

step 2: $\underline{{}^{210}_{83}} \rightarrow {}^{210}_{84}Po + {}^{0}_{-1}\beta$

step 3: $^{210}_{84}Po \rightarrow \underline{{}^{206}_{82}} + {}^{4}_{2}\alpha$

11.8 The three types of naturally occurring radioactive emissions differ in their ability to penetrate matter and, therefore, in their biological effects. Complete the following table summarizing properties of the three types of radiation:

Type of radiation	Speed	Penetration	Biological damage
alpha particles	Slow	little paper	
beta particles	fast	wood	
gamma radiation	fastest	very lead	

Self-Test

True-false: Indicate whether the following statements are true or false. If the statement is false, give the word or phrase that may be substituted for the underlined portion to make the statement true.

1. Elements retain their identity during <u>nuclear reactions</u>.
2. <u>Alpha particles</u> are the same type of radiation as X rays.
3. When an atom loses an alpha particle, its atomic number is <u>decreased</u> by 2.
4. Beta particles are more penetrating than <u>alpha particles</u>.
5. The half-life of a radionuclide is the length of time needed for <u>all</u> of the nuclide to decay.
6. <u>Nuclear fission</u> is the reaction that provides the sun's energy.
7. The combination of two small nuclei to produce a larger nucleus is called <u>nuclear fusion</u>.
8. Ionizing radiation knocks <u>protons</u> off some atoms so that they become ions.
9. The energy involved in a nuclear reaction is <u>smaller</u> than the energy involved in a chemical reaction.
10. One <u>diagnostic application</u> of radionuclides is the use of ionizing radiation to kill cancer cells.
11. In a <u>Geiger counter,</u> radiation is detected by the ionization of argon gas in a metal tube.
12. A free radical is an atom or molecule with <u>a positive charge</u> whose formation can be caused by radiation.
13. Different isotopes of an element have practically identical <u>chemical properties</u>.
14. An average American is exposed to <u>more</u> radiation from natural sources than from human-made sources.
15. A transmutation process is a nuclear reaction in which a nuclide of one element is changed into a nuclide of <u>the same</u> element.

Multiple choice:

16. The nuclide produced by the emission of an alpha particle from the platinum nuclide $^{186}_{78}Pt$ is:

    a. $^{188}_{80}Hg$ b. $^{182}_{76}Os$ c. $^{184}_{75}Re$ d. $^{186}_{79}Au$ e. none of these

17. The nuclide produced by beta emission from the barium radionuclide $^{142}_{56}Ba$ is:

    a. $^{142}_{57}La$ b. $^{143}_{55}Cs$ c. $^{138}_{54}Xe$ d. $^{146}_{58}Ce$ e. none of these

18. The half-life for tritium $^{3}_{1}H$ is 12.26 years. After 36.78 years, how much of an original 0.40-g sample of tritium will remain?

a. 0.35 g b. 0.20 g c. 0.10 g d. 0.050 g e. none of these

19. Four radionuclides have the half-lives listed below. Which of the four has the most stable nucleus?

a. 22 days b. 4000 years c. 16 minutes
d. 56 seconds e. all are very unstable

20. An alpha particle is made up of:

a. two protons and four neutrons
b. two protons and two neutrons
c. two neutrons and two electrons
d. two protons and two electrons
e. none of these

21. The nuclear chain reaction of uranium-235 provides the energy in:

a. nuclear power plants
b. nuclear fusion reactors
c. the sun
d. hydroelectric generators
e. none of these

22. Radiation is harmful to the human body because:

a. it causes the formation of free radicals
b. it causes molecules to fragment
c. it produces ions in the body
d. a and b only
e. a, b, and c

23. Fusion reactions are not generally used to provide energy on Earth because:

a. they give off too little energy
b. a very high temperature is required to start the reactions
c. the reactions proceed only at very low pressures
d. the starting materials (reactants) are too expensive
e. none of these

24. Radiation therapy includes the following use of radiation:

a. radiographs of bone tissue
b. radiation of cancer cells
c. dental X rays
d. use of radioactive tracer
e. all of the above

Answers to Practice Exercises

11.1

Type of radiation	Mass number	Charge	Symbol
alpha particles	4	+2	$^{4}_{2}\alpha$
beta particles	0	–1	$^{0}_{-1}\beta$
gamma rays	0	0	$^{0}_{0}\gamma$

11.2 a. $^{149}_{65}Tb \rightarrow ^{145}_{63}Eu + ^{4}_{2}\alpha$

b. $^{231}_{91}Pa \rightarrow ^{227}_{89}Ac + ^{4}_{2}\alpha$

11.3 a. $^{31}_{14}Si \rightarrow ^{31}_{15}P + ^{0}_{-1}\beta$

b. $^{59}_{26}Fe \rightarrow ^{59}_{27}Co + ^{0}_{-1}\beta$

11.4 a. $\left(\begin{array}{l}\text{Amount of radionuclide}\\ \text{undecayed after 3 half-lives}\end{array}\right) = (0.43\text{ mg}) \times \left(\frac{1}{2^3}\right) = (0.43\text{ mg}) \times \left(\frac{1}{8}\right) = 0.054\text{ mg}$

b. $\text{Number of half-lives} = \frac{122{,}000\text{ years}}{24{,}400\text{ years / half-life}} = 5\text{ half-lives}$

$\left(\begin{array}{l}\text{Amount of radionuclide}\\ \text{undecayed after 5 half-lives}\end{array}\right) = (0.43\text{ mg}) \times \left(\frac{1}{2^5}\right) = (0.43\text{ mg}) \times \left(\frac{1}{32}\right) = 0.013\text{ mg}$

11.5 $\left(\frac{1}{2}\right) \times \left(\frac{1}{2}\right) \times \left(\frac{1}{2}\right) = \frac{1}{8} = \frac{1}{2^3} = \frac{1}{2^n}$; $n = 3$ = number of half-lives

$\frac{135\text{ days}}{3\text{ half-lives}} = 45\text{ days} = 1$ half-life of iron-59

11.6 a. $^{10}_{5}B + ^{1}_{0}n \rightarrow ^{7}_{3}Li + ^{4}_{2}\alpha$

b. $^{7}_{3}Li + ^{1}_{1}p \rightarrow ^{4}_{2}He + ^{4}_{2}\alpha$

c. $^{10}_{5}B + ^{4}_{2}\alpha \rightarrow ^{13}_{7}N + ^{1}_{0}n$

d. $^{238}_{92}U + ^{14}_{7}N \rightarrow ^{247}_{99}Es + 5\,^{1}_{0}n$

11.7 step 1: $^{210}_{82}Pb \rightarrow ^{210}_{83}Bi + ^{0}_{-1}\beta$

step 2: $^{210}_{83}Bi \rightarrow ^{210}_{84}Po + ^{0}_{-1}\beta$

step 3: $^{210}_{84}Po \rightarrow ^{206}_{82}Pb + ^{4}_{2}\alpha$

11.8

Type of radiation	Speed	Penetration	Biological damage
alpha particles	slow	very little, stopped by skin or paper	ingestion – damages internal organs
beta particles	faster	penetrating, stopped by wood or aluminum foil	skin burns, ingestion – damages internal organs
gamma rays	fastest (speed of light)	very penetrating, stopped by lead or concrete	ionizing radiation causes serious damage to all tissues

Answers to Self-Test

The numbers in parentheses refer to sections in your textbook.
1. F; chemical reactions (11.13) **2**. F; gamma rays (11.2) **3**. T (11.3) **4**. T (11.8)
5. F; one-half (11.4) **6**. F; nuclear fusion (11.12) **7**. T (11.12) **8**. F; electrons (11.7)
9. F; larger (11.13) **10**. F; therapeutic application (11.11) **11**. T (11.9)
12. F; an unpaired electron (11.7) **13**. T (11.13) **14**. T (11.10) **15**. F; another (11.5)
16. b (11.3) **17**. a (11.3) **18**. d (11.4) **19**. b (11.4) **20**. b (11.2) **21**. a (11.12)
22. e (11.8) **23**. b (11.12) **24**. b (11.11)

Saturated Hydrocarbons — Chapter 12

Chapter Overview

Carbon compounds are the basis of life on Earth; all organic materials are carbon-based. The hydrocarbons, which you will study in this chapter, make up petroleum and are thus an important part of the industrial world, as fuel and in the manufacture of synthetic materials.

In this chapter you will find out how carbon forms such a vast variety of compounds. You will write structural and condensed structural formulas for alkanes and cycloalkanes and name them according to the IUPAC rules. You will identify and draw structural isomers and *cis-trans* isomers. You will write equations for the two major reactions of hydrocarbons and name halogenated hydrocarbons.

Practice Exercises

12.1 **Organic chemistry** (Sec. 12.1) is the study of **hydrocarbons** (compounds of hydrogen and carbon, Sec. 12.3) and **hydrocarbon derivatives** (compounds that contain carbon and hydrogen and one or more additional elements, Sec. 12.3).

Hydrocarbons may be **saturated** (containing only single carbon-to-carbon bonds, Sec. 12.3) or **unsaturated** (containing only single carbon-to-carbon multiple bonds, Sec. 12.3). The simplest saturated hydrocarbons are the **alkanes** (Sec. 12.4). Alkanes may be cyclic (carbon atoms arranged in a ring) or acyclic (not cyclic).

Methane, ethane, propane, butane, and pentane are acyclic alkanes containing one, two, three, four, and five carbon atoms, respectively. Keeping in mind that each carbon atom forms four covalent bonds, complete the following table.

Alkane	Molecular formula	Total number of atoms	Number of C–H bonds	Number of C–C bonds
methane				
ethane				
propane				
butane				
pentane				

12.2 The structures of alkanes and other organic compounds are usually represented in two dimensions, rather than three. The common types of representations are the **expanded structural formula** (Sec. 12.5), which shows all atoms and all bonds, the **condensed structural formula** (Sec. 12.5), which shows groupings of atoms, and the **skeletal formula** (Sec. 12.5), which shows carbon atoms and bonds but omits hydrogen atoms.

Complete the following table of structural representations. All carbons are connected in a straight chain.

Molecular formula	Condensed structural formula	Expanded structural formula	Skeletal formula
C_2H_6		H H | | H–C–C–H | | H H	
C_4H_{10}			
	$CH_3-CH_2-CH_2-CH_2-CH_3$		
			C–C–C–C–C–C

12.3 **Structural isomers** (Sec. 12.6) are compounds with the same molecular formula but different structural formulas. Draw and name three structural isomers of heptane, C_7H_{16}. Use condensed structural formulas. Rules for naming alkanes can be found in Section 12.8 of your textbook. (There are nine possible structural isomers for this molecular formula.)

C_7H_{16}	C_7H_{16}	C_7H_{16}

12.4 **Conformations** (Sec. 12.7) are differing orientations of a molecule made possible by rotation about a single bond. They are not isomers, because one form can change to another without breaking or forming bonds.

Using skeletal formulas, draw two conformations of the straight-chain alkane, heptane C_7H_{16}.

12.5 The IUPAC rules for naming organic compounds make it possible to give each compound a name that uniquely identifies it and also to draw its structural formula from that name.

Using the rules for nomenclature found in Section 12.8 in your textbook, give the IUPAC name for each of the following structures:

a.
```
CH3       CH3
|         |
CH2-CH2-CH-CH3
```

b.
```
                    CH3 CH3
                    |   |
CH3-CH2-CH2-CH-CH-CH3
```

c.
```
                CH3
                |
CH3             CH2 CH3
|               |   |
CH2-CH2-CH2-CH-CH-CH3
```

d.
```
    CH3
    |
    CH2
    |
CH3 CH2         CH3 CH3
|   |           |   |
CH2-CH-CH2-CH-CH-CH2-CH3
```

12.6 Once you have learned the IUPAC rules for naming organic compounds, you can translate the name of an alkane into a structural formula.

Give the condensed structural formulas for the following alkanes:

a. 3-methylhexane	b. 2,2-dimethylpentane
c. 3,5-dimethylheptane	d. 2,2,4,4-tetramethylhexane

12.7 Each carbon atom in a **hydrocarbon** (Sec. 12.3) can be classified according to the number of other carbon atoms to which it forms bonds: a primary carbon atom is bonded to one other carbon atom, a secondary to two, a tertiary to three, and a quaternary to four other carbon atoms.

Using the structural formulas that you drew in Practice Exercise 12.6, determine the total number of **primary, secondary, tertiary,** and **quaternary carbons** (Sec. 12.9) in each compound.

Compound	Primary carbons	Secondary carbons	Tertiary carbons	Quaternary carbons
3-methylhexane				
2,2-dimethylpentane				
3,5-dimethylheptane				
2,2,4,4-tetramethylhexane				

12.8 There are four common **alkyl groups** (Sec. 12.8) containing **branched chains** (Sec. 12.6, 12.10) that you should learn to name and draw.

Complete the following table by drawing the structural formulas for these four branched-chain alkyl groups:

a. isopropyl	b. *tert*-butyl	c. isobutyl	d. *sec*-butyl

12.9 In a **cycloalkane** (Sec. 12.11), the carbon atoms are attached to one another in a ringlike arrangement. The general formula for cycloalkanes is C_nH_{2n}. IUPAC naming procedures for cycloalkanes are found in Section 12.12 of your textbook. **Line-angle drawings** (Sec. 12.11) are often used to represent cycloalkane structures. A line-angle drawing can also denote a chain of carbon atoms using a sawtooth pattern of lines.

Give the IUPAC names for the following cycloalkanes:

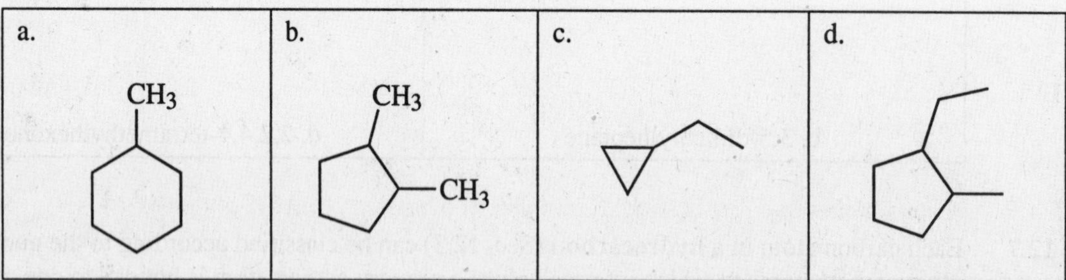

12.10 ***Cis-trans* isomers** (Sec. 12.13) are compounds that have the same molecular and structural formulas but that have different spatial arrangements of their atoms because rotation is restricted around bonds.

Give the IUPAC name for the *cis-trans* isomers in parts a and b.

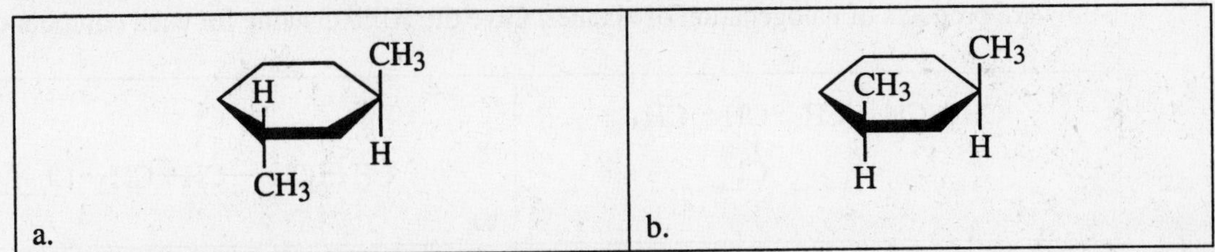

c. Draw a cyclohexane ring with two methyl groups, one on carbon 1 and one on carbon 2.
d. Draw a second cyclohexane ring with two methyl groups on carbon-1.
Determine whether *cis-trans* isomerism is possible for the structures you drew in parts c and d. Give the IUPAC name for the structures in parts c and d.

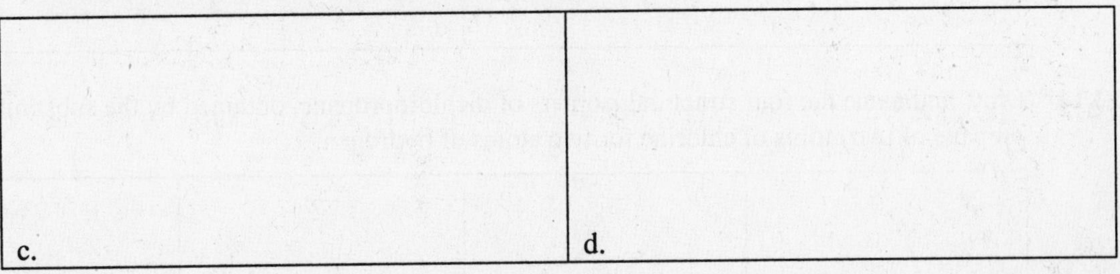

12.11 The problems in this exercise give a review of structural isomerism in alkanes. Compare the following pairs of molecules. Use line-angle notation to draw a structural formula for each molecule, and write its molecular formula. Tell whether the molecules in each pair are isomers of one another.

a. 1-methylcyclohexane and 1,3-dimethylcyclobutane

b. 2,3-dimethylpentane and 2,2,3-trimethylbutane

c. ethylcyclopropane and 2-methylbutane

12.12 One of the principal reactions of hydrocarbons is **combustion** (Sec. 12.16). Complete combustion is the reaction of a hydrocarbon with oxygen to produce carbon dioxide and water.

Write a balanced equation for the complete combustion of the following alkanes:

a. pentane

b. 2,3-dimethylhexane

12.13 The other important reaction of alkanes is **halogenation** (Sec. 12.16). The compounds shown below are products of halogenation of alkanes. Give the IUPAC name for each compound:

$CH_3-CH_2-CH_2-CH(Cl)-CH_3$ a.	$CH_3-CH_2-CH(Cl)-CH_2-Cl$ b.
$H_3C-CH_2-CH(Cl)-CH(CH_3)-CH_3$ c.	(cyclopentane ring with H, Br on one carbon and Br, H on the adjacent carbon) d.

12.14 Draw and name the four structural isomers of dichloropropane, obtained by the substitution in propane of two atoms of chlorine for two atoms of hydrogen.

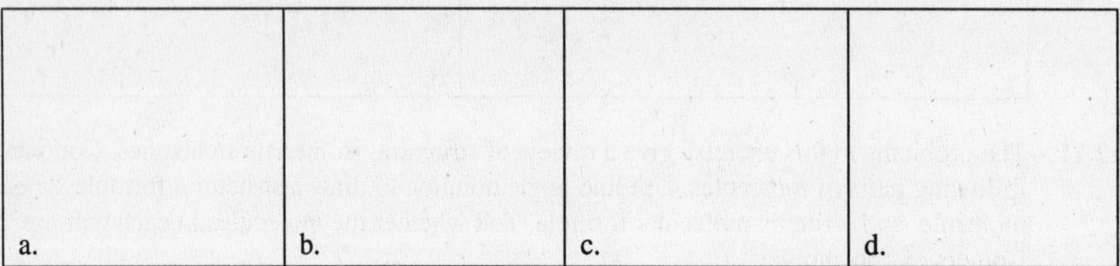

a.	b.	c.	d.

Self-Test

True-false: Indicate whether the following statements are true or false. If the statement is false, give the word or phrase that may be substituted for the underlined portion to make the statement true.

1. The smallest alkane is ethane.
2. Alkyl groups cannot rotate around the single bonds between the carbons in the ring of a cycloalkane.
3. Straight-chain pentane would have a higher boiling point than its structural isomer, 2,2-dimethylpropane.
4. Carbon atoms can form compounds containing chains and rings because carbon has six valence electrons.
5. A hydrocarbon contains only carbon, hydrogen, and oxygen atoms.
6. A saturated hydrocarbon contains at least one carbon-carbon double bond or triple bond.
7. The alkane 2-methylpentane contains one tertiary carbon atom.
8. An acyclic hydrocarbon contains carbon atoms arranged in a ring structure.
9. The alkyl group $-CH_2-CH_2-CH_3$ is named propanyl.
10. It takes a minimum of three carbon atoms to form a cyclic arrangement of carbon atoms.
11. The formula for a cycloalkane is C_nH_{2n}.
12. Natural gas consists mainly of ethane.

13. Alkanes are good preservatives for metals because they have low boiling points.
14. The reaction of alkanes with oxygen to form carbon dioxide and water is a substitution reaction.
15. The IUPAC name for isopropyl chloride is 2-chloropropane.
16. The compounds called CFCs are chlorofluorocarbons.

Multiple choice:

17. The straight-chain alkane having the molecular formula C_6H_{14} is called:

    a. hexane b. heptane c. pentane
    d. nonane e. none of these

18. Using IUPAC rules for naming, if the parent compound is pentane, an ethyl group could be attached to carbon:

    a. 1 b. 2 c. 3 d. 4 e. 5

19. How many structural isomers can butane have?

    a. two b. three c. four d. five e. six

20. Cyclopropane has the following molecular formula:

    a. CH_4 b. C_2H_6 c. C_3H_8 d. C_4H_{10} e. none of these

21. Which of the following compounds is an isomer of hexane?

    a. methylcyclopentane b. 2-methylbutane c. 3-ethylpentane
    d. 2-methylpentane e. none of these

22. Which of the following compounds forms *cis-trans* isomers?

    a. 2,3-dimethylpentane b. 2,2-dimethylpentane
    c. 1,1-dimethylcyclopentane d. 1,2-dimethylcyclopentane
    e. none of these

23. Choose the correct name for this compound:

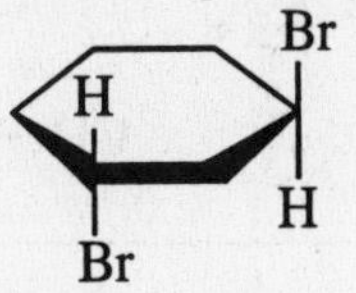

    a. *trans*-1,3-dibromocyclohexane
    b. *cis*-1,3-bromohexane
    c. *trans*-1,5-bromocyclohexane
    d. *cis*-1,2-bromocyclohexane
    e. none of these

24. Which of the following compounds can have structural isomers?

    a. CH_3Cl b. C_3H_7Cl c. C_3H_8 d. C_2H_5Cl e. none of these

25. Which of the following is a correct IUPAC name?

    a. 2-methylcyclobutane b. *cis*-2,3-dimethylpentane
    c. 1-methylbutane d. 3-ethylhexane
    e. *cis*-1,2-methylpropane

26. How many possible isomers can be written for dichloropropane?

    a. two b. three c. four d. five e. six

27. How many possible isomers can be written for dimethylcyclobutane?

a. two b. three c. four d. five e. six

28. Draw structures for the following compounds:

2-bromo-3-methylbutane	*trans*-1,2-dichlorocyclopropane

Answers to Practice Exercises

12.1

Alkane	Molecular formula	Total number of atoms	Number of C–H bonds	Number of C–C bonds
methane	CH_4	5	4	0
ethane	C_2H_6	8	6	1
propane	C_3H_8	11	8	2
butane	C_4H_{10}	14	10	3
pentane	C_5H_{12}	17	12	4

12.2

Molecular formula	Condensed structural formula	Expanded structural formula	Skeletal formula
C_2H_6	CH_3-CH_3	H H | | H−C−C−H | | H H	C−C
C_4H_{10}	$CH_3-CH_2-CH_2-CH_3$	H H H H | | | | H−C−C−C−C−H | | | | H H H H	C−C−C−C
C_5H_{12}	$CH_3-CH_2-CH_2-CH_2-CH_3$	H H H H H | | | | | H−C−C−C−C−C−H | | | | | H H H H H	C−C−C−C−C
C_6H_{14}	$CH_3-CH_2-CH_2-CH_2-CH_2-CH_3$	H H H H H H | | | | | | H−C−C−C−C−C−C−H | | | | | | H H H H H H	C−C−C−C−C−C

12.3

CH_3 $CH_3-CH_2-CH-CH-CH_3$ CH_3	$CH_3-CH-CH_2-CH-CH_3$ CH_3 CH_3	CH_3 CH_3 $CH_3-CH-C-CH_3$ CH_3
2,3-dimethylpentane	2,4-dimethylpentane	2,2,3-trimethylbutane

12.4

C—C—C—C C C—C	C C C—C—C C—C
heptane	heptane

There are many more conformations of heptane; these are just two of the representations.

12.5

a. 2-methylpentane

b. 2,3-dimethylhexane

c. 3-ethyl-2-methylheptane

d. 6-ethyl-3,4-dimethylnonane

12.6

CH_3 $CH_3-CH_2-CH-CH_2-CH_2-CH_3$ a. 3-methylhexane	CH_3 $CH_3-C-CH_2-CH_2-CH_3$ CH_3 b. 2,2-dimethylpentane
CH_3 $CH_3-CH_2-CH-CH_2-CH-CH_2-CH_3$ CH_3 c. 3,5-dimethylheptane	CH_3 CH_3 $CH_3-C-CH_2-C-CH_2-CH_3$ CH_3 CH_3 d. 2,2,4,4-tetramethylhexane

12.7

	Primary carbons	Secondary carbons	Tertiary carbons	Quaternary carbons
3-methylhexane	3	3	1	0
2,2-dimethylpentane	4	2	0	1
3,5-dimethylheptane	4	3	2	0
2,2,4,4-tetramethylhexane	6	2	0	2

12.8

a. isopropyl	b. *tert*-butyl	c. isobutyl	d. *sec*-butyl
$CH_3-CH(CH_3)-$	$CH_3-C(CH_3)_2-$	$CH_3-CH(CH_3)-CH_2-$	$CH_3-CH_2-CH(CH_3)-$

12.9 a. methylcyclohexane
b. 1,2-dimethylcyclopentane
c. ethylcyclopropane
d. 1-ethyl-2-methylcyclopentane

12.10 a. *trans*-1,3-dimethylcyclohexane
b. *cis*-1,3-dimethylcyclohexane

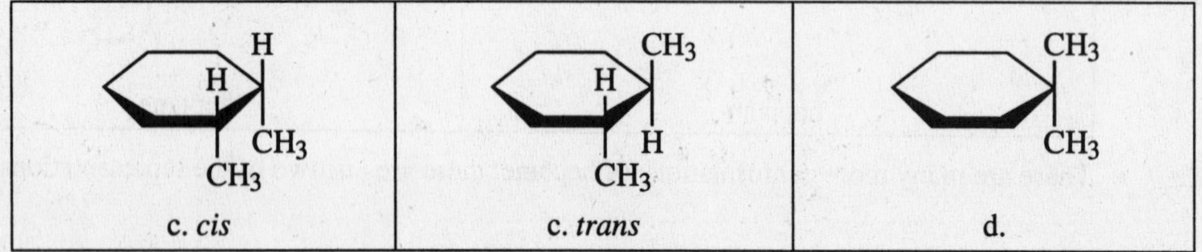

c. *cis* c. *trans* d.

c. Yes, *cis-trans* isomerism is possible ,because the methyl groups can be on the same side of the molecule or one methyl group can be above and one below. The molecules are *cis*-1,2-dimethylcyclohexane and *trans*-1,2-dimethylcyclohexane.
d. No, there are not *cis* and *trans* forms of this molecule because both methyl groups are attached to the same carbon. The name is 1,1-dimethylcyclohexane.

12.11 a.

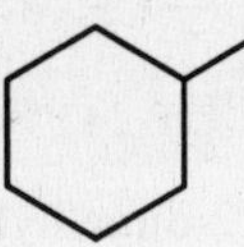

1-methylcyclohexane C_7H_{14}

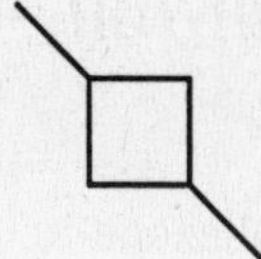

1,3-dimethylcyclobutane C_6H_{12}

not isomers (different numbers of carbons and hydrogens)

b.

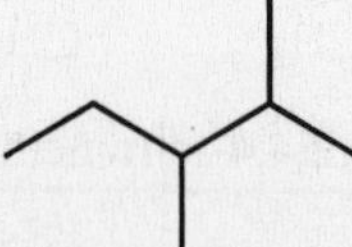

2,3-dimethylpentane C_7H_{16}

2,2,3-trimethylbutane C_7H_{16}

isomers (same molecular formula)

c.

ethylcyclopropane C_5H_{10}

2-methylbutane C_5H_{12}

not isomers (different numbers of hydrogens)

12.12 a. pentane

$$C_5H_{12} + 8O_2 \rightarrow 5CO_2 + 6H_2O$$

b. 2,3-dimethylhexane

$$2C_8H_{18} + 25O_2 \rightarrow 16CO_2 + 18H_2O$$

12.13
a. 2-chloropentane
b. 1,2-dichlorobutane
c. 3-chloro-2-methylpentane
d. *trans*-1,3-dibromocyclopentane

12.14

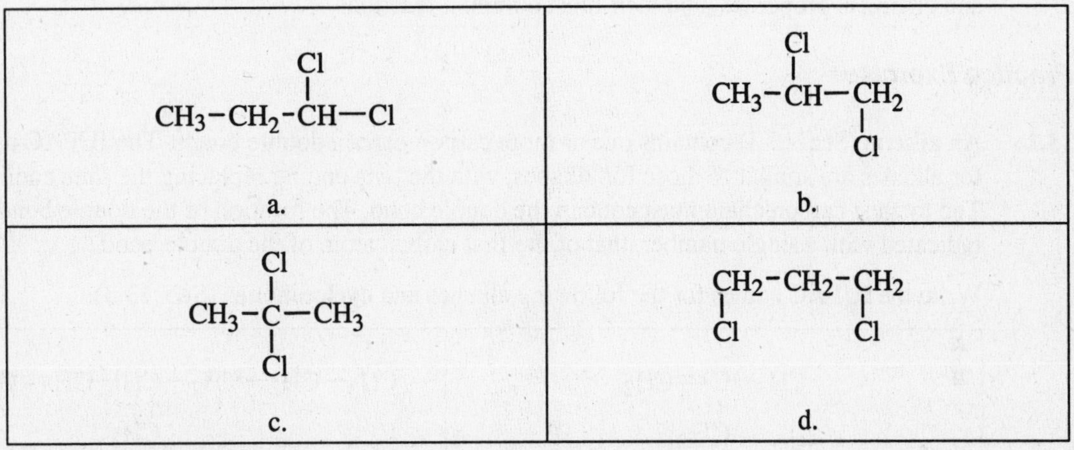

$CH_3-CH_2-CH(Cl)-Cl$ a.	$CH_3-CH(Cl)-CH_2(Cl)$ b.
$CH_3-C(Cl)_2-CH_3$ c.	$CH_2(Cl)-CH_2-CH_2(Cl)$ d.

a. 1,1-dichloropropane
b. 1,2-dichloropropane
c. 2,2-dichloropropane
d. 1,3-dichloropropane

Answers to Self-Test

The numbers in parentheses refer to sections in your textbook.
1. F; methane (12.4) **2.** T (12.13) **3.** T (12.15) **4.** F; four (12.2)
5. F; carbon and hydrogen (12.3) **6.** F; unsaturated (12.3) **7.** T (12.9)
8. F; cyclic (12.11) **9.** F; propyl (12.8) **10.** T (12.11) **11.** T (12.11)
12. F; methane (12.4) **13.** F; low water solubility (12.15) **14.** F; combustion (12.16)
15. T (12.17) **16.** T (12.17) **17.** a (12.8) **18.** c (12.8) **19.** a (12.6) **20.** e; C_3H_6 (12.11)
21. d (12.6) **22.** d (12.13) **23.** a (12.13) **24.** b (12.6, 12.17) **25.** d (12.8, 12.12, 12.13)
26. c (12.6, 12.17) **27.** d (12.13)
28.

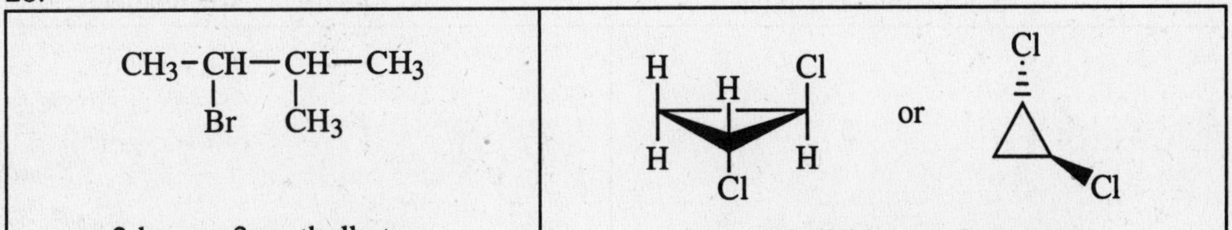

$CH_3-CH(Br)-CH(CH_3)-CH_3$ 2-bromo-3-methylbutane	H, Cl, H, H, H, Cl or Cl, Cl *trans*-1,2-dichlorocyclopropane

Unsaturated Hydrocarbons — Chapter 13

Chapter Overview

Unsaturated hydrocarbons contain fewer than the largest possible number of hydrogen atoms because they have one or more carbon-carbon double or triple bonds. Because a multiple bond is more easily broken than a single bond, unsaturated hydrocarbons are more reactive chemically than saturated hydrocarbons.

In this chapter you will name and write structural formulas for alkenes, alkynes, and aromatic hydrocarbons. You will learn the characteristics of these compounds, their physical and chemical properties, and their most common reactions.

Practice Exercises

13.1 An **alkene** (Sec. 13.3) contains one or more carbon-carbon double bonds. The IUPAC names for alkenes are similar to those for alkanes, with the *-ene* ending replacing the *-ane* ending. The longest carbon chain must contain the double bond. The location of the double bond is indicated with a single number, that of the first carbon atom of the double bond.

Write the IUPAC names for the following alkenes and **cycloalkenes** (Sec. 13.3).

$CH_3-\underset{\displaystyle CH_3}{\underset{\displaystyle\vert}{C}}=CH_2$ a.	$CH_3-CH_2-CH_2-\underset{\displaystyle CH_3}{\underset{\displaystyle\vert}{CH}}-CH=CH_2$ b.
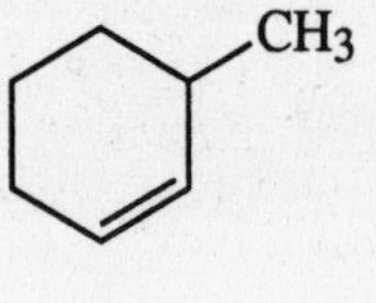 (cyclohexene ring with CH_3 substituent) c.	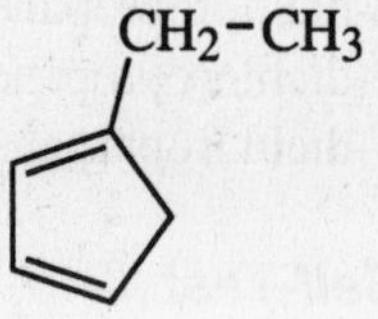 (cyclopentadiene ring with CH_2-CH_3 substituent) d.

13.2 Draw structural formulas for the following alkenes and cycloalkenes.

a. 3-methyl-1-pentene	b. 2-methyl-1,3-pentadiene
c. 1-methylcyclopentene	d. 1,4-cycloheptadiene

13.3 Using common names for the three most frequently encountered alkenyl groups listed in Section 13.3 in your textbook, give the common names for the following structures.

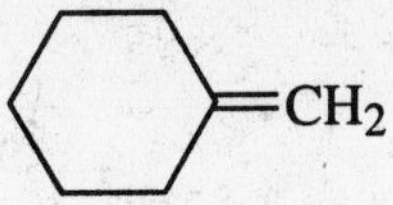 a. (cyclohexane)$=CH_2$	b. $CH_2=CH-Br$	c. $CH_2=CH-CH_2-I$

13.4 *Cis-trans* isomerism is possible for some alkenes, because the double bond is rigid, preventing rotation around its axis. For *cis-trans* isomers to exist, each of the two carbon atoms of the double bond must have two different groups attached to it.

Determine whether each of these alkenes can exist as *cis-trans* isomers. If isomers do exist, draw and give the IUPAC name for each isomer.

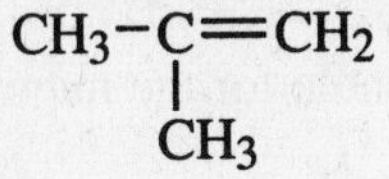 a. $CH_3-C(CH_3)=CH_2$	b. $CH_3-CH=CH-CH_2-CH_3$
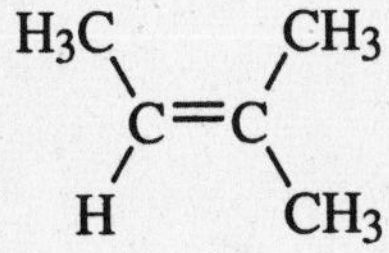 c. $(H_3C)(H)C=C(CH_3)(CH_3)$	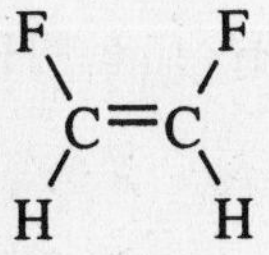 d. $(F)(H)C=C(F)(H)$

13.5 Draw structural formulas for the following alkenes.

a. *trans*-3-hexene	b. *cis*-2-hexene	c. *trans*-1,2-dibromoethene

13.6 The most important reactions of alkenes are **addition reactions** (Sec. 13.7). **Symmetrical addition reactions** (Sec. 13.7), such as hydrogenation and halogenation, involve adding identical atoms to each carbon of the double bond. In hydrogenation, a hydrogen atom is added to each carbon of the double bond by heating the alkene and H_2 in the presence of a catalyst. Halogenation involves the use of Br_2 or Cl_2 to add a halogen atom to each carbon of the double bond.

Complete the following reactions. Give the structural formula for the product.

a. $CH_3-CH_2-CH=CH_2 \quad + \quad H_2 \xrightarrow[\text{catalyst}]{\text{Ni}}$

b. $CH_3-CH_2-CH=CH_2 \quad + \quad Br_2 \longrightarrow$

c. (cyclopentene ring)$-CH_3 \quad + \quad H_2 \xrightarrow[\text{catalyst}]{\text{Ni}}$

13.7 Addition reactions may also be **unsymmetrical** (Sec. 13.7); different atoms or groups of atoms are added to the carbons of the double bond. Hydrohalogenation and hydration are important types of unsymmetrical addition. **Markovnikov's rule** (Sec. 13.7) states that, in an unsymmetrical addition, the hydrogen atom from the molecule being added becomes attached to the unsaturated carbon atom that already has the most hydrogen atoms.

Complete the following reactions. In each case give the structural formula for the major expected product.

a. $CH_3-CH_2-CH=CH_2 \quad + \quad HBr \longrightarrow$

b. $CH_3-CH_2-CH=CH_2 \quad + \quad H_2O \xrightarrow{H_2SO_4}$

c. (cyclopentene ring)$-CH_3 \quad + \quad HBr \longrightarrow$

d. (cyclopentene ring)$-CH_3 \quad + \quad H_2O \xrightarrow{H_2SO_4}$

13.8 Write the name of the alkene that could be used to prepare each of the following compounds. Remember Markovnikov's rule.

$CH_3-CH_2-CH_2-CH(Br)-CH_3$	$CH_3-CH(CH_3)-CH(OH)-CH_3$	1-bromo-1-methylcyclopentane (Br, CH_3 on ring carbon)
a.	b.	c.

13.9 How many molecules of hydrogen gas, H_2, would react with one molecule of each of the following compounds? Give the structure and IUPAC name for each product.

a. $CH_2{=}CH{-}CH_2{-}CH{=}CH_2$	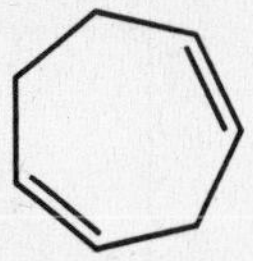 b. (cycloheptadiene structure)

13.10 A **polymer** (Sec. 13.8) is a very large molecule composed of many identical repeating units called **monomers** (Sec. 13.8). Alkenes can form **addition polymers** (Sec. 13.8) when the alkene monomers simply add together. The double bond of the alkene is broken, and single carbon-carbon bonds form between monomers.

Draw structural formulas for the monomer units from which these addition polymers were made and for the first three repeating units of the polymers.

General formula of polymer	Monomer formula	First three units of polymer structural formula
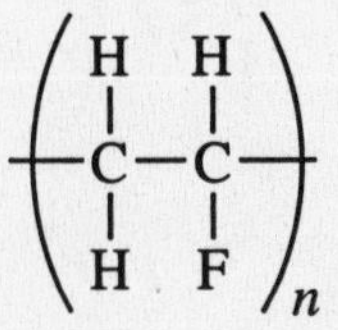 $\left(-\underset{H}{\overset{H}{C}}-\underset{F}{\overset{H}{C}}- \right)_n$		
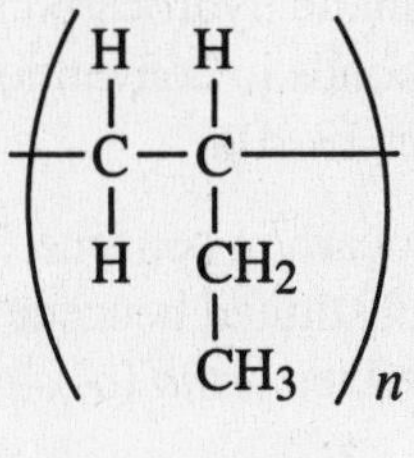 $\left(-\underset{H}{\overset{H}{C}}-\underset{CH_2-CH_3}{\overset{H}{C}}- \right)_n$		

13.11 An **alkyne** (Sec. 13.9) has one or more carbon-carbon triple bonds. The rules for naming alkynes are the same as those for naming alkenes, with the ending *-yne* instead of *-ene*.

Give the IUPAC names for the following alkynes:

a. $CH_3{-}CH_2{-}CH_2{-}C{\equiv}CH$	b. $\underset{Br}{CH_2}{-}CH_2{-}C{\equiv}CH$	c. $CH_3{-}CH_2{-}\underset{CH_3}{\overset{CH_3}{C}}{-}C{\equiv}C{-}CH_3$

13.12 Draw structural formulas for the following alkynes:

a. 3-methyl-1-pentyne	b. 4-chloro-2-pentyne	c. 4,5-dimethyl-2-hexyne

13.13 Addition reactions of alkynes are similar to those of alkenes. However, two molecules of a specific reactant can add to the triple bond.

Complete the following reactions. Give the structural formula for the major expected product.

a. $CH_3{-}CH_2{-}CH_2{-}C{\equiv}CH \quad + \quad 2Cl_2 \longrightarrow$

b. $CH_3{-}CH_2{-}CH_2{-}C{\equiv}CH \quad + \quad 2HBr \longrightarrow$

c. $CH_3{-}CH_2{-}CH_2{-}C{\equiv}CH \quad + \quad 1HCl \longrightarrow$

d. $CH_3{-}CH_2{-}CH_2{-}C{\equiv}CH \quad + \quad 2H_2 \xrightarrow[\text{catalyst}]{\text{Ni}}$

13.14 **Aromatic hydrocarbons** (Sec. 13.10) are unsaturated cyclic compounds that do not readily undergo addition reactions. The carbon-carbon bonds in simple aromatic hydrocarbons, such as benzene, are all equivalent, indicating the the double and single bonds in conventional structural diagrams of these compounds are **delocalized bonds** (Sec. 13.10).

The naming of the compounds below is based on the aromatic hydrocarbon benzene. The name of a substituent on the benzene ring is used as a prefix. A disubstituted benzene may be named using either numbered positions on the ring or one of the prefixes *ortho* (*o*-), *meta* (*m*-), and *para* (*p*-) (Sec. 13.11).

a. Give the IUPAC names for the following aromatic compounds, using numbered ring positions.

b. Name the compounds using prefixes for the positions of the substituents.

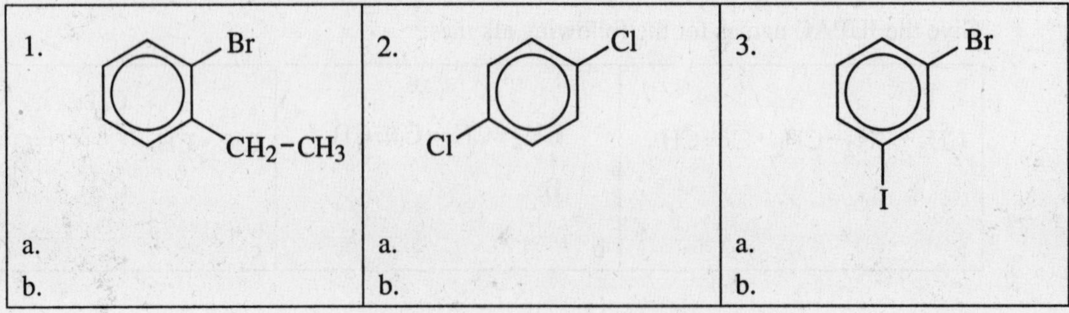

13.15 If the group attached to the benzene ring is not easily named as a substituent, the benzene ring is treated as the attached group and called a *phenyl* group. The compound is then named as an alkane, an alkene, or an alkyne.

Write the IUPAC name for the following phenyl-substituted compounds:

a. $CH_3-CH_2-CH_2-CH(C_6H_5)-CH_2Cl$	b. $CH_3-CH=CH-CH(C_6H_5)-CH_3$

13.16 Draw structural formulas for the following compounds:

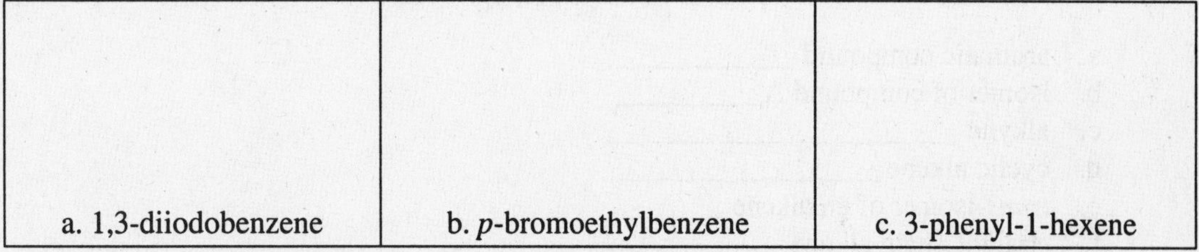

a. 1,3-diiodobenzene
b. *p*-bromoethylbenzene
c. 3-phenyl-1-hexene

13.17 Aromatic hydrocarbons do not readily undergo addition reactions. The most important reactions of aromatic compounds are substitution reactions. These include alkylation (using alkyl halides and the catalyst $AlCl_3$) and halogenation (using Br_2 or Cl_2 in the presence of a catalyst, $FeBr_3$ or $FeCl_3$).

Complete the following equations by supplying the missing information.

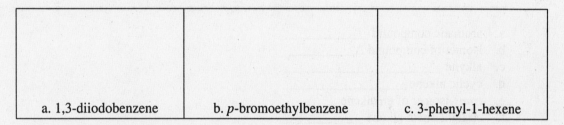

a. $C_6H_6 + CH_3Cl \xrightarrow{AlCl_3}$? + HCl

b. ? + ? $\xrightarrow{FeBr_3} C_6H_5Br + HBr$

c. $C_6H_6 + Cl_2 \xrightarrow{FeCl_3}$? + ?

d. ? + ? $\xrightarrow{?} C_6H_5-CH_2-CH_3 + HCl$

13.18 Use this identification exercise to review your knowledge of the structures introduced in this chapter. Using structures A through F, give the best choice for each of the terms below.

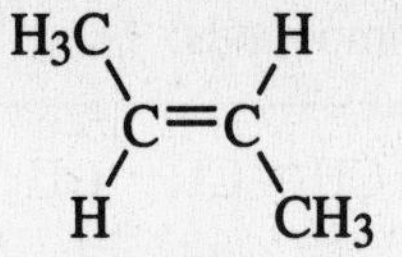 H_3C H C=C H CH_3 A.	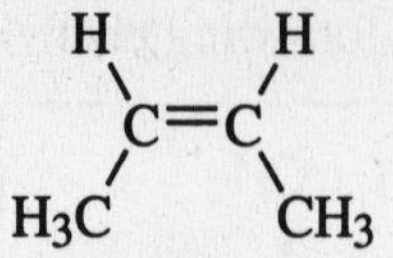 H H C=C H_3C CH_3 B.	$CH_3-CH_2-CH_2-CH_3$ C.
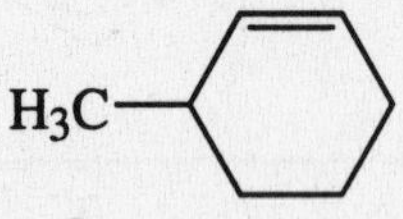 H_3C- (cyclohexene ring) D.	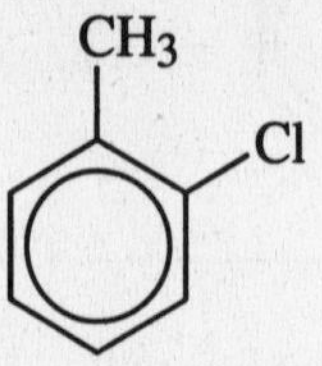 CH_3 Cl (benzene ring) E.	$CH_3-CH_2-C\equiv CH$ F.

a. aromatic compound ____________
b. isomer of compound A ________
c. alkyne ____________________
d. cyclic alkene ________________
e. *trans*-isomer of an alkene _______
f. straight-chain alkane ___________

Give the IUPAC names for structures A through F above:
g. Structure A ________________________
h. Structure B ________________________
i. Structure C ________________________
j. Structure D ________________________
k. Structure E ________________________
l. Structure F ________________________

Self-Test

True-false: Indicate whether the following statements are true or false. If the statement is false, give the word or phrase that may be substituted for the underlined portion to make the statement true.

1. The specific part of a molecule that governs the molecule's chemical properties is called a <u>functional</u> group.
2. The general formula for an <u>alkene</u> is C_nH_{2n}.
3. When naming alkenes by the IUPAC system, select as the parent carbon chain the longest chain that contains <u>at least one carbon atom</u> of the double bond.
4. The name of a cycloalkene <u>does not include</u> the numbered location of the double bond.
5. The carbon atoms at each end of a double bond have a <u>trigonal planar</u> arrangement of bonds.
6. The common name for the simplest alkyne is <u>ethylene</u>.
7. Benzene undergoes an <u>addition</u> reaction with bromine in the presence of a catalyst.
8. <u>Propene</u> is the simplest alkene that has *cis-trans* isomerism.

9. If 2-butene is halogenated with bromine gas, the most probable product is 2-bromobutane.
10. Hydration of an alkene produces an alcohol.
11. Polyethylene contains many double bonds.

Multiple choice:

12. According to Markovnikov's rule, addition of HBr to the double bond of 1-hexene would produce:
    a. 1-bromohexane b. 1,2-dibromohexane c. 2-bromo-1-hexene
    d. 2-bromohexane e. none of these

13. The geometry of the carbon-carbon triple bond of an alkyne is:
    a. linear b. trigonal planar c. tetrahedral
    d. angular e. none of these

14. Which of the following is a correct name according to IUPAC rules?
    a. 2-methylbenzene b. 1-chlorobenzene c. 1-bromo-2-chlorobenzene
    d. 2,4-dichlorobenzene e. none of these

15. Another name for *meta*-dichlorobenzene is:
    a. 1,2-dichlorobenzene b. 2,3-dichlorobenzene c. 1,3-dichlorobenzene
    d. 1,4-dichlorobenzene e. none of these

16. How many mono-chloro isomers of 1-butene are possible?
    a. four b. one c. five d. three e. none of these

17. Aromatic compounds have structures based on what parent molecule?
    a. benzene b. cyclopropane c. cyclohexane
    d. hexane e. none of these

18. Which of the following compounds does *not* have the formula C_5H_8?
    a. 1-methylcyclobutene b. 3-methyl-1-butyne c. 2-methyl-1-butene
    d. 1,3-pentadiene e. none of these

19. If you wished to prepare 2-bromo-3-methylbutane by the addition of HBr to an alkene, which of these alkenes would you use?
    a. 2-methyl-2-butene b. 2-methyl-1-butene c. 3-methyl-1-butene
    d. 3-methyl-2-butene e. none of these

20. Which of the following compounds would be the most likely to undergo an addition reaction with HCl?
    a. toluene b. cyclohexene c. benzene
    d. heptane e. none of these

Answers to Practice Exercises

13.1
a. 2-methylpropene
b. 3-methyl-1-hexene
c. 3-methylcyclohexene
d. 1-ethyl-1,3-cyclopentadiene

13.2

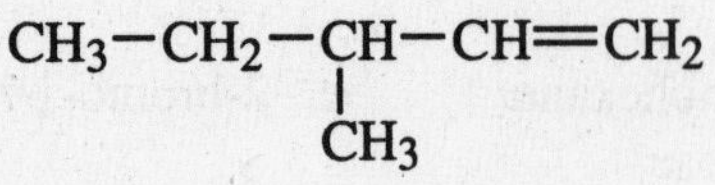 $CH_3-CH_2-CH(CH_3)-CH=CH_2$ a. 3-methyl-1-pentene	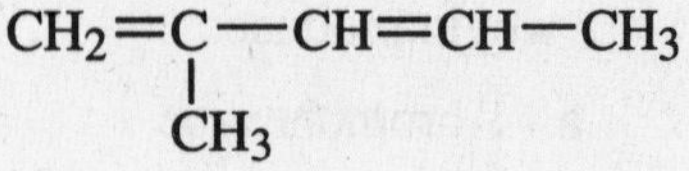 $CH_2=C(CH_3)-CH=CH-CH_3$ b. 2-methyl-1,3-pentadiene
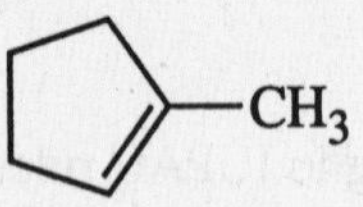 (cyclopentene ring)$-CH_3$ c. 1-methylcyclopentene	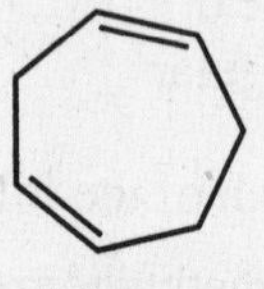 (cycloheptadiene ring) d. 1,4-cycloheptadiene

13.3
a. methylene cyclohexane
b. vinyl bromide
c. allyl iodide

13.4
a. no

b. yes; *trans*-2-pentene and *cis*-2-pentene

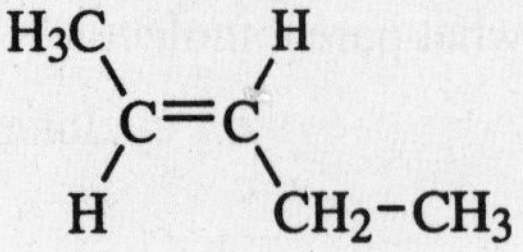 H_3C, H / C=C / H, CH_2-CH_3 *trans*-2-pentene	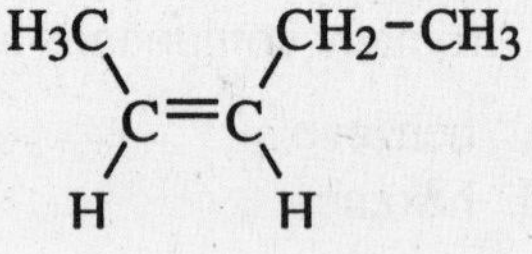 H_3C, CH_2-CH_3 / C=C / H, H *cis*-2-pentene

c. no

d. yes; *cis*-1,2-difluoroethene and *trans*-1,2-difluoroethene

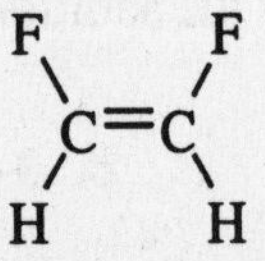 F, F / C=C / H, H *cis*-1,2-difluoroethene	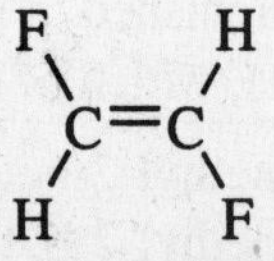 F, H / C=C / H, F *trans*-1,2-difluoroethene

13.5

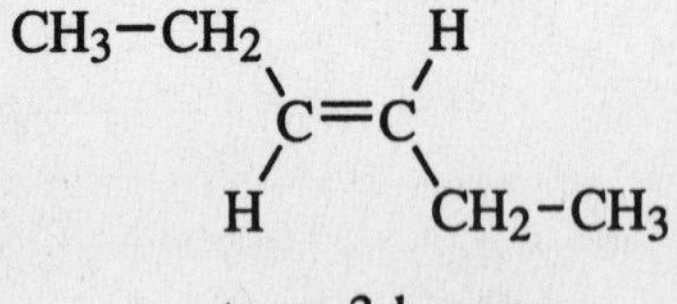 CH_3-CH_2, H / C=C / H, CH_2-CH_3 a. *trans*-3-hexene	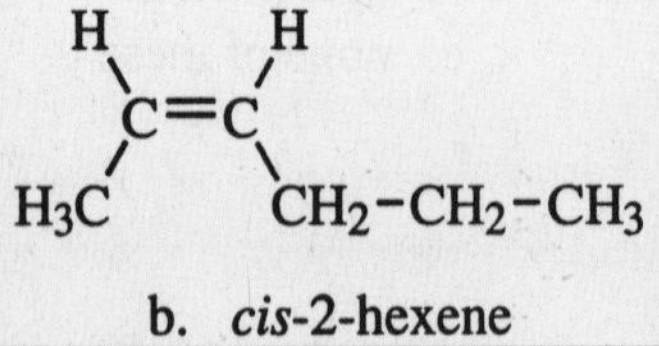 H, H / C=C / H_3C, $CH_2-CH_2-CH_3$ b. *cis*-2-hexene	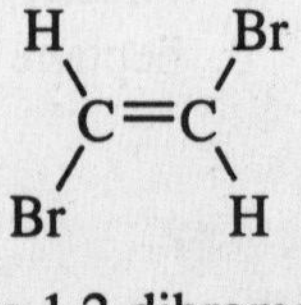 H, Br / C=C / Br, H c. *trans*-1,2-dibromoethene

13.6 a. $CH_3-CH_2-CH{=}CH_2 + H_2 \xrightarrow[\text{catalyst}]{\text{Ni}} CH_3-CH_2-CH_2-CH_3$

b. $CH_3-CH_2-CH{=}CH_2 + Br_2 \longrightarrow CH_3-CH_2-CH(Br)-CH_2(Br)$

c. 3-methylcyclopentene ($C_5H_7-CH_3$) $+ H_2 \xrightarrow[\text{catalyst}]{\text{Ni}}$ methylcyclopentane ($C_5H_9-CH_3$)

13.7 a. $CH_3-CH_2-CH{=}CH_2 + HBr \longrightarrow CH_3-CH_2-CH(Br)-CH_3$

b. $CH_3-CH_2-CH{=}CH_2 + H_2O \xrightarrow{H_2SO_4} CH_3-CH_2-CH(OH)-CH_3$

c. 1-methylcyclopentene ($C_5H_7-CH_3$) $+ HBr \longrightarrow$ 1-bromo-1-methylcyclopentane (ring carbon bearing CH_3 and Br)

d. 1-methylcyclopentene ($C_5H_7-CH_3$) $+ H_2O \xrightarrow{H_2SO_4}$ 1-methylcyclopentanol (ring carbon bearing CH_3 and OH)

13.8 a. 1-pentene
b. 3-methyl-1-butene
c. 1-methylcyclopentene

13.9 a. 2 molecules of hydrogen gas
b. 2 molecules of hydrogen gas

$CH_3-CH_2-CH_2-CH_2-CH_3$	(cycloheptane ring)
a. pentane	b. cycloheptane

13.10

General formula of polymer	Monomer formula	First three units of polymer structural formula
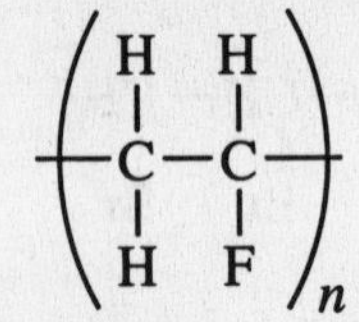 $\left(-CH_2-CHF-\right)_n$	$CH_2{=}CH{-}F$	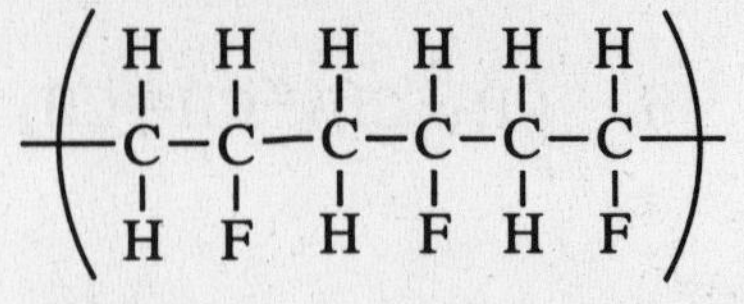 $-CH_2-CHF-CH_2-CHF-CH_2-CHF-$
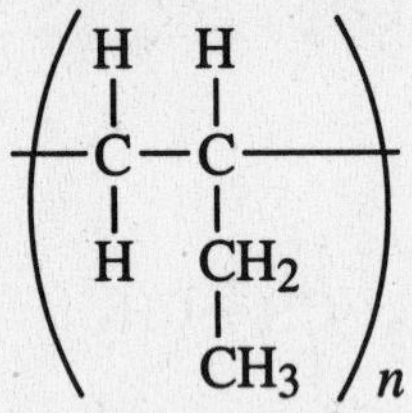 $\left(-CH_2-CH(CH_2CH_3)-\right)_n$	$CH_2{=}CH{-}CH_2{-}CH_3$	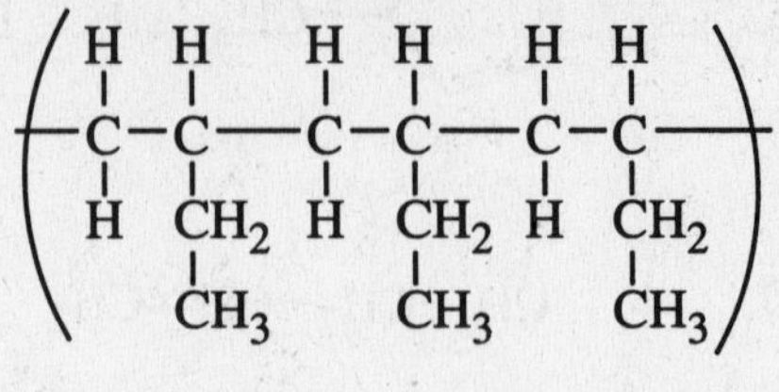 $-CH_2-CH(CH_2CH_3)-CH_2-CH(CH_2CH_3)-CH_2-CH(CH_2CH_3)-$

13.11 a. 1-pentyne
b. 4-bromo-1-butyne
c. 4,4-dimethyl-2-hexyne

13.12

$CH_3-CH_2-CH(CH_3)-C\equiv CH$	$CH_3-CH(Cl)-C\equiv C-CH_3$	$CH_3-CH(CH_3)-CH(CH_3)-C\equiv C-CH_3$
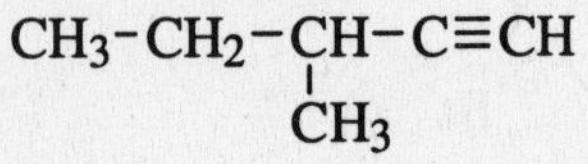 a. 3-methyl-1-pentyne	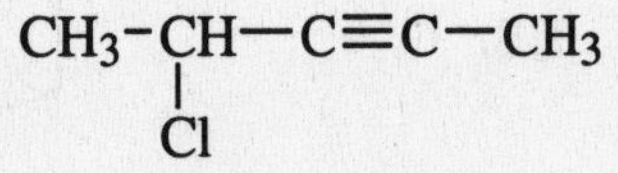 b. 4-chloro-2-pentyne	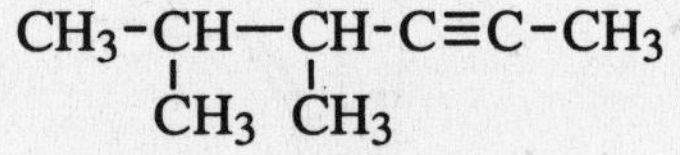 c. 4,5-dimethyl-2-hexyne

13.13

a. $CH_3-CH_2-CH_2-C\equiv CH + 2Cl_2 \longrightarrow CH_3-CH_2-CH_2-CCl_2-CHCl_2$

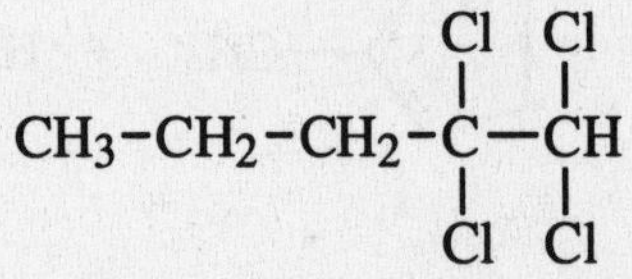

b. $CH_3-CH_2-CH_2-C\equiv CH + 2HBr \longrightarrow CH_3-CH_2-CH_2-CBr_2-CH_3$

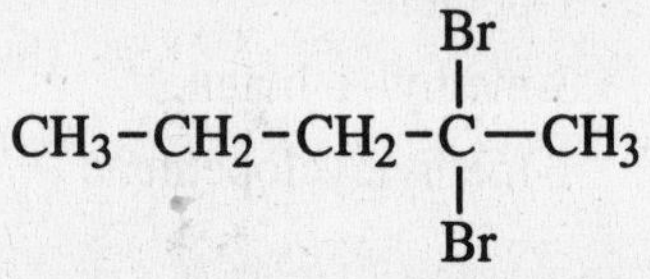

c. $CH_3-CH_2-CH_2-C\equiv CH + 1HCl \longrightarrow CH_3-CH_2-CH_2-C(Cl){=}CH_2$

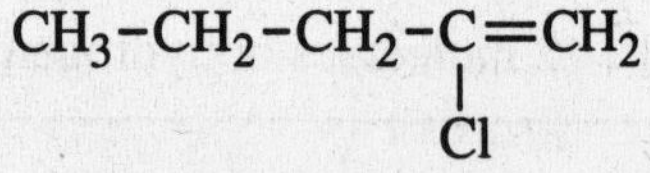

d. $CH_3-CH_2-CH_2-C\equiv CH + 2H_2 \xrightarrow[\text{catalyst}]{\text{Ni}} CH_3-CH_2-CH_2-CH_2-CH_3$

13.14 1. a. 1-bromo-2-ethylbenzene b. *o*-bromoethylbenzene
2. a. 1,4-dichlorobenzene b. *p*-dichlorobenzene
3. a. 1,3-bromoiodobenzene b. *m*-bromoiodobenzene

13.15 a. 1-chloro-2-phenylpentane
b. 4-phenyl-2-pentene

13.16

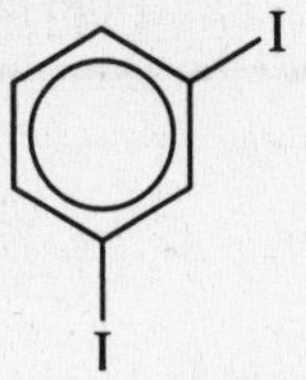 a. 1,3-diiodobenzene	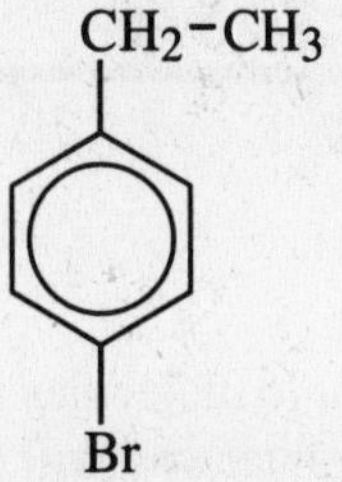 b. *p*-bromoethylbenzene	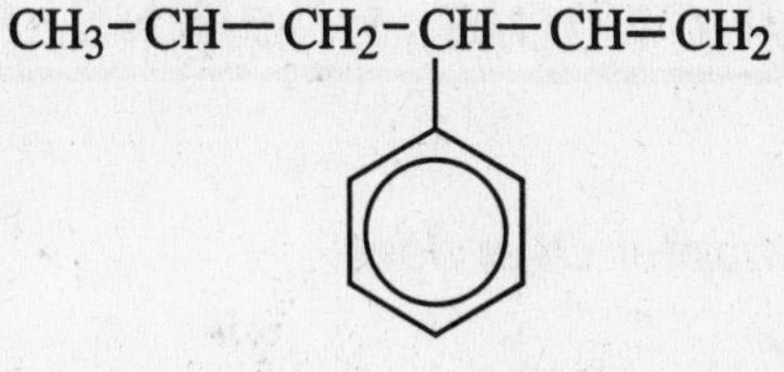 c. 3-phenyl-1-hexene
(benzene ring with I at positions 1 and 3)	(benzene ring with CH_2-CH_3 and Br para)	$CH_3-CH-CH_2-CH-CH=CH_2$ with phenyl on C3

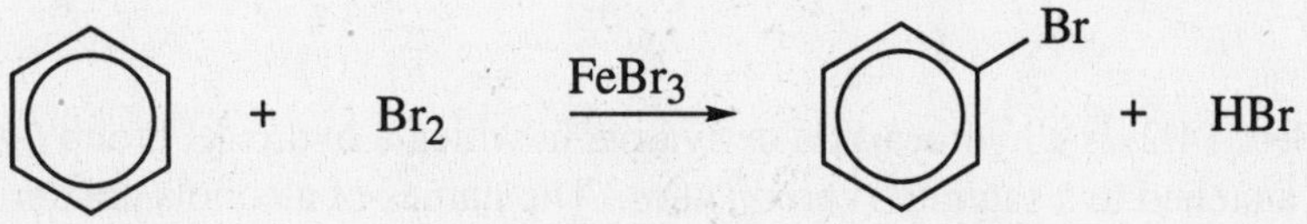

13.17 a. 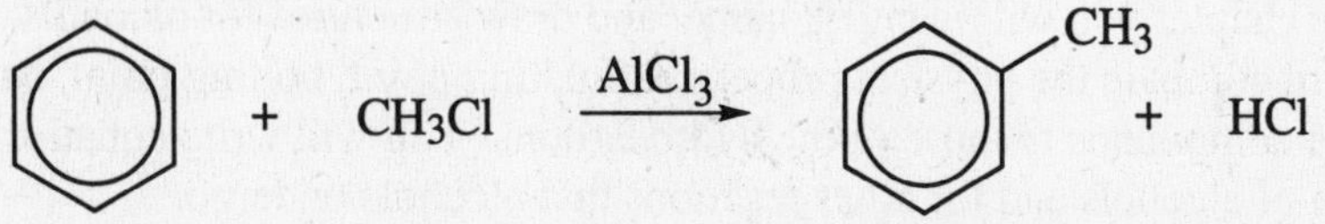 C_6H_6 + CH_3Cl $\xrightarrow{AlCl_3}$ $C_6H_5CH_3$ + HCl

b. C_6H_6 + Br_2 $\xrightarrow{FeBr_3}$ C_6H_5Br + HBr

c. C_6H_6 + Cl_2 $\xrightarrow{FeCl_3}$ C_6H_5Cl + HCl

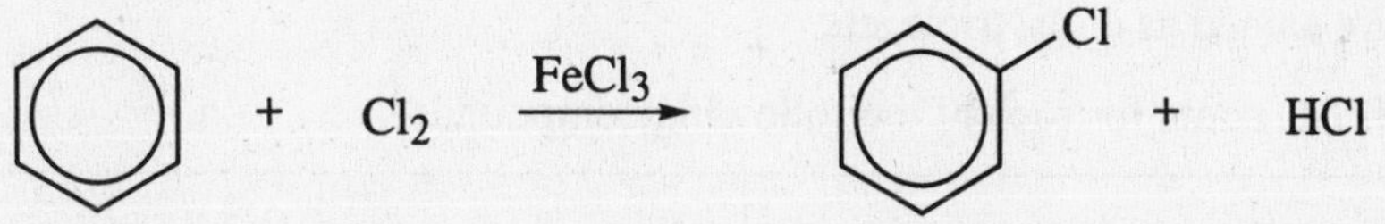

d. 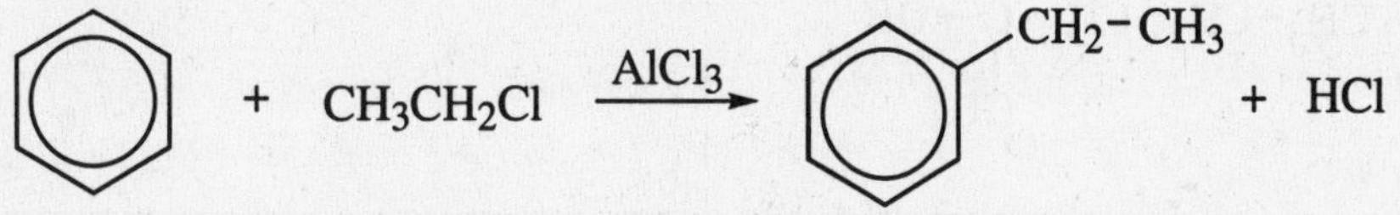 C_6H_6 + CH_3CH_2Cl $\xrightarrow{AlCl_3}$ $C_6H_5CH_2-CH_3$ + HCl

13.18 a. E; b. B; c. F; d. D; e. A; f. C;

g. Structure A *trans*-2-butene
h. Structure B *cis*-2-butene
i. Structure C butane
j. Structure D 3-methylcyclohexene
k. Structure E 2-chlorotoluene or 1-chloro-2-methylbenzene
l. Structure F 1-butyne

Answers to Self-Test

The numbers in parentheses refer to sections in your textbook.
1. T (13.1) **2**. T (13.2) **3**. F; both carbon atoms (13.3) **4**. T (13.3) **5**. T (13.2)
6. F; acetylene (13.9) **7**. F; substitution (13.13) **8**. F; 2-butene (13.5)
9. F; 2,3-dibromobutane (13.7) **10**. T (13.7) **11**. F; no (13.8) **12**. d (13.7)
13. a (13.9) **14**. c (13.11) **15**. c (13.11) **16**. c (13.5) **17**. a (13.10) **18**. c (13.2, 13.3)
19. c (13.7) **20**. b (13.7, 13.13)

Alcohols, Phenols, and Ethers — Chapter 14

Chapter Overview

Most of the chemical reactions of organic molecules involve the functional groups on the hydrocarbon chains or rings. In this chapter you will consider the properties, chemical and physical, of some hydrocarbon derivatives with oxygen- and sulfur-containing functional groups. The alcohols, phenols, and ethers are compounds that are commonly encountered in naturally occurring substances.

In this chapter you will identify, name, and draw structures for alcohols, phenols, and ethers. You will compare the physical properties (melting point, boiling point, solubility in water) of these compounds to those of the hydrocarbons. You will write equations for the mild oxidation of alcohols and for other reactions that alcohols undergo.

Practice Exercises

14.1 An **alcohol** (Sec. 14.2) is a hydrocarbon derivative in which a hydroxyl group (–OH) (Sec. 14.2) is attached to a saturated carbon atom. The names of alcohols are derived from the hydrocarbons and given the ending *-ol*. Polyhydroxy alcohols have more than one hydroxyl group and are named as diols, triols, etc.

Give the IUPAC name for each of the following compounds:

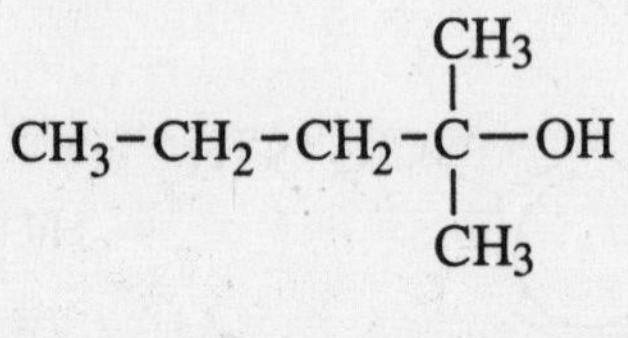 $CH_3-CH_2-CH_2-C(CH_3)_2-OH$ (C bearing CH_3 above and CH_3 below) a.	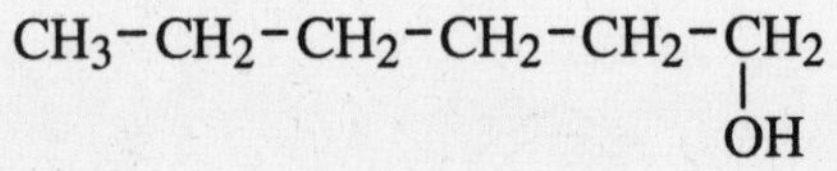 $CH_3-CH_2-CH_2-CH_2-CH_2-CH_2$ with OH on the last CH_2 b.
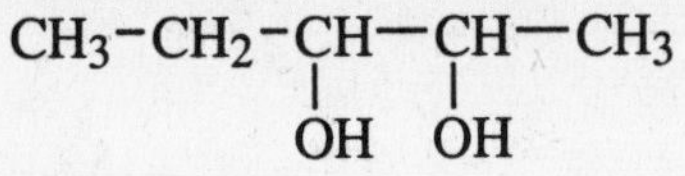 $CH_3-CH_2-CH(OH)-CH(OH)-CH_3$ c.	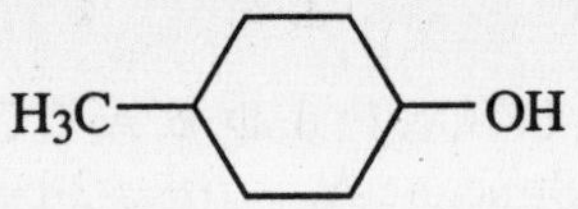 H_3C–(cyclohexane ring)–OH d.

14.2 Draw the structural formula for each of the following compounds:

a. 3-hexanol	b. 2,3-dimethyl-1-butanol
c. 1,2-butanediol	d. 2-methylcyclopentanol

14.3 One method for preparing alcohols is the hydration of alkenes in the presence of a sulfuric acid catalyst. The predominant alcohol product can be determined from Markovnikov's rule.

Write an equation for the preparation of each of the following alcohols from the appropriate alkene.

a. 4-methyl-2-pentanol

b. cyclopentanol

14.4 Alcohols undergo **dehydration** (Sec. 14.7), the removal of a water molecule, when they are heated in the presence of a sulfuric acid catalyst. The product formed is dependent on the temperature of the reaction. At 180°C, intramolecular dehydration, an **elimination reaction** (Sec. 14.7), produces an alkene. According to **Zaitsev's rule** (Sec. 14.7), the alkene with the greatest number of alkyl groups attached to the double bond will be formed. At a lower temperature (140°C), intermolecular dehydration, a **condensation reaction** (Sec. 14.7), produces an **ether** (Sec. 14.13).

Complete the following equations by supplying the missing information:

a. $CH_3-\underset{\displaystyle CH_3}{\underset{|}{CH}}-CH_2-OH \xrightarrow[H_2SO_4]{140^\circ C}$?

b. $CH_3-CH_2-CH_2-\underset{\displaystyle OH}{\underset{|}{CH}}-CH_3 \xrightarrow[H_2SO_4]{180^\circ C}$?

c. ? $\xrightarrow[H_2SO_4]{140^\circ C} CH_3-\underset{\displaystyle CH_3}{\underset{|}{CH}}-O-\underset{\displaystyle CH_3}{\underset{|}{CH}}-CH_3$

d. ? $\xrightarrow[H_2SO_4]{180^\circ C} CH_3-CH_2-CH_2-CH=\underset{\displaystyle CH_3}{\underset{|}{C}}-CH_3$

14.5 **Primary and secondary alcohols** (Sec. 14.7) undergo oxidation in the presence of a mild oxidizing agent to produce compounds that contain a carbon-oxygen double bond. Primary alcohols may oxidize even further to form the carboxylic acid. **Tertiary alcohols** (Sec. 14.7) cannot be oxidized in this way since the carbon attached to the hydroxyl group is attached to three other carbons and cannot form a carbon-oxygen double bond without breaking carbon-carbon bonds.

Alcohols also undergo substitution reactions. An example is the replacement of the hydroxyl group by a halogen atom using PCl_3 or PBr_3 as a reagent.

Complete the following equations by supplying the missing information:

a. ? $\xrightarrow[\text{agent}]{\text{mild oxidizing}}$ $CH_3-CH_2-CH_2-CH_2-\overset{H}{\overset{|}{C}}=O$

b. 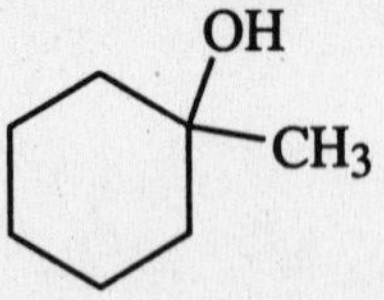 (1-methylcyclohexanol: cyclohexane ring with OH and CH_3 on the same carbon) $\xrightarrow[\text{agent}]{\text{mild oxidizing}}$?

c. $CH_3-CH_2-CH_2-\underset{OH}{\underset{|}{CH}}-CH_3$ $\xrightarrow[\text{agent}]{\text{mild oxidizing}}$?

d. ? + PCl_3 $\xrightarrow{\Delta}$ $CH_3-\underset{CH_3}{\underset{|}{CH}}-CH_2-\underset{Cl}{\underset{|}{CH}}-CH_3$

14.6 A **phenol** (Sec. 14.2) is a compound in which a hydroxyl group is attached to a carbon atom in an aromatic ring system. Phenols are named in the same way as benzene derivatives, but with the parent name *phenol.*

Write the IUPAC name for each of the following compounds:

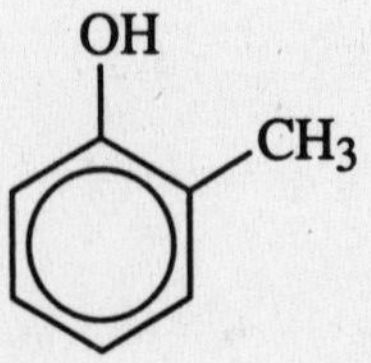 a. (benzene ring with OH and adjacent CH_3)	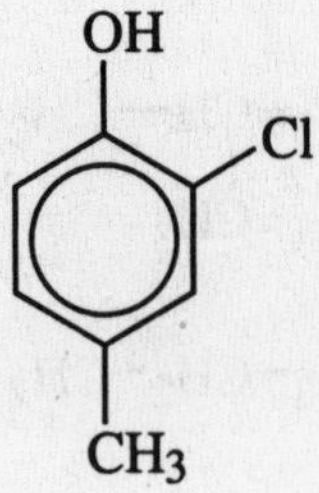 b. (benzene ring with OH, Cl adjacent to OH, and CH_3 para to OH)

14.7 Draw a structural formula for each of the following compounds:

a. 3-ethylphenol	b. 2,4-dibromophenol

14.8 An **ether** (Sec. 14.13) is a compound in which an oxygen atom is bonded to two carbon atoms by single bonds. According to the IUPAC system, ethers are named as substituted hydrocarbons. The smaller hydrocarbon and the oxygen atom are called an **alkoxy group** (Sec. 14.14) and considered as a substituent on the larger hydrocarbon.

Write the IUPAC name for each of the following compounds:

$CH_3{-}CH_2{-}CH_2{-}CH_2{-}O{-}CH_3$	(cyclohexane ring)$-O-CH_2-CH_3$
a.	b.

14.9 Ethers are often known by common names in which the hydrocarbon groups are named alphabetically followed by the word *ether*. Give the common name for each of the ethers in Practice Exercise 14.8.

a.

b.

14.10 Draw the structural formula and the line-angle drawing for each of the following ethers:

a. 2-ethoxypentane	b. 1-chloro-2-ethoxycyclopentane

14.11 Many of the physical properties of an organic compound that has an oxygen-containing functional group are determined by the molecule's ability to form hydrogen bonds with a like molecule or with water molecules.

Indicate which compound in each of the following pairs would have the higher boiling point and the greater solubility in water.

Compounds	Higher boiling point	Greater solubility in water
1-propanol and propane		
2-methyl-2-propanol and diethyl ether		
phenol and toluene		

14.12 Sulfur analogs of alcohols are called **thiols** (Sec. 14.17). These compounds contain a **sulfhydryl group** (Sec. 14.17) and are named by adding *-thiol* to the name of the parent alkane. Sulfur analogs of ethers are called sulfides and are named in the same way as ethers with *sulfide* in place of *ether*, or *alkylthio* in place of *alkoxy*.

Give the IUPAC name for the following thiols:

$CH_3-CH_2-CH_2-CH_2-SH$	(cyclohexane ring)–SH	CH_3-CH_2-S-(cyclopentane ring)
a.	b.	c.

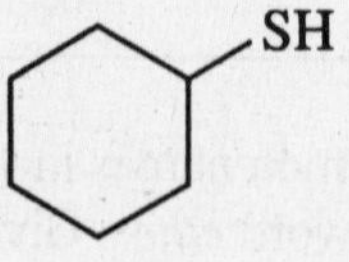

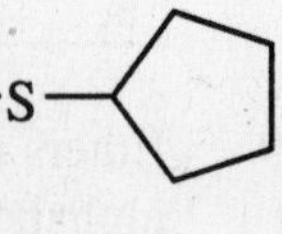

14.13 Write structural formulas for the following sulfur compounds:

a. 2-methyl-1-butanethiol	b. 2-bromocyclobutanethiol	c. diethylsulfide

14.14 Use this identification exercise to review your knowledge of the structures introduced in this chapter. Using structures A through I, give the best match for the terms that follow:

$HO-CH_2-CH_2-OH$ A.	$CH_3-O-CH_2-CH_3$ B.	CH_3-S-CH_3 C.
$CH_3-CH_2-C(CH_3)_2-OH$ D.	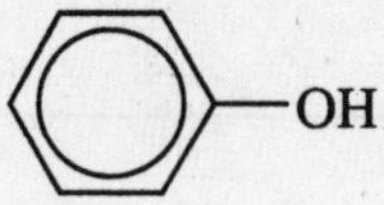 (benzene ring)–OH E.	$CH_3-CH_2-CH_2-OH$ F.
CH_3-CH_2-SH G.	$CH_3-O-O-CH_3$ H.	$CH_3-S-S-CH_3$ I.

Give the letter of the best match for the structures above:

a. 3° alcohol ____________

b. glycol ____________

c. peroxide ____________

d. ether ____________

e. phenol ____________

f. disulfide ____________

g. thiol ____________

h. sulfide ____________

i. 1° alcohol ____________

Give the IUPAC names for structures A through G:

j. Structure A ____________

k. Structure B ____________

l. Structure C ____________

m. Structure D ____________

n. Structure E ____________

o. Structure F ____________

p. Structure G ____________

Self-Test

True-false: Indicate whether the following statements are true or false. If the statement is false, give the word or phrase that may be substituted for the underlined portion to make the statement true.

1. A compound in which a hydroxyl group is attached to a carbon atom in an aromatic ring system is called a <u>thiol</u>.
2. A secondary alcohol has <u>two</u> hydrogen atoms attached to the hydroxyl carbon.
3. Ether molecules <u>cannot</u> form hydrogen bonds with other ether molecules.
4. Diethyl ether and <u>1-butanol</u> are structural isomers.
5. Oxygen has six valence electrons, which include <u>two nonbonding electron pairs</u>.
6. 2-Butanol is an example of a <u>tertiary</u> alcohol.
7. Alcohols are <u>more soluble</u> in water than alkanes of similar molecular mass.
8. A symmetrical ether can be prepared by the <u>intramolecular</u> dehydration of an alcohol.
9. Phenols are used as antioxidants because they are <u>low-melting solids</u>.
10. Furan is an example of a <u>heterocyclic</u> organic compound.
11. Storage of ethers can be hazardous if unstable <u>peroxides</u> form.

Multiple choice:

12. An alcohol may be prepared by:

a. hydrogenation of an alkene
b. hydration of an alkene
c. dehydration of a carbonyl group
d. dehydrogenation of an alkane
e. none of these

13. Mercaptans is an older term for:

a. thiols
b. phenols
c. glycols
d. ethers
e. none of these

14. The mild oxidation of a secondary alcohol can result in the production of a(n):

a. acid
b. aldehyde
c. ketone
d. alkene
e. none of these

15. The IUPAC name for isopropyl alcohol is:

a. 1-propyl alcohol
b. 2-propanol
c. 2-propyl alcohol
d. 1-propanol
e. none of these

16. The common name for 1,2-ethanediol is:

a. ethylene glycol
b. glycerol
c. diethyl ether
d. 1,2-ethyl alcohol
e. cresol

17. According to Zaitsev's rule, the main product of the dehydration of 2-methyl-2-butanol would be:

a. 2-methyl-2-butene
b. 2-methyl-1-butene
c. 1-methyl-2-butene
d. 1-methyl-1-butene
e. none of these

18. Which of the following names is correct according to IUPAC rules?

a. 2-ethyl-1-butene-2-ol b. 3-propylene-2-ol c. 3,4-butanediol
d. 1,2-dimethylphenol e. none of these

19. Rank butane, butanol, and diethyl ether by boiling points, lowest boiling point first and highest boiling point last:

a. butane, butanol, diethyl ether b. butanol, butane, diethyl ether
c. butane, diethyl ether, butanol d. butanol, diethyl ether, butane
e. diethyl ether, butane, butanol

20. The intramolecular dehydration of 2-pentanol produces:

a. 2-methoxypentane b. 2-pentene c. 1-pentene
d. 2-methylpentene e. none of these

21. An example of a cyclic ether is:

a. phenol b. methoxybenzene c. catechol
d. pyran e. methyl *tert*-butyl ether

22. Mild reduction of a disulfide produces:

a. a thiol b. a peroxide c. a thioether
d. a cyclic ether e. none of these

Answers to Practice Exercises

14.1 a. 2-methyl-2-pentanol
b. 1-hexanol
c. 2,3-pentanediol
d. 4-methylcyclohexanol

14.2

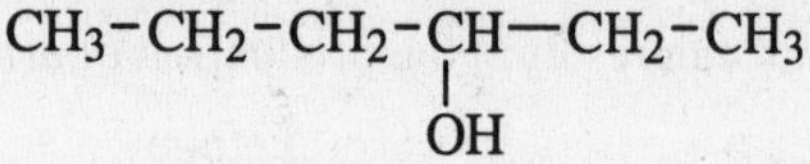 $CH_3-CH_2-CH_2-CH(OH)-CH_2-CH_3$ a. 3-hexanol	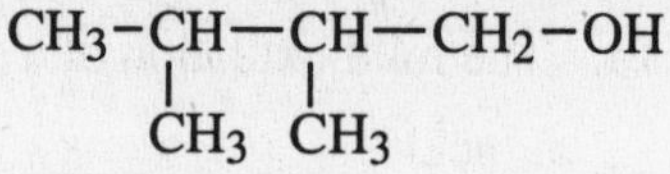 $CH_3-CH(CH_3)-CH(CH_3)-CH_2-OH$ b. 2,3-dimethyl-1-butanol
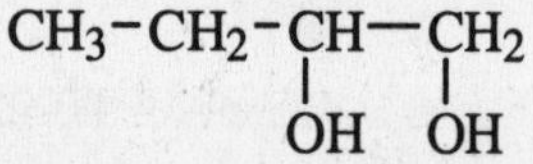 $CH_3-CH_2-CH(OH)-CH_2(OH)$ c. 1,2-butanediol	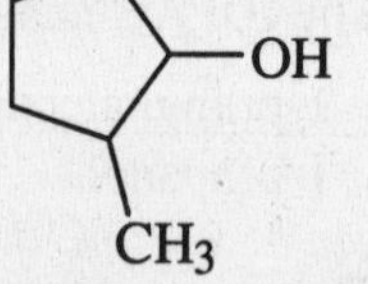 (cyclopentane ring with –OH and CH_3) d. 2-methylcyclopentanol

14.3 a. 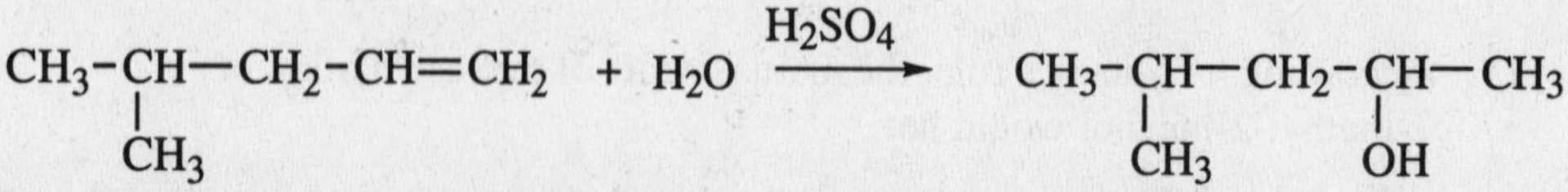 $CH_3-CH(CH_3)-CH_2-CH=CH_2 + H_2O \xrightarrow{H_2SO_4} CH_3-CH(CH_3)-CH_2-CH(OH)-CH_3$

b. 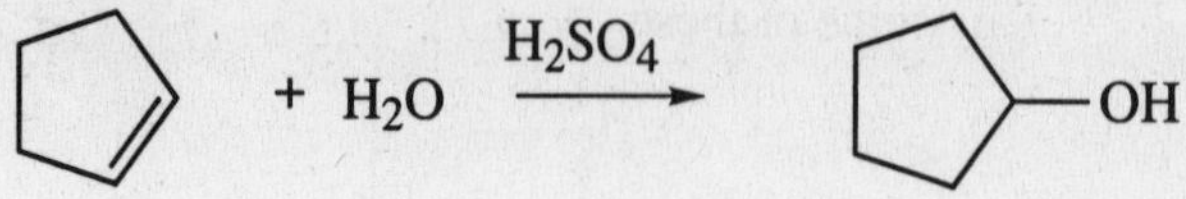 cyclopentene $+ H_2O \xrightarrow{H_2SO_4}$ cyclopentanol (cyclopentyl–OH)

14.4 a. $CH_3{-}\underset{\displaystyle CH_3}{\underset{|}{CH}}{-}CH_2{-}OH \xrightarrow[H_2SO_4]{140^\circ C} CH_3{-}\underset{\displaystyle CH_3}{\underset{|}{CH}}{-}CH_2{-}O{-}CH_2{-}\underset{\displaystyle CH_3}{\underset{|}{CH}}{-}CH_3$

b. $CH_3{-}CH_2{-}CH_2{-}\underset{\displaystyle OH}{\underset{|}{CH}}{-}CH_3 \xrightarrow[H_2SO_4]{180^\circ C} CH_3{-}CH_2{-}CH{=}CH{-}CH_3$

c. $CH_3{-}\underset{\displaystyle CH_3}{\underset{|}{CH}}{-}OH \xrightarrow[H_2SO_4]{140^\circ C} CH_3{-}\underset{\displaystyle CH_3}{\underset{|}{CH}}{-}O{-}\underset{\displaystyle CH_3}{\underset{|}{CH}}{-}CH_3$

d. $CH_3{-}CH_2{-}CH_2{-}CH_2{-}\overset{\displaystyle OH}{\overset{|}{\underset{\displaystyle CH_3}{\underset{|}{C}}}}{-}CH_3 \xrightarrow[H_2SO_4]{180^\circ C} CH_3{-}CH_2{-}CH_2{-}CH{=}\underset{\displaystyle CH_3}{\underset{|}{C}}{-}CH_3$

14.5 a. $CH_3{-}CH_2{-}CH_2{-}CH_2{-}\underset{\displaystyle OH}{\underset{|}{CH_2}} \xrightarrow[\text{agent}]{\text{mild oxidizing}} CH_3{-}CH_2{-}CH_2{-}CH_2{-}\overset{\displaystyle H}{\overset{|}{C}}{=}O$

b. 1-methylcyclohexanol (cyclohexane ring bearing OH and CH_3 on the same carbon) $\xrightarrow[\text{agent}]{\text{mild oxidizing}}$ No reaction, tertiary alcohol

c. $CH_3{-}CH_2{-}CH_2{-}\underset{\displaystyle OH}{\underset{|}{CH}}{-}CH_3 \xrightarrow[\text{agent}]{\text{mild oxidizing}} CH_3{-}CH_2{-}CH_2{-}\underset{\displaystyle O}{\underset{||}{C}}{-}CH_3$

d. $CH_3{-}\underset{\displaystyle CH_3}{\underset{|}{CH}}{-}CH_2{-}\underset{\displaystyle OH}{\underset{|}{CH}}{-}CH_3 + PCl_3 \xrightarrow{\Delta} CH_3{-}\underset{\displaystyle CH_3}{\underset{|}{CH}}{-}CH_2{-}\underset{\displaystyle Cl}{\underset{|}{CH}}{-}CH_3$

14.6 a. 2-methylphenol
b. 2-chloro-4-methylphenol

14.7

Structure: benzene ring bearing OH, with $CH_2{-}CH_3$ meta to the OH	Structure: benzene ring bearing OH, with Br ortho and Br para to the OH
a. 3-ethylphenol	b. 2,4-dibromophenol

14.8 a. 1-methoxybutane b. ethoxycyclohexane

14.9 a. butylmethyl ether b. ethylcyclohexyl ether

14.10

a. 2-ethoxypentane	b. 1-chloro-2-ethoxycyclopentane
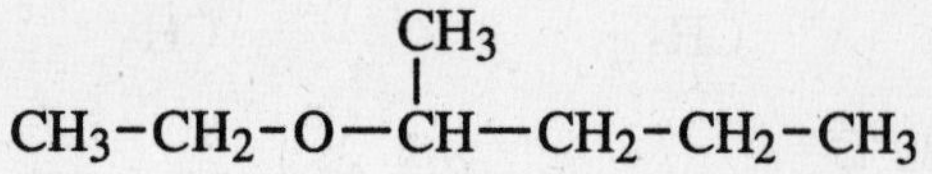 $CH_3-CH_2-O-CH(CH_3)-CH_2-CH_2-CH_3$	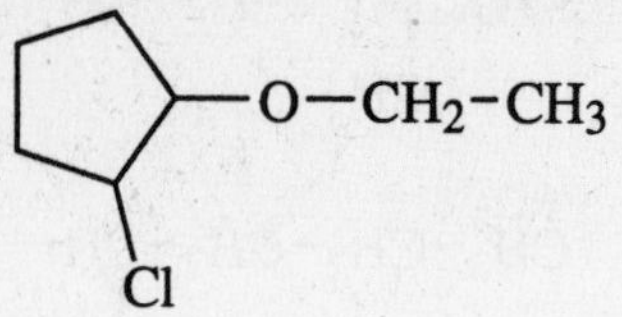 (cyclopentane ring) $-O-CH_2-CH_3$, Cl
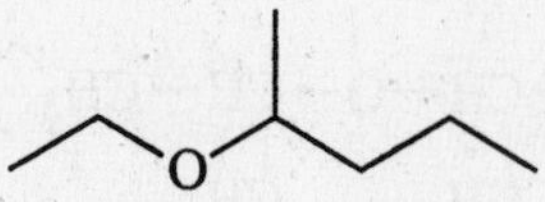 O	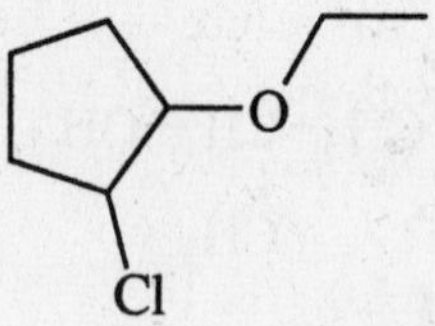 O, Cl

14.11

Compounds	Higher boiling point	Greater solubility in water
1-propanol and propane	1-propanol	1-propanol
2-methyl-2-propanol and diethyl ether	2-methyl-2-propanol	2-methyl-2-propanol
phenol and toluene	phenol	phenol

14.12 a. 1-butanethiol b. cyclohexanethiol c. ethylthiocyclopentane

14.13

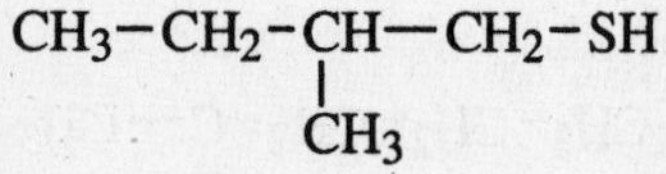 $CH_3-CH_2-CH(CH_3)-CH_2-SH$	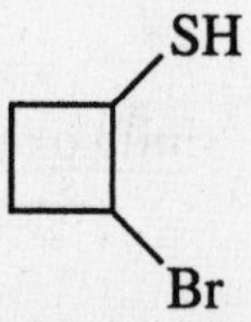 SH, Br	$CH_3-CH_2-S-S-CH_2-CH_3$
a. 2-methyl-1-butanethiol	b. 2-bromocyclobutanethiol	c. diethylsulfide

14.14 a. D; b. A; c. H; d. B; e. E; f. I; g. G; h. C; i. F

j. Structure A 1,2-ethanediol
k. Structure B methoxyethane
l. Structure C methylthiomethane
m. Structure D 2-methyl-2-butanol
n. Structure E phenol
o. Structure F 1-propanol
p. Structure G ethanethiol
q. Structure H methylthiomethane
r. Structure I ethanethiol

Answers to Self-Test

The numbers in parentheses refer to sections in your textbook.
1. F; phenol (14.2) **2.** F; one (14.7) **3.** T (14.11) **4.** T (14.10) **5.** T (14.1) **6.** F; secondary (14.7) **7.** T (14.5) **8.** F; intermolecular (14.7) **9.** F; easily oxidized (14.9) **10.** T (14.12) **11.** T (14.11) **12.** b (14.6) **13.** a (14.13) **14.** c (14.7) **15.** b (14.4) **16.** a (14.4) **17.** a (14.7) **18.** e (14.3) **19.** c (14.5, 14.11) **20.** b (14.7) **21.** d (14.3, 14.10) **22.** a (14.13)

Aldehydes and Ketones — Chapter 15

Chapter Overview

The carbonyl functional group is very commonly found in nature. It contains an oxygen atom joined to a carbon atom by a double bond. Aldehydes and ketones contain the carbonyl functional group.

In this chapter you will learn to recognize, name, and write structural formulas for aldehydes and ketones and write equations for their preparation by oxidation of alcohols. You will compare the physical properties of aldehydes and ketones with those of other organic compounds. You will learn some tests used to distinguish aldehydes from ketones and will write equations for reactions involving addition to the carbonyl group of aldehydes and ketones.

Practice Exercises

15.1 The functional group that identifies **aldehydes and ketones** (Sec. 15.2) is the **carbonyl group** (Sec. 15.1), a carbon atom and an oxygen atom joined by a double bond. Aldehydes are compounds in which the carbonyl carbon is bonded to at least one hydrogen. Ketones are compounds in which the carbonyl carbon is bonded to two other carbons.

Classify each of the following structural formulas as an aldehyde, a ketone, or neither.

$CH_3-CH_2-CH_2-O-CH_3$ a.	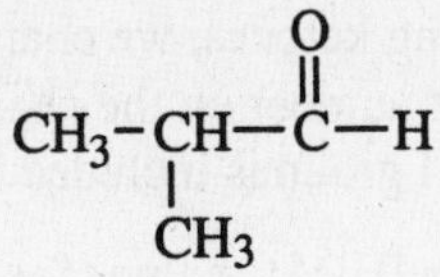 $\overset{O}{\overset{\|\|}{}}$ $CH_3-\underset{\underset{CH_3}{\|}}{CH}-\overset{\overset{O}{\|\|}}{C}-H$ b.
$H_3C-\bigcirc=O$ c.	$CH_3-\underset{\underset{O}{\|\|}}{C}-\overset{\overset{Cl}{\|}}{CH}-\underset{\underset{CH_3}{\|}}{CH}-CH_3$ d.
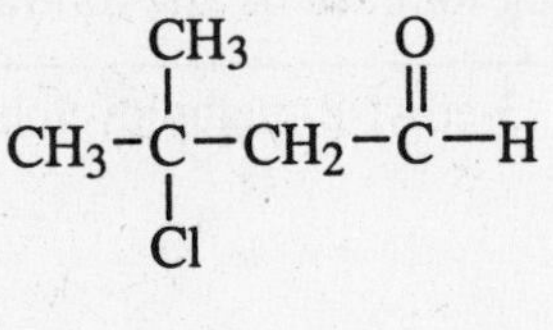 $CH_3-\overset{\overset{CH_3}{\|}}{\underset{\underset{Cl}{\|}}{C}}-CH_2-\overset{\overset{O}{\|\|}}{C}-H$ e.	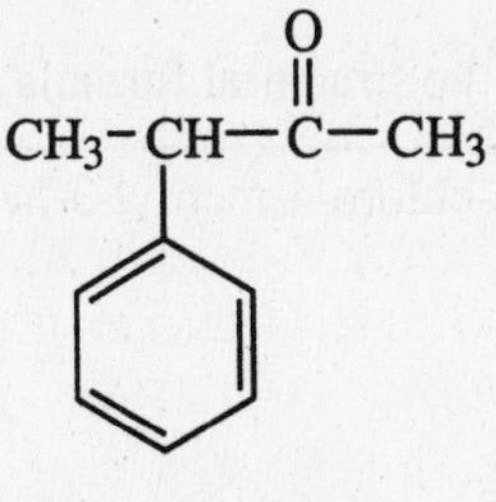 $CH_3-\underset{\underset{C_6H_5}{\|}}{CH}-\overset{\overset{O}{\|\|}}{C}-CH_3$ f.

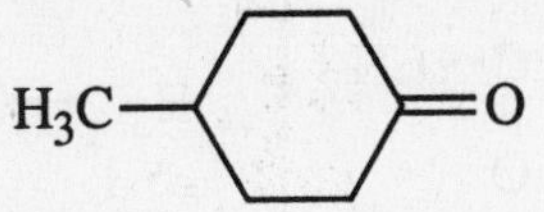

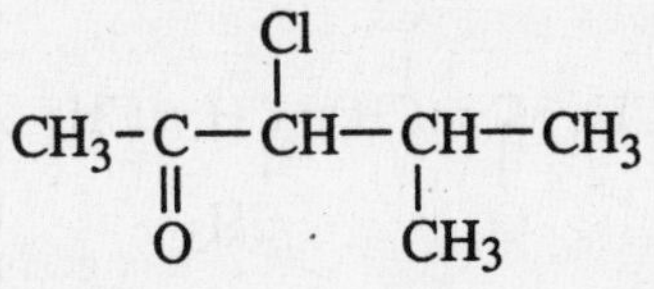

15.2 In naming aldehydes, we change the *-e* ending of the hydrocarbon to *-al*, and name any substituents, starting the counting from the carbonyl carbon. No number is specified for the carbonyl group.

Give the IUPAC name for each of the following aldehydes:

$CH_3-CH(Br)-C(=O)-H$	$CH_3-C(CH_3)(Cl)-CH_2-CHO$	$H-C(=O)-CH_2-C(Cl)_3$
a.	b.	c.

15.3 Draw the structural formula and the line-angle drawing for each of the following aldehydes.

a. 4-chloro-3,3-dimethylbutanal	b. 4-bromo-3-methylpentanal

15.4 In naming ketones, we change the *-e* ending to *-one*. We give the carbonyl group the lowest possible number on the chain and then name the substituents. The numbered position of the carbonyl group is included in the name.

Give the IUPAC names for the following ketones.

$CH_3-C(=O)-CH(Cl)-CH(CH_3)-CH_3$	(benzene ring)$-CH(CH_3)-C(=O)-CH_3$	H_3C-(cyclohexane ring)$=O$
a.	b.	c.

15.5 Draw the structural formula and the line-angle drawing for each of the following ketones.

a. 6-chloro-4-methyl-3-hexanone	b. 3-ethyl-2-methylcyclohexanone

15.6 Common names are frequently used for some of the simpler aldehydes and ketones. Give the common name(s) for each of the following structures.

$CH_3-\overset{O}{\overset{\|\|}{C}}-CH_3$	$C_6H_5-\overset{O}{\overset{\|\|}{C}}-CH_3$	$H-\overset{O}{\overset{\|\|}{C}}-H$	$CH_3-\overset{O}{\overset{\|\|}{C}}-H$
a.	b.	c.	d.

15.7 Hydrogen bonding cannot occur between the molecules of an aldehyde or between the molecules of a ketone. However, dipole-dipole attractions occur between molecules.

Determine which one of each pair of compounds would have a higher boiling point and explain your choice.

Compounds	Higher-boiling compound	Explanation
pentanal and hexane		
2-hexanone and 2-octene		
pentanal and 1-pentanol		

15.8 Aldehydes and ketones can be produced by the oxidation of primary and secondary alcohols, respectively. Write an equation for the preparation of each of the following compounds by oxidation of an alcohol. (Assume that no further oxidation of aldehydes to carboxylic acids occurs.)

a. 3-methyl-2-phenylpentanal

b. 3-ethyl-2-methylcyclohexanone

15.9 Aldehydes are easily oxidized to carboxylic acids; ketones are resistant to oxidation. The Tollens test and the Benedict's test, which can distinguish between the two types of compounds, use metal ions (Ag^+ and Cu^{2+}) as oxidizing agents. The appearance of the reduced metal (Ag) or metal oxide (Cu_2O) indicates that an aldehyde is present.

Indicate whether each of the following compounds will give a positive test with Tollens or Benedict's reagent.

Compound	Benedict's reagent	Tollens reagent
a. 2-methyl-2-propanol		
b. 2-methylhexanal		
c. 4-chloro-3-methyl-2-pentanone		
d. 3,3-dimethylpentane		
e. 4-chlorocyclohexanone		

15.10 Aldehydes and ketones are easily reduced by hydrogen gas (H_2), in the presence of a catalyst (Ni, Pt, or Cu), to form alcohols.

Write an equation for the preparation of each of the following alcohols by the reduction of the appropriate aldehyde or ketone.

a. 3-chloro-1-butanol

b. 1-bromo-3-methyl-2-pentanol

15.11 Aldehydes and ketones easily undergo addition of an alcohol molecule to the double bond of the carbonyl group, forming a **hemiacetal** (Sec. 15.9).

Draw the structural formula of the hemiacetal formed when one molecule of methanol, CH_3OH, reacts with one molecule of each of the following carbonyl compounds.

a. 2-bromopropanal + methanol	b. 4-methylcyclohexanone + methanol	c. 3-chloro-2-pentanone + methanol

15.12 If a hemiacetal molecule reacts with a second molecule of alcohol, an **acetal** (Sec. 15.9) is produced.

Draw the structural formula of the acetal formed when one additional molecule of methanol is added to each of the hemiacetal molecules in Practice Exercise 15.11.

a.	b.	c.

15.13 The hydrolysis of an acetal or a hemiacetal produces the aldehyde or ketone and the alcohol that originally reacted to form the acetal or hemiacetal. Complete the following acid hydrolysis reactions.

a. 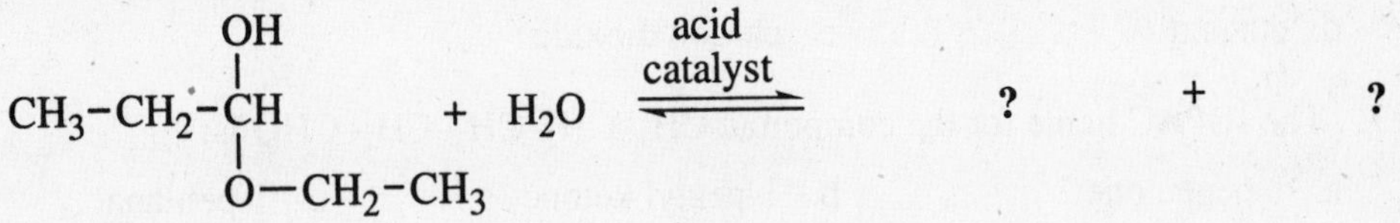 $CH_3{-}CH_2{-}CH(OH){-}O{-}CH_2{-}CH_3 + H_2O \xrightleftharpoons[\text{catalyst}]{\text{acid}}$? + ?

b. 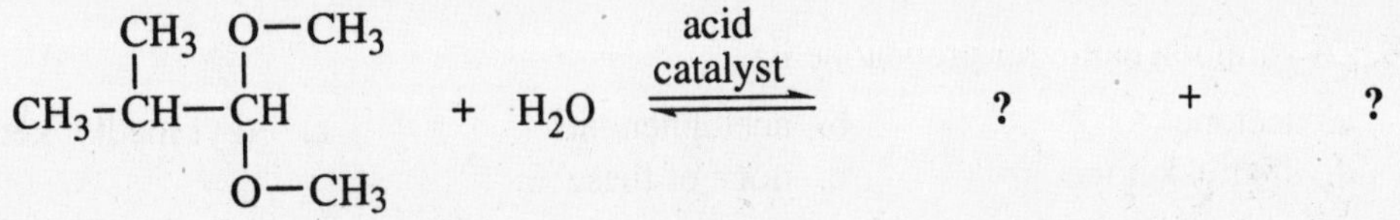 $CH_3{-}CH(CH_3){-}CH(O{-}CH_3){-}O{-}CH_3 + H_2O \xrightleftharpoons[\text{catalyst}]{\text{acid}}$? + ?

c. 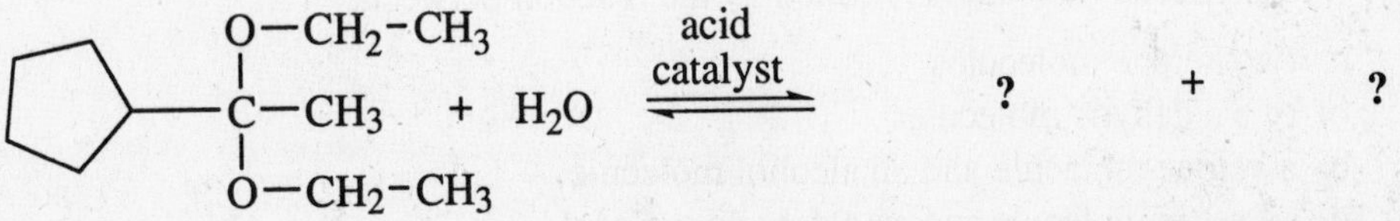 $C_5H_9{-}C(CH_3)(O{-}CH_2{-}CH_3){-}O{-}CH_2{-}CH_3 + H_2O \xrightleftharpoons[\text{catalyst}]{\text{acid}}$? + ? (cyclopentyl group)

Self-Test

True-false: Indicate whether the following statements are true or false. If the statement is false, give the word or phrase that may be substituted for the underlined portion to make the statement true.

1. The simplest ketone contains <u>two</u> carbon atoms.
2. The simplest aldehyde has the common name <u>formaldehyde</u>.
3. The oxidation of 2-butanol produces a <u>ketone</u>.
4. Many important steroid hormones are <u>aldehydes</u>.
5. Aldehydes and ketones have <u>higher</u> boiling points than the corresponding alcohols.
6. Low-molecular-mass aldehydes and ketones are water-soluble because water <u>forms hydrogen bonds</u> with them.
7. Ketones are often produced by oxidation of the corresponding <u>tertiary</u> alcohol.
8. If an aldehyde is produced by oxidation of an alcohol, further oxidation to <u>a ketone</u> may take place.
9. A positive Tollens test indicates that <u>an aldehyde</u> is present.
10. Benedict's test uses <u>Ag^+</u> ion as an oxidizing agent for aldehydes.
11. A ketone can be reduced by hydrogen gas in the presence of a <u>Ni catalyst</u>.
12. Addition to the carbon-oxygen double bond of a carbonyl group takes place <u>less easily</u> than addition to a carbon-carbon double bond.
13. Hemiacetals are <u>much more stable</u> than acetals.
14. An acetal molecule is the product of the addition of <u>two molecules</u> of an alcohol to one molecule of an aldehyde.
15. Replacement of the oxygen atom in a carbonyl compound with a sulfur atom produces a <u>sulfoxide</u>.

Multiple choice:

16. Oxidation of ethanol cannot yield:

    a. acetaldehyde b. acetone c. acetic acid
    d. ethanal e. carbon dioxide

17. The IUPAC name for the compound $CH_3–CH_2–CH_2–CH_2–CHO$ is:

    a. 1-pentanone b. 1-pentyl ketone c. 1-pentanal
    d. pentanal e. none of these

18. A common name for propanone is:

    a. acetone b. acetophenone c. ethyl methyl ketone
    d. diethyl ketone e. none of these

19. A hemiacetal molecule is formed by the reaction between:

    a. two ketone molecules
    b. two aldehyde molecules
    c. a ketone molecule and an alcohol molecule
    d. a ketone molecule and an aldehyde molecule
    e. none of these

20. Aldehydes and ketones have boiling points lower than the corresponding alcohols because aldehydes and ketones:

    a. have stronger dipole-dipole attractions
    b. form more hydrogen bonds than alcohols do
    c. cannot form hydrogen bonds between molecules
    d. form bonds with water molecules
    e. none of these

21. How many ketones are isomeric with butanal?

    a. one b. two c. three d. four e. none of these

22. Acetaldehyde can be produced by the oxidation of:

    a. acetone b. ethanol c. dimethyl ether
    d. ethane e. none of these

Answers to Practice Exercises

15.1 a. neither (an ether) b. aldehyde c. ketone
d. ketone e. aldehyde f. ketone

15.2 a. 2-bromopropanal
b. 3-chloro-3-methylbutanal
c. 3,3,3-trichloropropanal

15.3

a. 4-chloro-3,3-dimethylbutanal	b. 4-bromo-3-methylpentanal
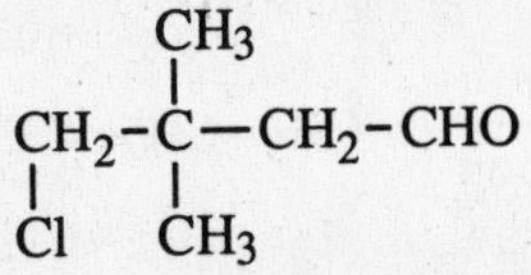 $\underset{\displaystyle Cl}{\underset{\vert}{CH_2}}-\overset{\displaystyle CH_3}{\overset{\vert}{\underset{\underset{\displaystyle CH_3}{\vert}}{C}}}-CH_2-CHO$	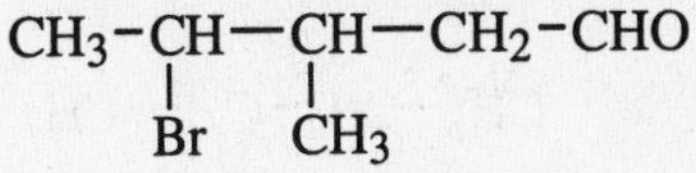 $CH_3-\underset{\underset{\displaystyle Br}{\vert}}{CH}-\underset{\underset{\displaystyle CH_3}{\vert}}{CH}-CH_2-CHO$
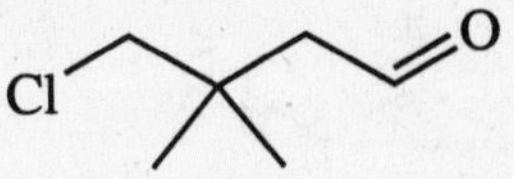 Cl, O	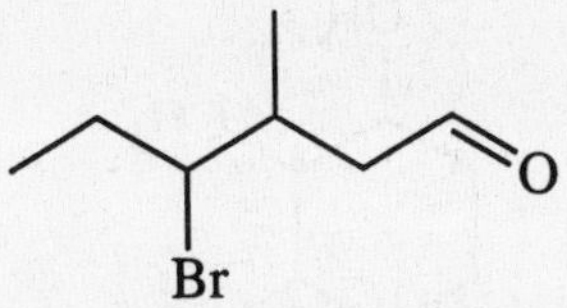 O, Br

15.4
a. 3-chloro-4-methyl-2-pentanone;
b. 3-phenyl-2-butanone
c. 4-methylcyclohexanone

15.5

a. 6-chloro-4-methyl-3-hexanone	b. 3-ethyl-2-methylcyclohexanone
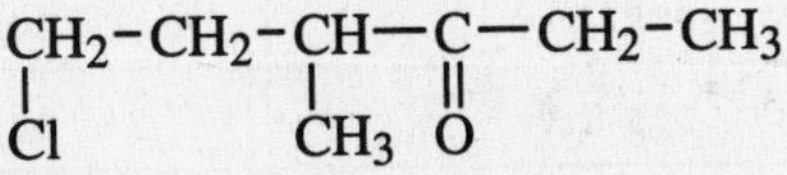 $\underset{\underset{\displaystyle Cl}{\vert}}{CH_2}-CH_2-\underset{\underset{\displaystyle CH_3}{\vert}}{CH}-\underset{\underset{\displaystyle O}{\Vert}}{C}-CH_2-CH_3$	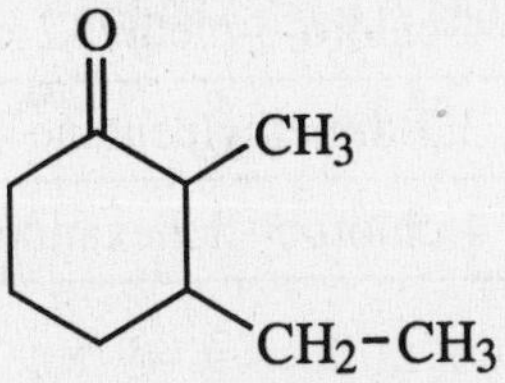 O, CH_3, CH_2-CH_3
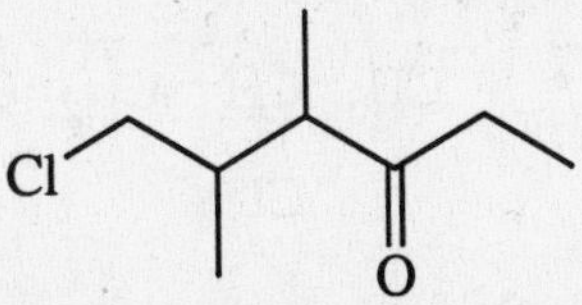 Cl, O	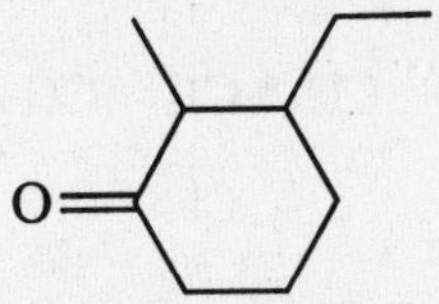 O

15.6
a. acetone or dimethyl ketone
b. methyl phenyl ketone
c. formaldehyde
d. acetaldehyde

15.7

Compounds	Higher-boiling compound	Explanation
pentanal and hexane	pentanal	Pentanal has dipole-dipole interactions between molecules.
2-hexanone and 2-octene	2-hexanone	2-Hexanone has dipole-dipole interactions between molecules.
pentanal and 1-pentanol	1-pentanol	1-Pentanol has hydrogen bonding between molecules.

15.8

a. 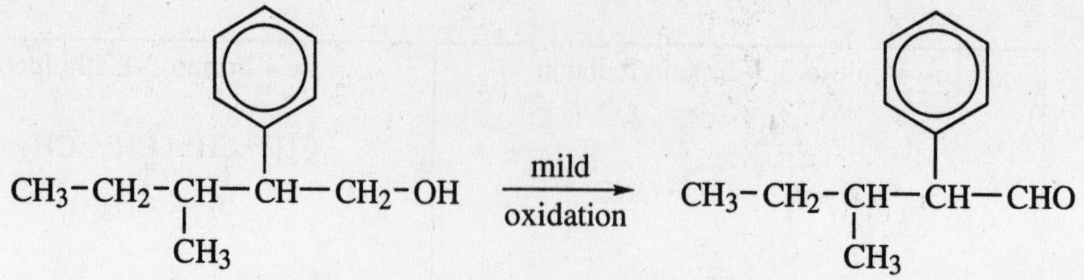 $CH_3-CH_2-CH(CH_3)-CH(C_6H_5)-CH_2-OH \xrightarrow{\text{mild oxidation}} CH_3-CH_2-CH(CH_3)-CH(C_6H_5)-CHO$

b. 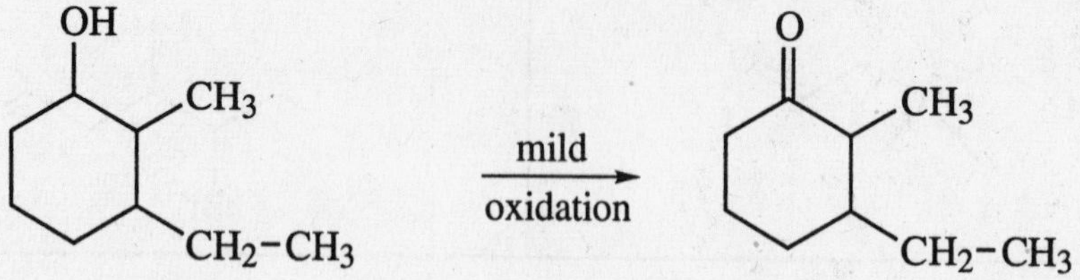 2-ethyl-6-methylcyclohexanol (ring with OH, CH_3, CH_2-CH_3) $\xrightarrow{\text{mild oxidation}}$ 3-ethyl-2-methylcyclohexanone (ring with O, CH_3, CH_2-CH_3)

15.9

Compound	Benedict's reagent	Tollens reagent
a. 2-methyl-2-propanol	negative	negative
b. 2-methylhexanal	positive	positive
c. 4-chloro-3-methyl-2-pentanone	negative	negative
d. 3,3-dimethylpentane	negative	negative
e. 4-chlorocyclohexanone	negative	negative

15.10

a. 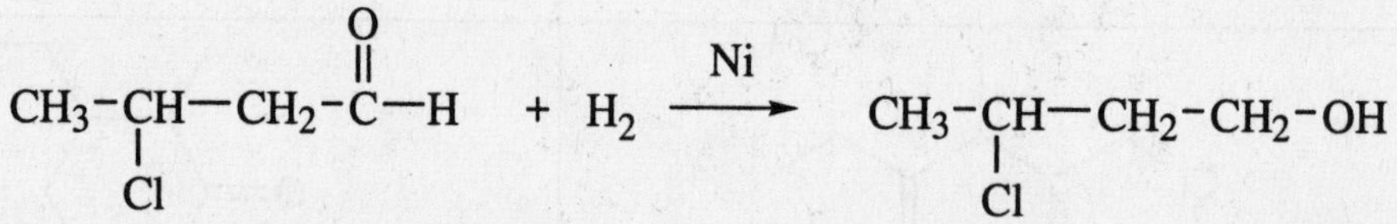 $CH_3-CH(Cl)-CH_2-C(=O)-H + H_2 \xrightarrow{Ni} CH_3-CH(Cl)-CH_2-CH_2-OH$

b. 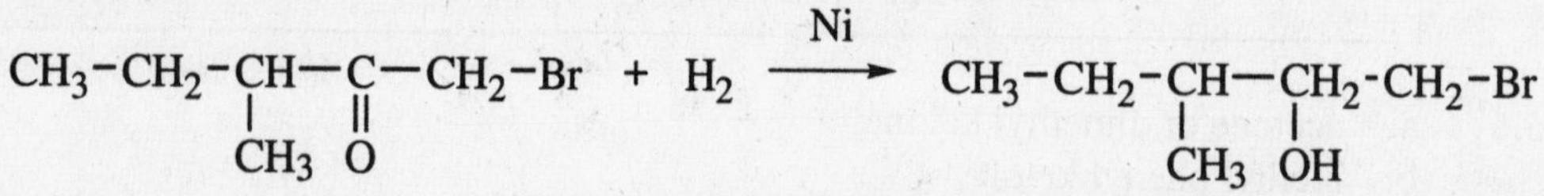 $CH_3-CH_2-CH(CH_3)-C(=O)-CH_2-Br + H_2 \xrightarrow{Ni} CH_3-CH_2-CH(CH_3)-CH(OH)-CH_2-Br$

15.11

$CH_3-CH(Br)-CH(OH)(O-CH_3)$	H_3C–cyclohexane–C(O–CH_3)(OH)	$CH_3-C(OH)(O-CH_3)-CH(Cl)-CH_2-CH_3$
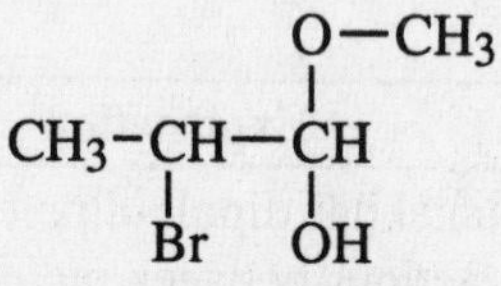 a. 2-bromopropanal + methanol	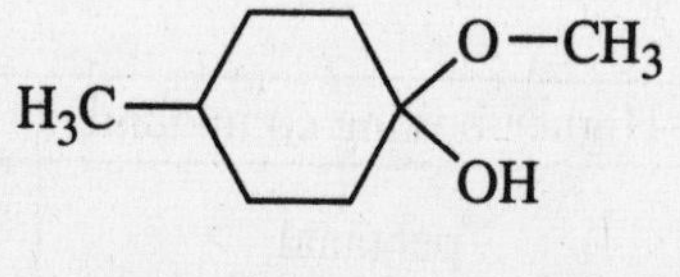 b. 4-methylcyclohexanone + methanol	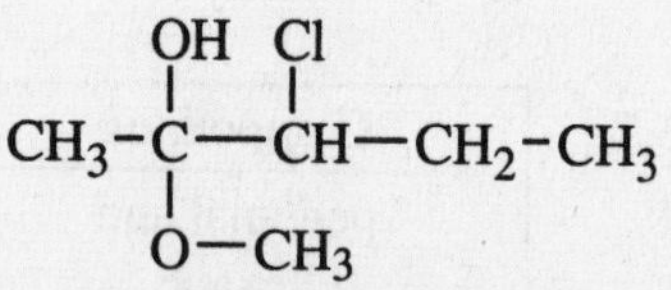 c. 3-chloro-2-pentanone + methanol

15.12

$CH_3-CH(Br)-CH(O-CH_3)(O-CH_3)$	H_3C–cyclohexane–C(O–CH_3)(O–CH_3)	$CH_3-C(O-CH_3)(H_3C-O)-CH(Cl)-CH_2-CH_3$
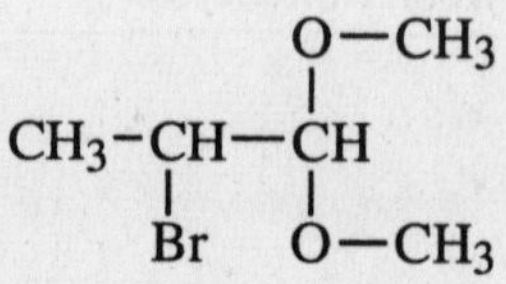 a.	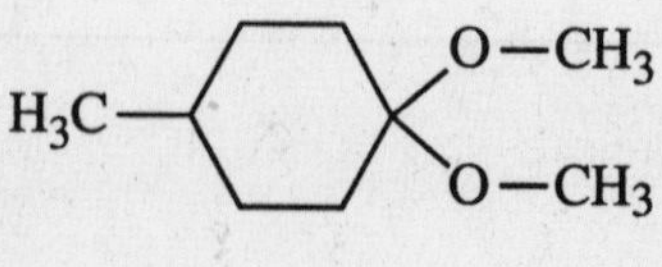 b.	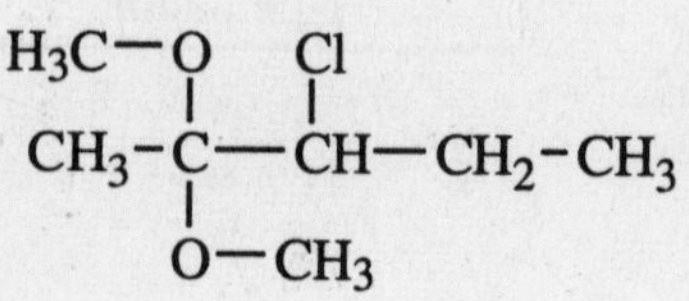 c.

15.13 a. $CH_3-CH_2-CH(OH)(O-CH_2-CH_3) + H_2O \xrightleftharpoons{\text{acid catalyst}} CH_3-CH_2-C(=O)-H + CH_3-CH_2-OH$

b. $CH_3-CH(CH_3)-CH(O-CH_3)_2 + H_2O \xrightleftharpoons{\text{acid catalyst}} CH_3-CH(CH_3)-C(=O)-H + 2\ CH_3-OH$

c. cyclopentyl$-C(CH_3)(O-CH_2-CH_3)_2 + H_2O \xrightleftharpoons{\text{acid catalyst}}$ cyclopentyl$-C(=O)-CH_3 + 2\ CH_3-CH_2-OH$

Answers to Self-Test

The numbers in parentheses refer to sections in your textbook.

1. F; three (15.2) **2.** T (15.3) **3.** T (15.2) **4.** F; ketones (15.5) **5.** F; lower (15.6) **6.** T (15.6) **7.** F; secondary (15.7) **8.** F; a carboxylic acid (15.7) **9.** T (15.8) **10.** F; Cu^{2+} (15.8) **11.** T (15.8) **12.** F; more easily (15.9) **13.** F; much less stable (15.9) **14.** T (15.9) **15.** F; thiocarbonyl compound (15.11) **16.** b (15.7, 15.8) **17.** d (15.3) **18.** a (15.4) **19.** c (15.9) **20.** c (15.6) **21.** a (15.2, 15.3) **22.** b (15.3, 15.7)

Carboxylic Acids and Esters — Chapter 16

Chapter Overview

Carboxylic acids participate in a variety of different reactions in organic chemistry and in biological systems. Some of their derivatives, soluble salts and esters, help to keep our world clean and sweet-smelling.

In this chapter you will learn to recognize and name carboxylic acids, carboxylic acid salts, and esters. You will write equations for preparation of these compounds. You will compare their physical properties and write equations for some of their chemical reactions. You will identify thioesters and esters of phosphoric acid.

Practice Exercises

16.1 **Carboxylic acids** (Sec. 16.1) contain the **carboxyl group** (Sec. 16.1), a carbonyl group with a hydroxyl group bonded to the carbonyl carbon atom. The naming of carboxylic acids is similar to that of aldehydes, with the *–al* ending replaced by *–oic acid.*

Give the IUPAC name for each of the following carboxylic acids.

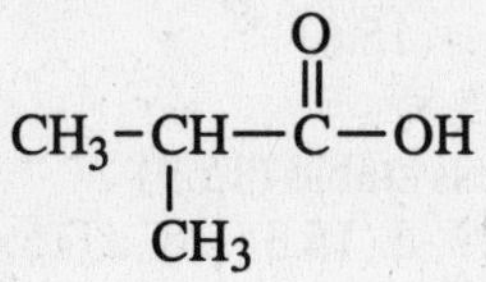 a. $CH_3-CH(CH_3)-C(=O)-OH$	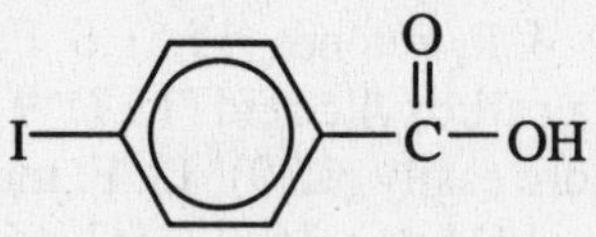 b. $I-C_6H_4-C(=O)-OH$ (para)	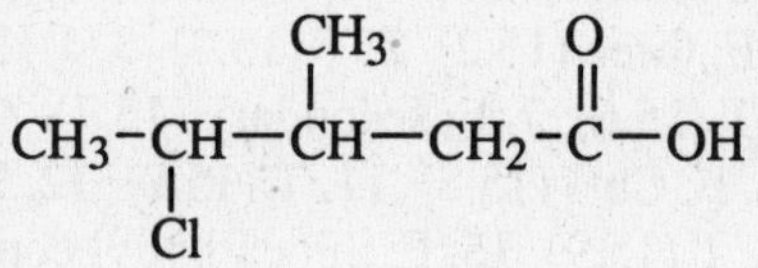 c. $CH_3-CH(Cl)-CH(CH_3)-CH_2-C(=O)-OH$

16.2 Draw the structural formulas for the following carboxylic acids:

a. 3-chloro-2-methylpropanoic acid	b. 2,2-dibromobutanoic acid

16.3 Common names are often used for many of the carboxylic acids and **dicarboxylic acids** (Sec. 16.2). Complete the following table to compare the common name and the IUPAC name for some acids.

IUPAC name	Common name	Structural formula
butanoic acid		
	formic acid	
		HOOC—COOH
	adipic acid	

16.4 A number of carboxylic acids are polyfunctional, and some of these are important in reactions that occur in the human body. Complete the following table giving structures and names for some common polyfunctional carboxylic acids.

Common name	Molecular formula	Structure of carboxylic acid
pyruvic acid		
		$CH_3-CH(OH)-C(=O)-OH$
fumaric acid		
		$HOOC-CH_2-C(OH)(COOH)-CH_2-COOH$

16.5 Carboxylic acids have very high boiling points because two molecules can form two hydrogen bonds with one another, from each of the double-bonded oxygens to the hydrogens of the –OH groups. Carboxylic acids also form hydrogen bonds with water and so are somewhat water-soluble, especially those with short hydrocarbon chains.

Draw diagrams of the hydrogen bonding between the following molecules. Use structural formulas and show hydrogen bonds as dotted lines.

a. two molecules of propanoic acid	b. one molecule of propanoic acid and water

16.6 Carboxylic acids can be prepared by oxidation of a primary alcohol or an aldehyde. Aromatic acids can be prepared by oxidation of an alkyl side chain on a benzene derivative.

Write equations for the preparation of carboxylic acids from the indicated reactants. Show the necessary reagents used in each reaction. Name the carboxylic acid formed.

a. 1-butanol

b. butanal

c. 4-bromo-1-ethylbenzene

16.7 Carboxylic acids react with strong bases to produce water and a **carboxylic acid salt** (Sec. 16.9). The negative ion of the salt is called a **carboxylate ion** (Sec. 16.8).

Complete the following table of carboxylic acids and their salts. (Use sodium salts.)

Name of acid	Structural formula of carboxylic acid salt	IUPAC name of carboxylic acid salt
propanoic acid		
	$CH_3-CH_2-CH_2-\overset{\overset{\displaystyle O}{\|\|}}{C}-O^-\ Na^+$	
benzoic acid		sodium benzoate

16.8 A carboxylic acid salt can be converted to the carboxylic acid by reacting the salt with a strong acid such as HCl or H_2SO_4.

Write an equation for the reaction of potassium ethanoate with hydrochloric acid to form the carboxylic acid.

16.9 The reaction of a carboxylic acid with an alcohol, in the presence of a strong-acid catalyst, produces an **ester** (Sec. 16.10), whose functional group is –COOR. The –OH from the carboxyl group combines with –H from the alcohol to form a molecule of water as a by-product.

Complete the following table showing ester structures, names, and the reactants that form the esters. In ester names, the "alcohol part" comes first, followed by the "acid part" with an *-ate* ending.

Acid and alcohol	Structure of ester formed	IUPAC name
ethanoic acid and methanol		
	$CH_3-CH_2-CH_2-\overset{\overset{\displaystyle O}{\|\|}}{C}-O-CH_3$	
		ethyl benzoate

16.10 Esters are often known by common names, which are based on the common names of the acid parts of the esters. Complete the following table comparing the IUPAC and common names of some esters.

IUPAC name	Common name
ethyl methanoate	
	methyl valerate
	propyl caproate

16.11 Ester molecules do not have a hydrogen atom bonded to an oxygen atom, so they cannot form hydrogen bonds to one another. They can, however, form hydrogen bonds with water molecules.

Indicate which compound in each of the following pairs would have a higher boiling point and explain your answer.

Compounds	Higher-boiling compound	Explanation
butanoic acid and methyl propanoate		
methyl propanoate and 1-butanol		
methyl hexanoate and decane		

16.12 Esters undergo hydrolysis with an acid catalyst to produce the acid and alcohol from which they were formed. They also undergo base-catalyzed hydrolysis, which is called **ester saponification** (Sec. 16.15). In this case the products are the alcohol and the salt of the carboxylic acid.

Give the names of the products formed in the following reactions.

a. Ethyl butanoate undergoes basic hydrolysis with sodium hydroxide (saponification).

b. Ethyl butanoate undergoes acidic hydrolysis.

16.13 Thiols react with carboxylic acids to form **thioesters** (Sec. 16.16). A molecule of water is formed as a by-product. Complete the following reactions showing thioester formation.

a. $HCOOH + CH_3-CH_2-SH \longrightarrow$?

b. ? + ? $\longrightarrow$ (cyclopentyl)$-\overset{O}{\overset{\|}{C}}-S-CH_3 + H_2O$

16.14 Inorganic acids react with alcohols to form esters in a manner similar to that for the formation of carboxylic acid esters. Phosphoric acid has three hydroxyl groups and so can form **phosphate esters** (Sec. 16.18) with one, two, or three molecules of an alcohol. Draw the structural formula of the ester formed from each set of reactants named below.

a. One molecule of ethanol and one molecule of phosphoric acid

b. Two molecules of ethanol and one molecule of phosphoric acid

16.15 Use this identification exercise to review your knowledge of the structures introduced in this chapter. Using structures A through I, give the best choice for each of the terms that follow.

$CH_3-C(Cl)(CH_3)-C(=O)-OH$ A.	$CH_3-CH_2-C(=O)-O^-\ K^+$ B.	$CH_3-CH_2-C(=O)-O-CH_3$ C.
$CH_3-CH(OH)-C(=O)-OH$ D.	$CH_3-CH_2-CH_2-OH$ E.	$HOOC-(CH_2)_2-COOH$ F.
O, =O G.	$CH_3-CH_2-CH_2-O-CH_3$ H.	$CH_3-C(=O)-CH_2-C(=O)-OH$ I.

a. dicarboxylic acid ________
b. carboxylic acid ________
c. β-keto carboxylic acid ________
d. lactone ________
e. potassium salt ________
f. ester ________
g. α-hydroxy carboxylic acid ________
h. alcohol ________
i. ether ________

Give the IUPAC names for compounds A through H.
A. ________
B. ________
C. ________
D. ________
E. ________
F. ________
G. ________
H. ________

Self-Test

True-false: Indicate whether the following statements are true or false. If the statement is false, give the word or phrase that may be substituted for the underlined portion to make the statement true.

1. Formic acid is the simplest carboxylic acid.
2. A carboxylic acid with six carbons in a straight chain is named 1-hexanoic acid.
3. Vinegar is made of glacial acetic acid.
4. Another name for 2-methylbutanoic acid is β-methylbutyric acid.
5. Glycolic acid is a dicarboxylic acid.
6. Oxidation of 2-butanol produces butanoic acid.
7. Because of their extensive hydrogen bonding, carboxylic acids have low boiling points.
8. Benzoic acid cannot be prepared by the oxidation of ethylbenzene.
9. A carboxylate ion is formed by the loss of an acidic hydrogen atom.
10. The reaction between sodium hydroxide and benzoic acid would produce sodium benzoate.
11. According to Le Châtelier's principle, adding an excess of alcohol to a carboxylic acid will decrease the amount of ester that is formed.
12. The pleasant fragrances of many flowers and fruits are produced by mixtures of carboxylic acids.
13. Aspirin has an acid functional group and an ester functional group.
14. The boiling points of esters are lower than those of alcohols and acids with comparable molecular mass.
15. Ester saponification is the base-catalyzed hydrolysis of an ester.
16. A polyester is an addition polymer with ester linkages.

Multiple choice:

17. The reaction between methanol and ethanoic acid will produce:

    a. methyl acetate b. ethyl formate c. methyl butyrate
    d. ethyl ethanoate e. none of these

18. Rank these three types of compounds – alcohol, carboxylic acid, ester of carboxylic acid of comparable molecular mass – in order of boiling point, highest to lowest:

    a. alcohol, acid, ester b. ester, acid, alcohol c. acid, alcohol, ester
    d. ester, alcohol, acid e. acid, ester, alcohol

19. The products of the acid-catalyzed hydrolysis of an ester include:

    a. a carboxylate ion b. a ketone c. an alcohol
    d. an ether e. none of these

20. A thioester is formed by the reaction between:

    a. sulfuric acid and a carboxylate b. an alcohol and sulfuric acid
    c. a carboxylic acid and a thiol d. a carboxylic acid and a sulfate
    e. none of these

21. When a carboxylic acid and an alcohol react to form an ester:
    a. a molecule of water is incorporated in the ester
    b. a molecule of water is produced
    c. a molecule of oxygen is incorporated in the ester
    d. a molecule of hydrogen is produced
    e. none of these

22. A carboxylate ion is produced in which of these reactions?
    a. esterification
    b. saponification
    c. acid hydrolysis of an ester
    d. reaction of an alcohol with an inorganic acid
    e. none of these

23. An example of a dicarboxylic acid is:
    a. succinic acid
    b. butyric acid
    c. caproic acid
    d. lactic acid
    e. none of these

24. The hydrolysis of methyl butanoate in the presence of strong acid would yield:
    a. methanoic acid and butyric acid
    b. methanol and butanoic acid
    c. methanoic acid and 1-butanol
    d. sodium butanoate and methanol
    e. none of these

Answers to Practice Exercises

16.1 a. 2-methylpropanoic acid
b. 4-iodobenzoic acid
c. 4-chloro-3-methylpentanoic acid

16.2

a. 3-chloro-2-methylpropanoic acid	b. 2,2-dibromobutanoic acid
$\underset{\text{Cl}}{CH_2}-\underset{CH_3}{CH}-\overset{O}{\overset{\Vert}{C}}-OH$	$CH_3-CH_2-\overset{Br}{\underset{Br}{C}}-\overset{O}{\overset{\Vert}{C}}-OH$

16.3

IUPAC name	Common name	Structural formula
butanoic acid	butyric acid	$CH_3-(CH_2)_2-COOH$
methanoic acid	formic acid	$H-COOH$
ethanedioic acid	oxalic acid	$HOOC-COOH$
hexanedioic acid	adipic acid	$HOOC-(CH_2)_4-COOH$

16.4

Common name	Molecular formula	Structure of carboxylic acid
pyruvic acid	$C_3H_4O_3$	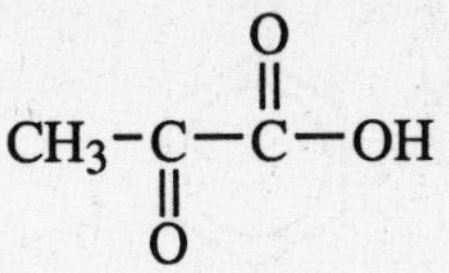 $CH_3-C(=O)-C(=O)-OH$
lactic acid	$C_3H_6O_3$	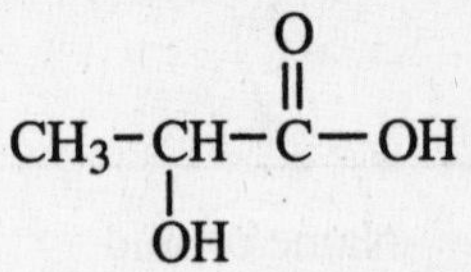 $CH_3-CH(OH)-C(=O)-OH$
fumaric acid	$C_4H_4O_4$	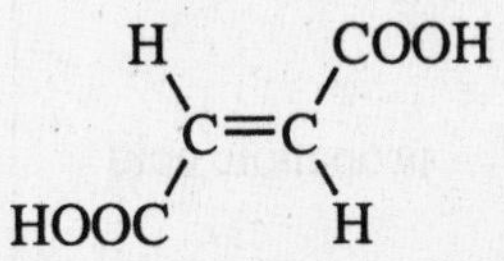 $(H)(HOOC)C=C(COOH)(H)$
citric acid	$C_6H_8O_7$	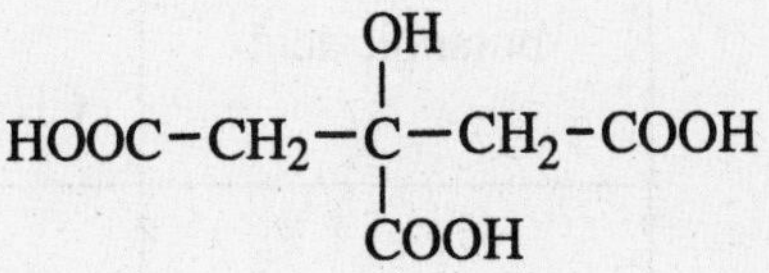 $HOOC-CH_2-C(OH)(COOH)-CH_2-COOH$

16.5 a. two molecules of propanoic acid

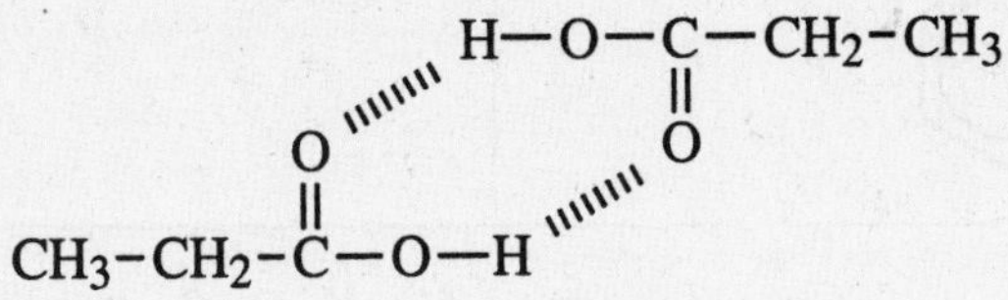

$$H-O-C(=O)-CH_2-CH_3$$

$$CH_3-CH_2-C(=O)-O-H$$

b. one molecule of propanoic acid and water

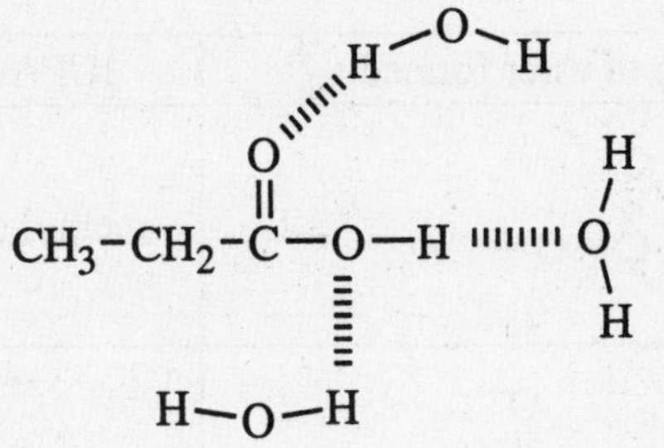

$$CH_3-CH_2-C(=O)-O-H \cdots OH_2$$

H–O–H

16.6 a. $CH_3-CH_2-CH_2-CH_2-OH \xrightarrow[\text{oxidizing agent}]{CrO_3} CH_3-CH_2-CH_2-COOH$

b. $CH_3-CH_2-CH_2-C(=O)-H \xrightarrow[\text{oxidizing agent}]{K_2Cr_2O_7} CH_3-CH_2-CH_2-COOH$

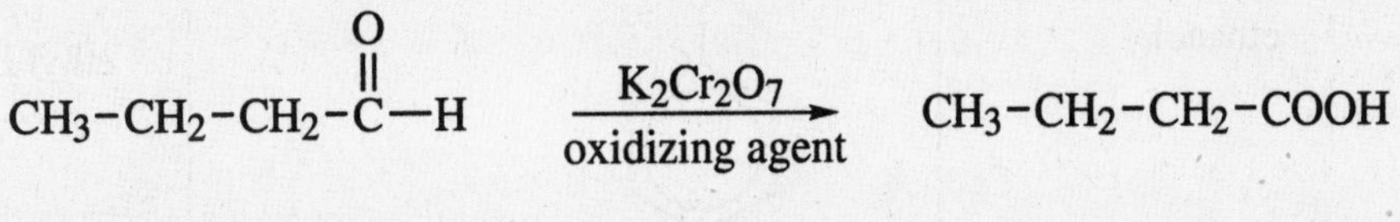

c. $BrC_6H_4CH_2{-}CH_3 \xrightarrow[H_2SO_4]{K_2Cr_2O_7} BrC_6H_4COOH$ (4-bromo substituted benzene ring: $CH_2{-}CH_3$ oxidized to COOH, Br para)

16.7

Name of acid	Structural formula of carboxylic acid salt	IUPAC name of carboxylic acid salt
propanoic acid	$CH_3{-}CH_2{-}\overset{O}{\overset{\parallel}{C}}{-}O^-\ Na^+$	sodium propanoate
butanoic acid	$CH_3{-}CH_2{-}CH_2{-}\overset{O}{\overset{\parallel}{C}}{-}O^-\ Na^+$	sodium butanoate
benzoic acid	$C_6H_5{-}\overset{O}{\overset{\parallel}{C}}{-}O^-\ Na^+$	sodium benzoate

16.8 $CH_3{-}\overset{O}{\overset{\parallel}{C}}{-}O^-\ K^+ + HCl \longrightarrow CH_3{-}\overset{O}{\overset{\parallel}{C}}{-}OH + KCl$

16.9

Acid and alcohol	Structure of ester formed	IUPAC name
ethanoic acid and methanol	$CH_3{-}\overset{O}{\overset{\parallel}{C}}{-}O{-}CH_3$	methyl ethanoate
butanoic acid and methanol	$CH_3{-}CH_2{-}CH_2{-}\overset{O}{\overset{\parallel}{C}}{-}O{-}CH_3$	methyl butanoate
benzoic acid and ethanol	$C_6H_5{-}\overset{O}{\overset{\parallel}{C}}{-}O{-}CH_2{-}CH_3$	ethyl benzoate

16.10

IUPAC name	Common name
ethyl methanoate	ethyl formate
methyl pentanoate	methyl valerate
propyl hexanoate	propyl caproate

16.11

Compounds	Higher-boiling compound	Explanation
butanoic acid and methyl propanoate	butanoic acid	hydrogen bonding between two molecules of the carboxylic acid (produces a dimer)
methyl propanoate and 1-butanol	1-butanol	hydrogen bonding between the alcohol molecules
methyl hexanoate and decane	methyl hexanoate	dipole-dipole attraction between ester molecules

16.12 a. ethanol and sodium butanoate b. ethanol and butanoic acid

16.13 a. $HCOOH + CH_3-CH_2-SH \longrightarrow HC(=O)-S-CH_2-CH_3 + H_2O$

b. 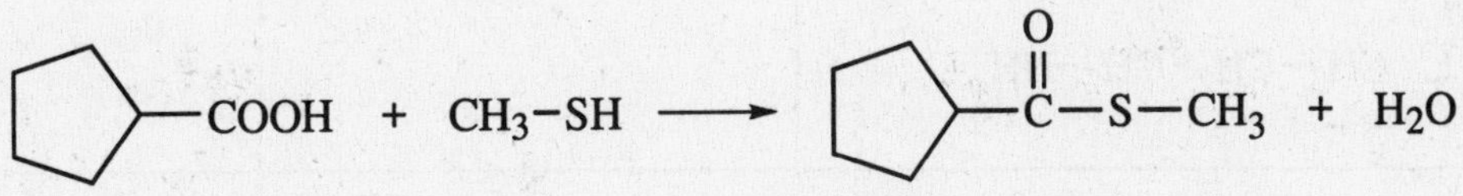 $C_5H_9-COOH + CH_3-SH \longrightarrow C_5H_9-C(=O)-S-CH_3 + H_2O$ (cyclopentyl)

16.14

a. 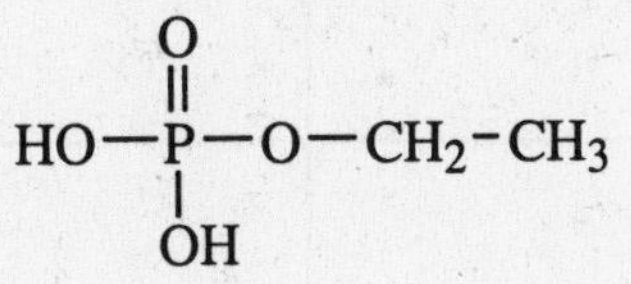 $HO-P(=O)(OH)-O-CH_2-CH_3$

b. 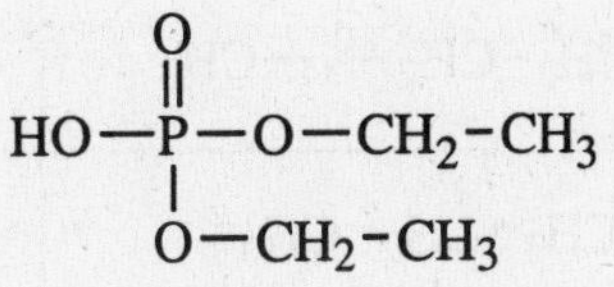 $HO-P(=O)(O-CH_2-CH_3)-O-CH_2-CH_3$

16.15 a. F; b. A; c. I; d. G; e. B; f. C; g. D; h. E; i. H

A. 2-chloro-2-methylpropanoic acid
B. potassium propanoate
C. methyl propanoate
D. 2-hydroxypropanoic acid
E. 1-propanol
F. butanedioic acid
G. 4-butanolide
H. 1-methoxypropane

Answers to Self-Test

The numbers in parentheses refer to sections in your textbook.
1. T (16.1) **2.** F; hexanoic acid (16.2) **3.** F; dilute aqueous acetic acid solution (16.3) **4.** F; α-methylbutyric acid (16.3) **5.** F; an α-hydroxy acid (16.4) **6.** F; butanone (16.6) **7.** F; high boiling points (16.5) **8.** F; can be prepared (16.6) **9.** T (16.7) **10.** T (16.8) **11.** F; increase (16.9) **12.** F; esters (16.11) **13.** T (16.11) **14.** T (16.12) **15.** T (16.13) **16.** F; a condensation polymer **17.** a (16.10) **18.** c (16.5, 16.12) **19.** c (16.13) **20.** c (16.14) **21.** b (16.9) **22.** b (16.9, 16.13) **23.** a (16.3, 16.4) **24.** b (16.13)

Amines and Amides — Chapter 17

Chapter Overview

Amines and amides are nitrogen-containing organic compounds that include many substances of importance in the human body as well as many natural and synthetic drugs.

In this chapter you will learn to recognize, name, and draw structural formulas for amines and amides. You will write equations for the preparation of amines and amides and for the hydrolysis of amides. You will learn the names and functions of some biologically important amines and amides.

Practice Exercises

17.1 An amine (Sec. 17.2) is an organic compound containing a nitrogen atom to which one, two or three hydrocarbon groups are attached. Both common and IUPAC names are used for amines.

Complete the following table. Follow the rules for naming given in Section 17.3 of your textbook.

Structural formula	IUPAC name	Common name
a. $CH_3-CH_2-CH_2-NH_2$		
b. $CH_3-CH_2-CH_2-N(CH_2-CH_3)-CH_3$		
c. $HN(CH_2-CH_3)-CH_2-CH_2-CH_3$		
d. benzene ring with NH_2 and F (meta)		
e. benzene ring with NH_2 and CH_2-CH_3 (ortho)		

17.2 Amines are classified as primary, secondary, or tertiary on the basis of the number of alkyl groups attached to the nitrogen atom. An **amino group** (Sec. 17.2) is the $-NH_2$ functional group; secondary and tertiary amines have substituted amino groups.

Classify each of the amines in Practice Exercise 17.1 as a primary, secondary, or tertiary amine.

a.

b.

c.

d.

e.

17.3 Amines are capable of forming hydrogen bonds between molecules; however, because nitrogen is less electronegative than oxygen, hydrogen bonds between molecules of amines are weaker than those between alcohol molecules.

Determine which compound of each pair would have a higher boiling point and explain your choice.

Compounds	Higher-boiling compound	Explanation
1-pentanamine and 1-pentanol		
1-pentanamine and hexane		

17.4 Because amines are weak bases, an amine's reaction with an acid produces an **amine salt** (Sec. 17.6). The amine may be regenerated by reaction of the salt with a strong base. Complete the following equations.

a. $CH_3{-}CH_2{-}\underset{\displaystyle NH_2}{\underset{|}{CH_2}}$ + HCl ⟶ ?

b. ? + ? ⟶ $C_6H_5{-}NH_2$ + NaBr + H_2O

17.5 Amine salts are named in the same way as other ionic compounds are: the positive ion, the **substituted ammonium ion** (Sec. 17.5), is named first, followed by the name of the negative ion. Give the name of each of the amine salts in Practice Exercise 17.4.

a.

b.

17.6 Amines can be prepared by the reaction of ammonia with an alkyl halide in the presence of a strong base. Primary amines formed by this method react further with more molecules of the alkyl halide to form secondary and tertiary amines.

Write the reaction and write names for the primary, secondary, and tertiary amines and the **quaternary ammonium salt** (Sec. 17.7) formed when ammonia reacts with chloroethane in the presence of sodium hydroxide, a strong base. There are four reactions.

a.

b.

c.

d.

17.7 **Heterocyclic amines** (Sec. 17.8) and their derivatives are common in naturally occurring compounds. Name the following heterocyclic amines:

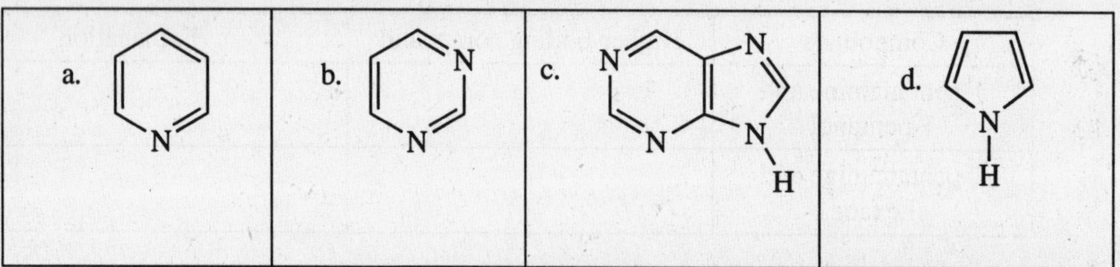

17.8 An **amide** (Sec. 17.11) is a carboxylic acid derivative in which the carboxyl –OH group is replaced by an amino or substituted amino group. Amides can be prepared by the condensation reaction at high temperature between a carboxylic acid and ammonia or a primary or secondary amine. (At room temperature, an acid-base reaction takes place, and there is no amide formation.)

Give the structure of the amide formed by the reaction of the following acids and amines.

a. $CH_3-CH(Cl)-C(=O)-OH + NH_2-CH_2-CH_3 \xrightarrow{\text{high T}}$?

b. (3-bromophenyl)–COOH + $CH_3-NH-CH_3 \xrightarrow{\text{high T}}$?

c. $CH_3-CH_2-CH_2-C(=O)-OH + NH_3 \xrightarrow{\text{high T}}$?

17.9 Amides can be classified as primary, secondary, or tertiary on the basis of the number of carbon atoms attached to the nitrogen atom. Amide names in the IUPAC system use the name of the parent acid with the ending *-amide*. We included alkyl groups attached to the nitrogen atom as prefixes, using *N-* to locate them.

Classify each of the amides in Practice Exercise 17.8 as a primary, secondary, or tertiary amide, and give the IUPAC name for the amide produced.

a.

b.

c.

17.10 Common names for amides are similar to IUPAC names, but the common name of the parent acid is used. Give the common names for the amides in Practice Exercise 17.8.

a.

b.

c.

17.11 Draw structural formulas for the following substituted amides.

a. *N*-ethyl-*N*-propylpentanamide	b. *N*-methylbenzamide

17.12 Amides undergo hydrolysis in a manner similar to that of esters. The products depend on the catalyst that is used: an acid catalyst produces a carboxylic acid and an amine salt, and a basic catalyst produces a carboxylic acid salt and an amine.

Complete the following equations for the hydrolysis of each of these amides:

a. 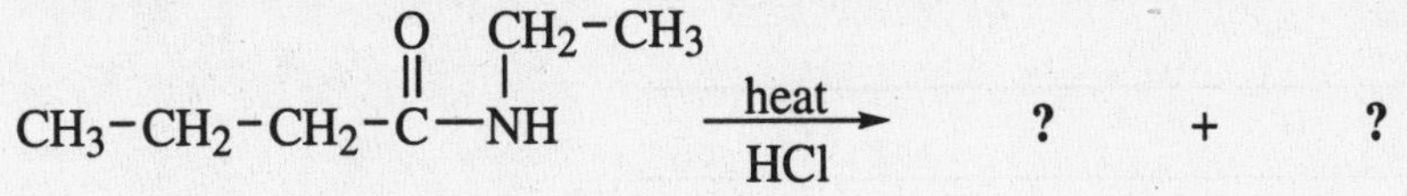 $CH_3-CH_2-CH_2-\overset{\overset{O}{\|}}{C}-\underset{}{\overset{\overset{CH_2-CH_3}{|}}{NH}} \xrightarrow[HCl]{heat} ? + ?$

b. $C_6H_5-\overset{\overset{O}{\|}}{C}-\underset{\underset{CH_2-CH_3}{|}}{\overset{\overset{CH_3}{|}}{N}} \xrightarrow[HCl]{heat} ? + ?$

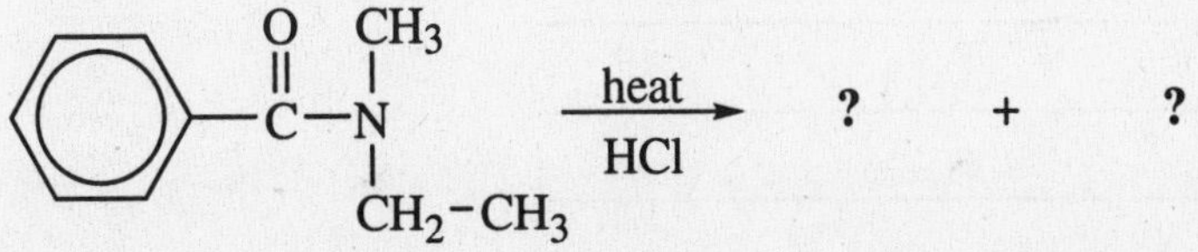

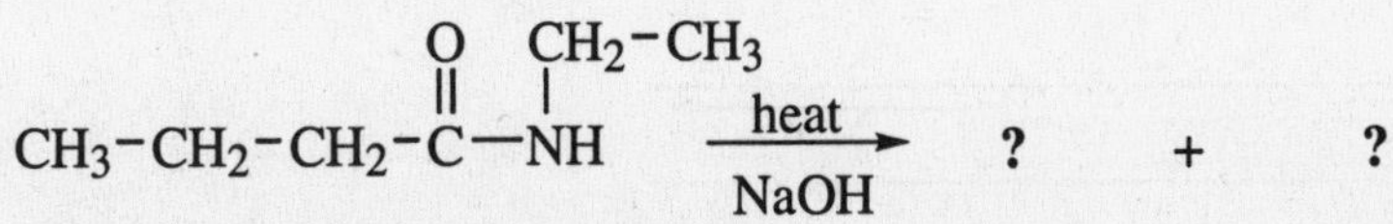

c. $CH_3-CH_2-CH_2-\overset{\overset{O}{\|}}{C}-\overset{\overset{CH_2-CH_3}{|}}{NH} \xrightarrow[NaOH]{heat} ? + ?$

17.13 Give the IUPAC names for the organic products formed in Practice Exercise 17.12.

a.

b.

c.

17.14 Use this identification exercise to review your knowledge of common structures. Using structures A through I, give the best choice for each of the terms that follow.

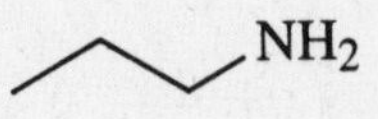 $CH_3CH_2CH_2-NH_2$ (line structure) A.	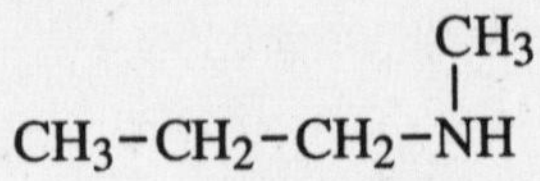 $CH_3-CH_2-CH_2-NH-CH_3$ B.	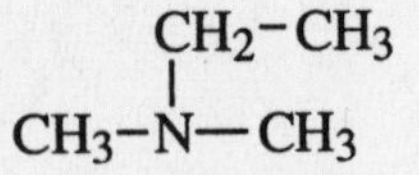 $CH_3-N(CH_2-CH_3)-CH_3$ C.
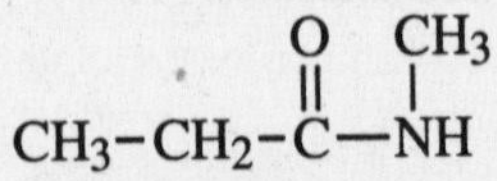 $CH_3-CH_2-C(=O)-NH-CH_3$ D.	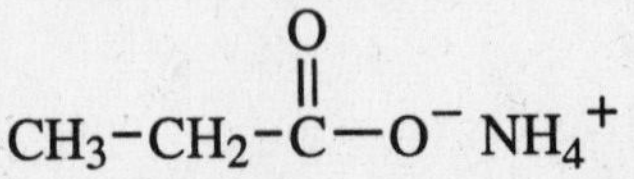 $CH_3-CH_2-C(=O)-O^-\ NH_4^+$ E.	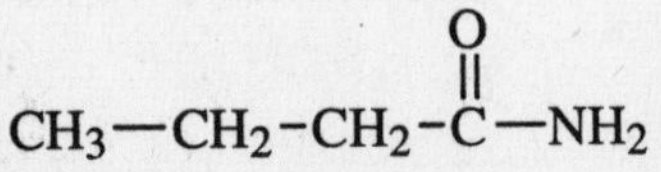 $CH_3-CH_2-CH_2-C(=O)-NH_2$ F.
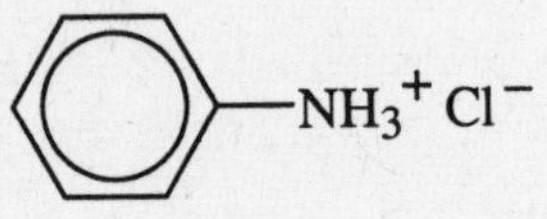 $C_6H_5-NH_3^+\ Cl^-$ G.	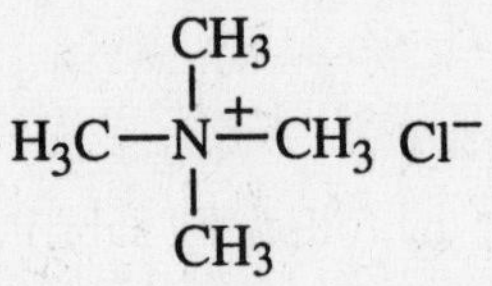 $H_3C-N^+(CH_3)_2-CH_3\ Cl^-$ H.	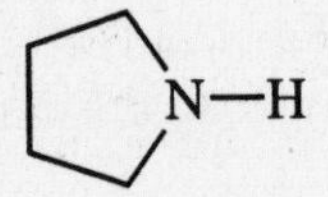 (cyclic) N—H I.

a. heterocyclic amine ________
b. salt of a carboxylic acid ________
c. quaternary ammonium salt ________
d. amine salt ________
e. primary amine ________
f. secondary amide ________
g. secondary amine ________
h. tertiary amine ________
i. primary amide ________

Give the IUPAC names of compounds A through I.

A. ________

B. ________

C. ________

D. ________

E. ________

F. ________

G. ________

H. ________

I. ________

Self-Test

True-false: Indicate whether the following statements are true or false. If the statement is false, give the word or phrase that may be substituted for the underlined portion to make the statement true.

1. Nitrogen forms two covalent bonds to complete its octet of electrons.
2. *tert*-Butyl amine is a primary amine.
3. The IUPAC name for an amine having two methyl groups and one ethyl group attached to a nitrogen atom is ethyldimethylamine.
4. A benzene ring with an attached amino group is called aniline.
5. Amines are noted for their pleasant odors.
6. Hydrogen bonding between the molecules of an amine is stronger than hydrogen bonding between alcohol molecules.
7. Reaction of excess alkyl halide with ammonia produces an amine salt.
8. The nitrogen atom of a heterocyclic amine cannot be part of an aromatic system.
9. Urea is a naturally occurring diamine with only one carbon atom.
10. The various types of nylon are synthesized by the reactions of diamines with dicarboxylic acids.
11. Kevlar is a very tough polyamide that owes its strength to the presence of aromatic rings in its "backbone."
12. Amides do not exhibit basic properties in water as amines do.
13. Disubstituted amides have no hydrogens attached to nitrogen for hydrogen bonding and so have low melting points.
14. The symptoms of Parkinson's disease are caused by a deficiency of serotonin.

Multiple choice:

15. An amine can be formed from its amine salt by treating the salt with:

    a. NaOH b. an alcohol c. H_2SO_4
    d. a carboxylic acid e. none of these

16. Which of the following is *not* a heterocyclic amine derivative?

    a. caffeine b. nicotine c. heme
    d. choline e. none of these

17. Alkaloids, a group of nitrogen-containing compounds obtained from plants, include all of the following compounds except:

    a. quinine b. nicotine c. atropine
    d. heroin e. morphine

18. Which of the following compounds is a secondary amine?

    a. 2-butanamine b. ethylmethylamine c. *N,N*-dimethylaniline
    d. 2-methylaniline e. none of these

19. *N*-methylpropanamide could be prepared as a product of the reaction between:

a. *N*-propanamine and methanol
b. propylamine and acetic acid
c. methylamine and propanoic acid
d. acetic acid and methylamine
e. none of these

20. Basic hydrolysis of an amide produces:

a. an amine salt and a carboxylic acid
b. an amine salt and a carboxylic acid salt
c. an amine and a carboxylic acid
d. an amine and a carboxylic acid salt
e. none of these

21. Mental depression may be caused by a deficiency of:

a. histamine
b. serotonin
c. atropine
d. porphyrin
e. none of these

22. A correct name for the compound containing nitrogen bonded to two ethyl groups and one hydrogen is:

a. diethylamine
b. 2-ethylamine
c. 2-ethyl amine
d. diethyl amine
e. none of these

Answers to Practice Exercises

17.1
a. 1-propanamine; propylamine
b. *N*-ethyl-*N*-methyl-1-propanamine; ethylmethylpropylamine
c. *N*-ethyl-1-propanamine; ethylpropylamine
d. 3-fluoroaniline; 3-fluoroaniline
e. 2-ethylaniline or *o*-ethylaniline

17.2
a. primary
b. tertiary
c. secondary
d. primary
e. primary

17.3

Compounds	Higher-boiling compound	Explanation
1-pentanamine and 1-pentanol	1-pentanol	-OH hydrogen bonds are stronger than -NH hydrogen bonds because oxygen is more electronegative than nitrogen.
1-pentanamine and hexane	1-pentanamine	hydrogen bonding between amine molecules

17.4 a. $CH_3-CH_2-CH_2(-NH_2) + HCl \longrightarrow CH_3-CH_2-CH_2(-{}^{+}NH_3)\ Cl^-$

b. $C_6H_5-\overset{+}{N}H_3\ Br^- + NaOH \longrightarrow C_6H_5-NH_2 + NaBr + H_2O$

17.5 a. propylammonium chloride
b. anilinium bromide

17.6 a. primary ethylamine $NH_3 + CH_3-CH_2-Cl \xrightarrow{OH^-} CH_3-CH_2-NH_2$

b. secondary diethylamine $CH_3-CH_2-NH_2 + CH_3-CH_2-Cl \xrightarrow{OH^-} CH_3-CH_2-NH(-CH_2-CH_3)$

c. tertiary triethylamine $CH_3-CH_2-NH(-CH_2-CH_3) + CH_3-CH_2-Cl \xrightarrow{OH^-} CH_3-CH_2-N(-CH_2-CH_3)_2$

d. quaternary tetraethylammonium iodide $CH_3-CH_2-N(-CH_2-CH_3)_2 + CH_3-CH_2-Cl \xrightarrow{OH^-} CH_3-CH_2-\overset{+}{N}(-CH_2-CH_3)_3\ Cl^-$

17.7 a. pyridine b. pyrimidine
c. purine d. pyrrole

17.8 a. $CH_3-CH(-Cl)-\overset{O}{\overset{\|}{C}}-OH + NH_2-CH_2-CH_3 \xrightarrow{\text{high T}} CH_3-CH(-Cl)-\overset{O}{\overset{\|}{C}}-NH-CH_2-CH_3 + H_2O$

b. $(3\text{-}Br)C_6H_4-COOH + CH_3-NH-CH_3 \xrightarrow{\text{high T}} (3\text{-}Br)C_6H_4-\overset{O}{\overset{\|}{C}}-N(-CH_3)-CH_3 + H_2O$

c. $CH_3-CH_2-CH_2-\overset{O}{\overset{\|}{C}}-OH + NH_3 \xrightarrow{\text{high T}} CH_3-CH_2-CH_2-\overset{O}{\overset{\|}{C}}-NH_2 + H_2O$

17.9 a secondary amide, *N*-ethyl-2-chloropropanamide
b. tertiary amide, *N,N*-dimethyl-*m*-bromobenzamide
c. primary, butanamide

17.10 a *N*-ethyl-2-chloropropionamide
b. *N,N*-dimethyl-*m*-bromobenzamide
c. butyramide

17.11

a. $H_3C-CH_2-CH_2-CH_2-C(=O)-N(CH_2-CH_3)(CH_2-CH_2-CH_3)$

a. *N*-ethyl-*N*-propylpentanamide

b. $C_6H_5-C(=O)-NH(CH_3)$

b. *N*-methylbenzamide

17.12 a. $CH_3-CH_2-CH_2-C(=O)-NH(CH_2-CH_3) \xrightarrow[HCl]{heat} CH_3-CH_2-CH_2-C(=O)-OH + CH_3-CH_2-NH_3^+\ Cl^-$

b. $C_6H_5-C(=O)-N(CH_3)(CH_2-CH_3) \xrightarrow[HCl]{heat} C_6H_5-C(=O)-OH + CH_3-NH_2^+(CH_2-CH_3)\ Cl^-$

c. $CH_3-CH_2-CH_2-C(=O)-NH(CH_2-CH_3) \xrightarrow[NaOH]{heat} CH_3-CH_2-CH_2-C(=O)-O^-\ Na^+ + CH_3-CH_2-NH_2$

17.13 a. butanoic acid and ethylammonium chloride
b. benzoic acid and ethylmethylammonium chloride
c. sodium butanoate and ethanamine

17.14 a. I; b. E; c. H; d. G; e. A; f. D; g. B; h. C; i. F

A. 1-propanamine
B. *N*-methyl-1-propanamine
C. *N,N*-dimethylethanamine
D. *N*-methylpropanamide
E. ammonium propanoate
F. butanamide
G. anilinium chloride
H. tetramethylammonium chloride
I. pyrrolidine

Answers to Self-Test

The numbers in parentheses refer to sections in your textbook.
1. F; three covalent bonds (17.1) **2.** T (17.2) **3.** F; N,N-dimethylethanamine (17.3) **4.** T (17.3) **5.** F; unpleasant or fishlike odors (17.4) **6.** F; weaker (17.4) **7.** T (17.6) **8.** F; can (17.8) **9.** F; diamide (17.12) **10.** T (17.16) **11.** T (17.16) **12.** T (17.13) **13.** T (17.13) **14.** F; dopamine (17.9) **15.** a (17.6) **16.** d (17.8) **17.** d (17.10) **18.** b (17.3) **19.** c (17.14) **20.** d (17.15) **21.** b (17.9) **22.** a (17.3)

Carbohydrates — Chapter 18

Chapter Overview

The remaining chapters in the book will be concerned with compounds that are important in biological systems. Carbohydrates are important energy-storage compounds. Their oxidation provides energy for the activities of living organisms.

In this chapter you will learn to identify various types of carbohydrates and write structural diagrams for some of the most common ones. You will be able to explain stereoisomerism in terms of chiral carbons and draw figures that represent the three-dimensional structures of compounds containing chiral carbons. You will identify the structural features of some important disaccharides and polysaccharides and will learn where in nature they are found and what functions they have.

Practice Exercises

18.1 A **monosaccharide** (Sec. 18.3) is a **carbohydrate** (Sec. 18.3) that contains a single polyhydroxy aldehyde or polyhydroxy ketone unit. Carbohydrates can be classified in terms of the number of monosaccharide units they contain. Give the names for carbohydrates having the following numbers of monosaccharide units.

a. 2 monosaccharide units ____________________

b. 2 to 10 monosaccharide units ____________________

c. 3 monosaccharide units ____________________

d. many (greater than 10) units ____________________

18.2 A molecule that cannot be superimposed on its **mirror image** (Sec. 18.4) is a **chiral molecule** (Sec. 18.4). An organic molecule is chiral if it contains a **chiral center** (Sec. 18.4), a single atom with four different atoms or groups of atoms attached to it. Indicate whether the circled carbon in each structure below is chiral or **achiral** (Sec. 18.4):

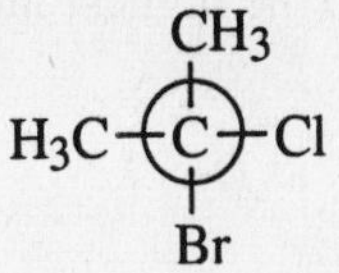

a.	b.	c.
CH_3 (top), H_3C–(C)–Cl, Br (bottom)	CH_3 (top), H–(C)–Br, I (bottom)	CH_3 (top), CH_3-CH_2–(C)–$CH_2-CH_2-CH_3$, H (bottom)

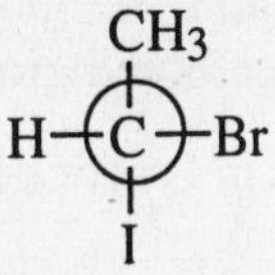

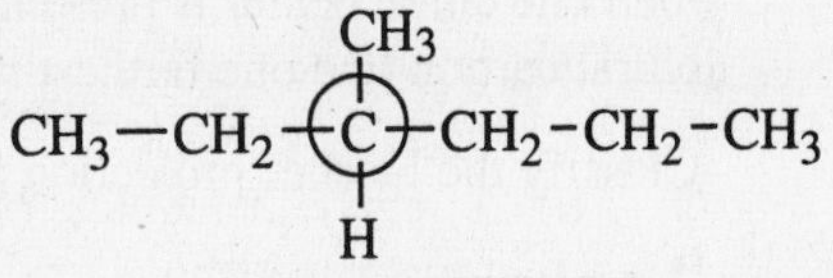

18.3 Organic molecules may contain more than one chiral center. Circle the chiral centers in the following condensed structures.

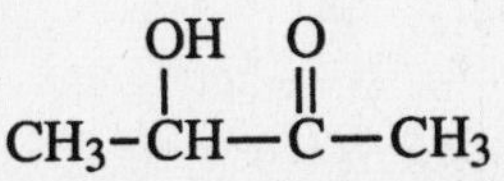

a.	b.	c.
$CH_3-CH(OH)-C(=O)-CH_3$	$CH_3-CH(OH)-CH(OH)-CH_3$	cyclohexane ring with H_3C and Br substituents

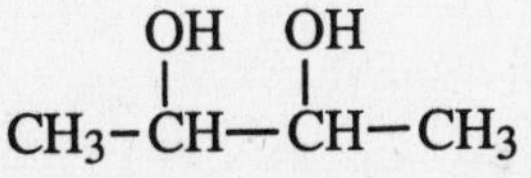

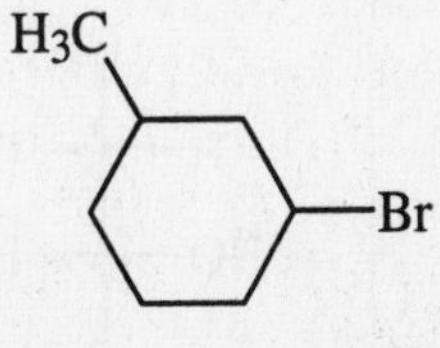

18.4 The three-dimensional nature of chiral molecules can be represented by using **Fischer projections** (18.6). A Fischer projection shows the chiral carbon as intersecting horizontal and vertical lines, with vertical lines representing bonds directed into the page and horizontal lines representing bonds directed out of the page.

Convert the following Fischer projection formulas into structural formulas.

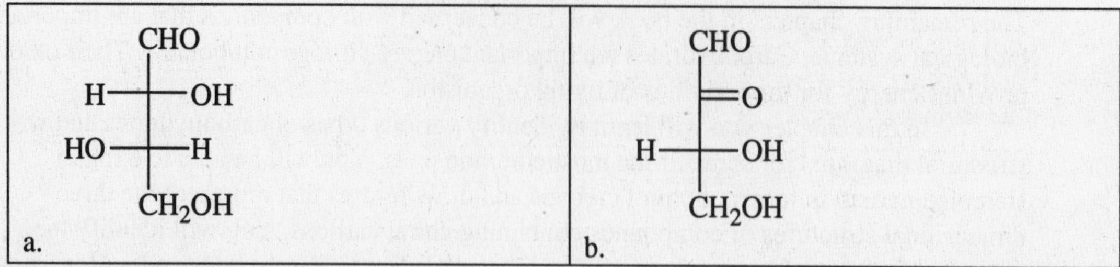

18.5 **Stereoisomers** (Sec. 18.5) are isomers whose atoms are connected in the same way but differ in their arrangement in space. **Enantiomers** (Sec. 18.5) are stereoisomers whose molecules are mirror images of one another.

Using Fischer projections, draw the enantiomers of the Fischer projection formulas in Practice Exercise 18.4.

a. b.

18.6 By definition, the enantiomer with the –OH on the right in the Fischer projection of a monosaccharide is the right-handed enantiomer and has the designation D. The enantiomer with the –OH on the left is the left-handed enantiomer and has the designation L. When more than one chiral center is present, the D and L designations refer to the highest-numbered chiral carbon (the one farthest from the carbonyl group).

Classify the Fischer projection formulas in Practice Exercise 18.4 as D or L.

a. __________ b. __________

18.7 **Diastereomers** (Sec. 18.5) are stereoisomers whose molecules are not mirror images of each other. For the structure shown in part a.: b. draw the enantiomer and c. draw one diasteromer.

CH_3

H—OH

HO—H

CH_3

a. b. c.

18.8 A monosaccharide is classified as an **aldose** or a **ketose** (Sec. 18.8) on the basis of the carbonyl group: an aldose contains an aldehyde group and a ketose contains a ketone group. Monosaccharides can be further classified by the number of carbons in each molecule (for example, aldopentose). Monosaccharides and **disaccharides** (Sec. 18.3) are often called **sugars** (Sec. 18.8). A monosaccharide that has n chiral centers may exist in a maximum of 2^n steroisomeric forms.

Fill in information for each of the following carbohydrates: a. classify using carbon number and type of carbonyl group present; b. calculate the total number of stereoisomers; c. name the isomer shown.

CHO H—OH H—OH H—OH CH_2OH	CH_2OH =O HO—H H—OH H—OH CH_2OH	CHO H—OH HO—H HO—H H—OH CH_2OH
a. b. c.	a. b. c.	a. b. c.

18.9 For monosaccharides containing five or more carbons, open-chain structures are in equilibrium with cyclic structures formed by the intramolecular reaction of the carbonyl group with a hydroxyl group to form a cyclic hemiacetal. The structural representations of these cyclic forms are called **Haworth projections** (Sec. 18.11). See Section 18.11 in your textbook for conventions observed in drawing Haworth projections.

Complete the following equations by giving the structural formula for either the cyclic or the open-chain form.

a.
CHO
HO—H
H—OH
H—OH
H—OH
CH_2OH
⟶

b.
CH_2OH
O
~OH
OH
OH OH

18.10 In naming the cyclic form of a monosaccharide, the α or β-configuration is determined by the position of the –OH on carbon 1 relative to the $–CH_2OH$ group on carbon 5: the –OH group is α when it is drawn in the "up" position and β when it is drawn in the "down" position.

For the following structures: a. Circle carbon1 and note whether the –OH group is α or β. b. Name each compound.

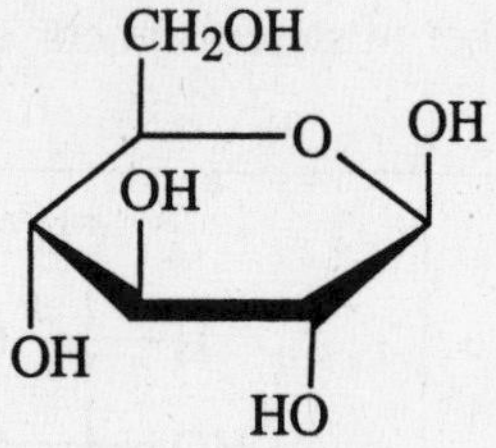

a.	b.	c.
CH_2OH, O, OH, OH, OH, HO	CH_2OH, OH, O, OH, OH, OH	CH_2OH, O, OH, OH, OH

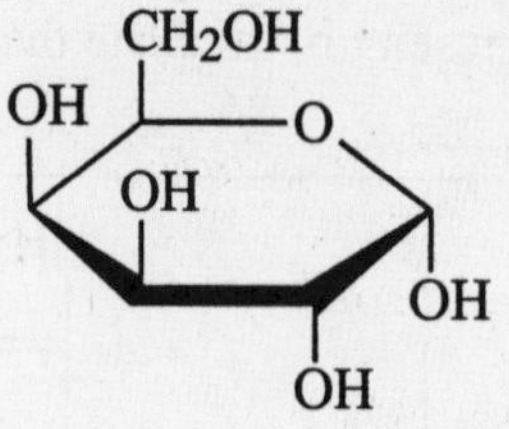

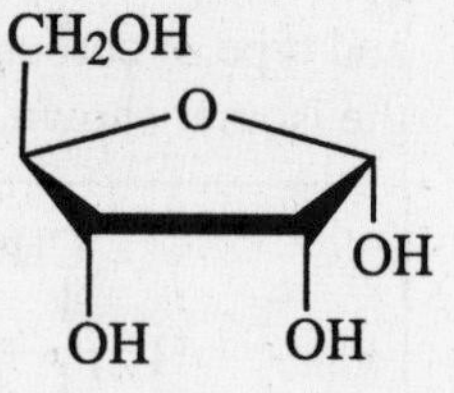

18.11 Weak oxidizing agents, such as Fehling's solution, oxidize the carbonyl group end of a monsaccharide to give an acid. Sugars that can be oxidized by these agents are called **reducing sugars** (Sec. 18.12).

The carbonyl group of sugars can also be reduced to a hydroxyl group using H_2 as the reducing agent. Another important reaction of monosaccharides is **glycoside** (Sec. 18.12) formation, reaction of the cyclic hemiacetal with an alcohol to form an acetal.

Complete the following equations for some common reactions of sugars.

a. CHO; H–C–OH; H–C–OH; H–C–OH; CH_2OH $\xrightarrow[\text{(weak oxidation)}]{\text{Fehling's solution}}$?

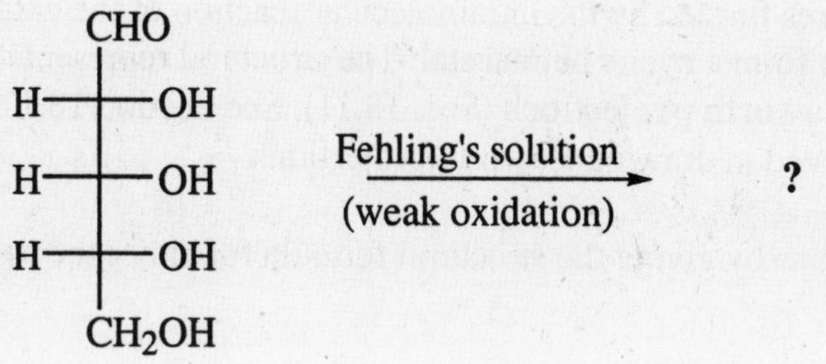

b. CHO; H–C–OH; H–C–OH; H–C–OH; CH_2OH $\xrightarrow[\text{(reduction)}]{H_2\text{/catalyst}}$?

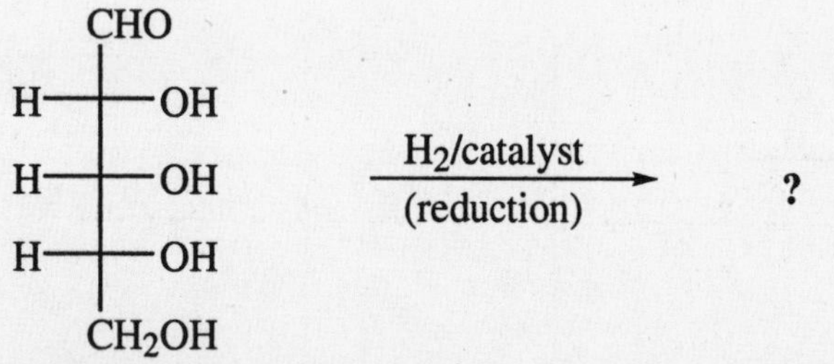

c. (ring: CH_2OH, OH, O, OH, OH, OH) + CH_3OH $\overset{H^+}{\rightleftharpoons}$?

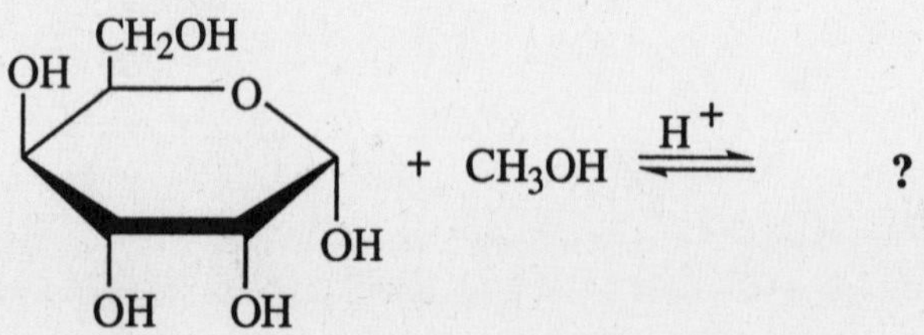

18.12 The two monosaccharides that form a disaccharide are joined by a **glycosidic linkage** (Sec. 18.13). The configuration (α or β) of the hemiacetal carbon atom of the cyclic form is often very important.

For the following disaccharides: a. give the name; b. circle the glycosidic bond and name the linkage (*α* or *β*); c. give the names of the monosaccharides that make up the disaccharide.

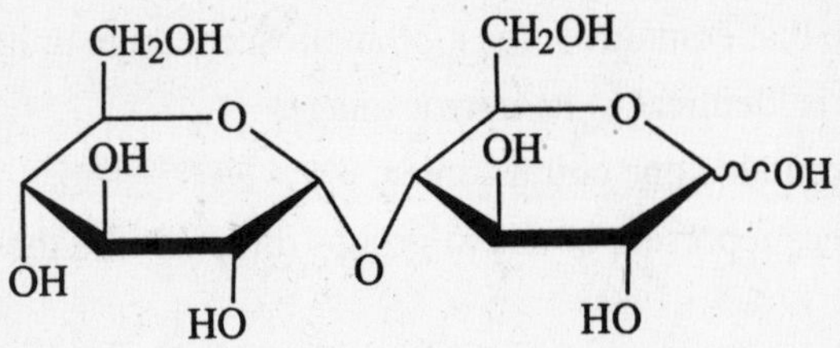 a. b. c.	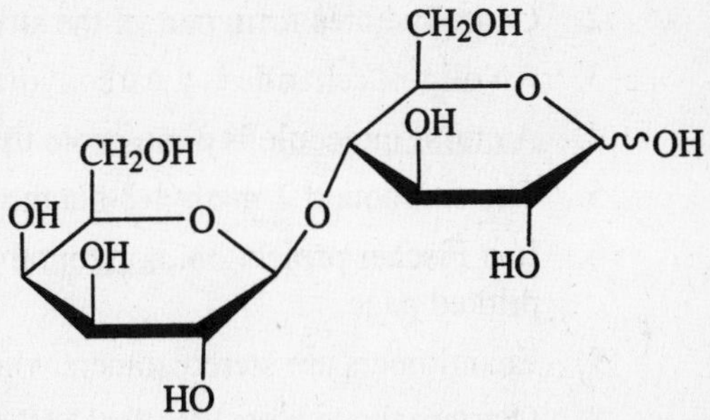 a. b. c.

18.13 Are the disaccharides in Practice Exercise 18.12 reducing or nonreducing sugars with Benedict's solution? ____________________ Place a box around the functional group (in each of the structural formulas above) that is responsible for giving a positive test with Benedict's solution.

18.14 There are a number of important **polysaccharides** (Sec. 18.14) that are made up of D-glucose units. They differ in the types of linkages (*α* or *β*) between monomers, the size, and the function. Complete the following table that summarizes these properties. Use information found in Section 18.14 of your textbook.

Polysaccharide	Linkage	Size	Branching	Function
glycogen				
amylose				
amylopectin				
cellulose				

Self-Test

True-false: Indicate whether the following statements are true or false. If the statement is false, give the word or phrase that may be substituted for the underlined portion to make the statement true.

1. Bioorganic substances include carbohydrates, lipids, proteins, and nucleic acids.
2. Carbohydrates form part of the structural framework of DNA and RNA.
3. An oligosaccharide is a carbohydrate that contains at least 20 monosaccharide units.
4. A chiral molecule is a molecule that is identical to its mirror image.
5. The compound 2-methyl-2-butanol contains one chiral center.
6. In a Fischer projection, horizontal lines represent bonds to groups directed into the printed page.
7. Enantiomers are stereoisomers whose molecules are not mirror images of each other.
8. Diastereomers have identical boiling and freezing points.
9. A dextrorotatory compound rotates plane-polarized light in a clockwise direction.
10. The direction of rotation of plane-polarized light by sugar molecules is designated in the compound's name as D or L.
11. The responses of the human body to the two enantiomeric forms of a chiral molecule are identical.
12. A six-carbon monosaccharide with a ketone functional group is called an aldopentose.
13. Glycogen is the glucose storage polysaccharide known as animal starch.
14. The carbonyl group of a monosaccharide can be reduced to a hydroxyl group using Tollens solution as a reducing agent.
15. The bond between two monosaccharide units in a disaccharide is a glycosidic linkage.
16. Humans cannot digest cellulose because they lack an enzyme to catalyze hydrolysis of the $\alpha(1 \rightarrow 4)$ linkage.
17. Amylopectin molecules are more highly branched than amylose molecules are.
18. A homopolysaccharide is a polysaccharide in which more than one type of monosaccharide unit is present.

Multiple choice:

19. Which of the following is *not* required for the production of carbohydrates by photosynthesis?

    a. carbon dioxide b. oxygen c. water
    d. sunlight e. chlorophyl

20. Which of the following compounds contains only three chiral centers?

    a. glyceraldehyde b. glucose c. fructose
    d. galactose e. none of these

21. Ribose is an example of a(n):

    a. ketohexose b. aldohexose c. aldotetrose
    d. aldopentose e. none of these

22. Dextrose, or blood sugar, is:

    a. D-glucose
    b. D-galactose
    c. D-fructose
    d. D-ribose
    e. none of these

23. The sugar sometimes known as levulose is:

    a. D-ribose
    b. L-glucose
    c. L-galactose
    d. D-fructose
    e. none of these

24. Which of the following compounds is *not* made up solely of D-glucose units?

    a. maltose
    b. lactose
    c. cellulose
    d. starch
    e. glycogen

25. Which of the following sugars is *not* a reducing sugar?

    a. lactose
    b. maltose
    c. sucrose
    d. galactose
    e. all are reducing sugars

26. The main storage form of D-glucose in animal cells is:

    a. glycogen
    b. amylopectin
    c. amylose
    d. sucrose
    e. none of these

27. Which of the following molecules has a chiral center?

    a. ethanol
    b. 1-chloro-1-bromoethane
    c. 1-chloro-2-bromoethane
    d. 1,2-dichloroethane
    e. none of these

28. Hydrolysis of sucrose yields:

    a. glucose and fructose
    b. glucose and galactose
    c. ribose and fructose
    d. ribose and glucose
    e. none of these

Answers to Practice Exercises

18.1
a. disaccharide
b. oligosaccharides
c. trisaccharide
d. polysaccharide

18.2
a. achiral
b. chiral
c. chiral

18.3

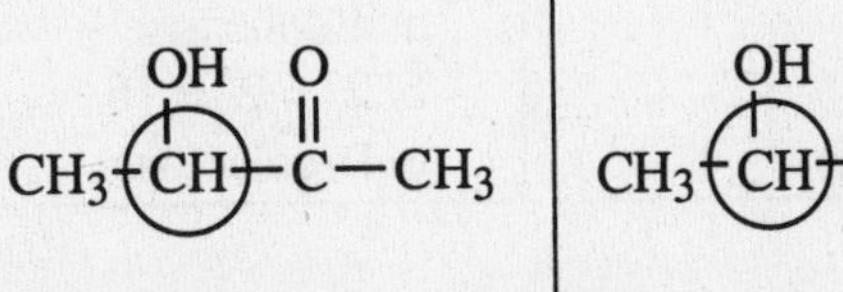

a.

b.

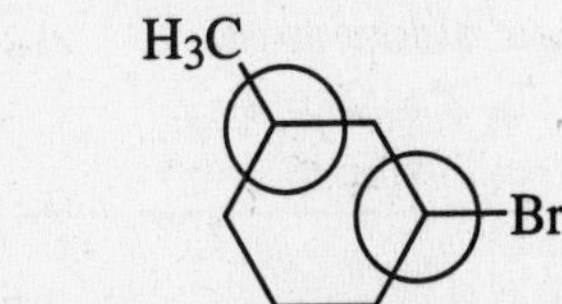

c.

18.4

a.

CHO
H—OH
HO—H
CH_2OH

CHO
CHOH
CHOH
CH_2OH

b.

CHO
=O
H—OH
CH_2OH

CHO
C=O
CHOH
CH_2OH

18.5

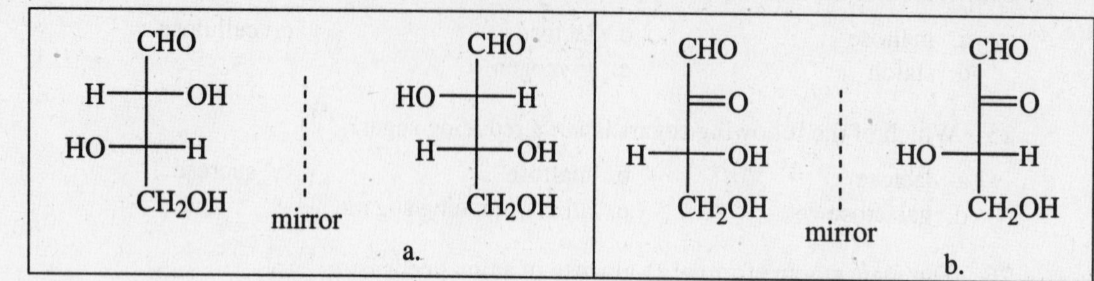

18.6 a. L b. D

18.7

a.

CH_3
H—OH
HO—H
CH_3

b. enantiomer

CH_3
HO—H
H—OH
CH_3

c. diastereomer

CH_3
HO—H
HO—H
CH_3

or

CH_3
H—OH
H—OH
CH_3

18.8

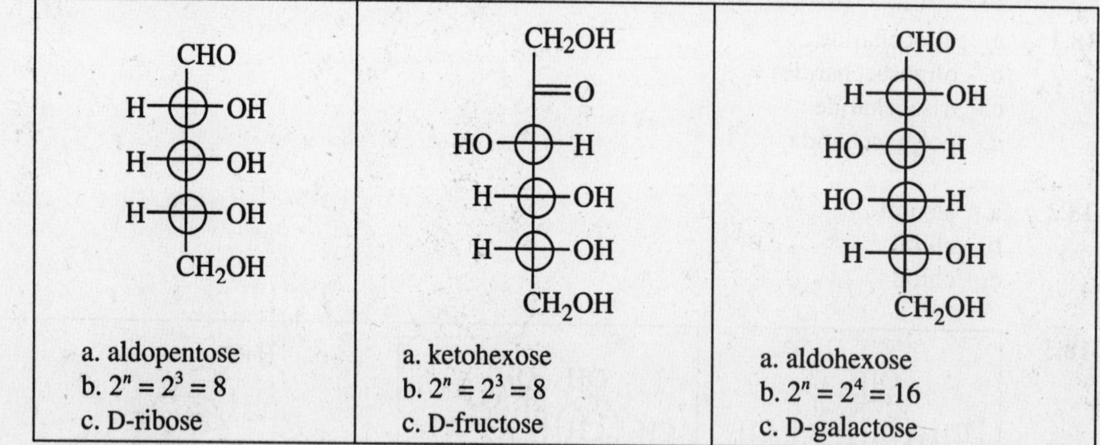

18.9 a. [Fischer projection: CHO; HO–H; H–OH; H–OH; H–OH; CH_2OH] → [Haworth ring: CH_2OH, O, OH, OH, OH, OH]

b. [Haworth ring: CH_2OH, O, OH, OH, OH, OH] → [Fischer projection: CHO; H–OH; H–OH; H–OH; H–OH; CH_2OH]

18.10

a. β-D-glucose	b. α-D-galactose	c. α-D-ribose

18.11 a. [CHO; H–OH; H–OH; H–OH; CH_2OH] —Fehling's solution (weak oxidation)→ [COOH; H–OH; H–OH; H–OH; CH_2OH]

b. [CHO; H–OH; H–OH; H–OH; CH_2OH] —H_2/catalyst (reduction)→ [CH_2OH; H–OH; H–OH; H–OH; CH_2OH]

c. (glucopyranose: CH_2OH, OH, OH, OH, OH) + CH_3OH $\overset{H^+}{\rightleftharpoons}$ (methyl glucoside: CH_2OH, OH, OH, OH, OCH_3) + H_2O

18.12

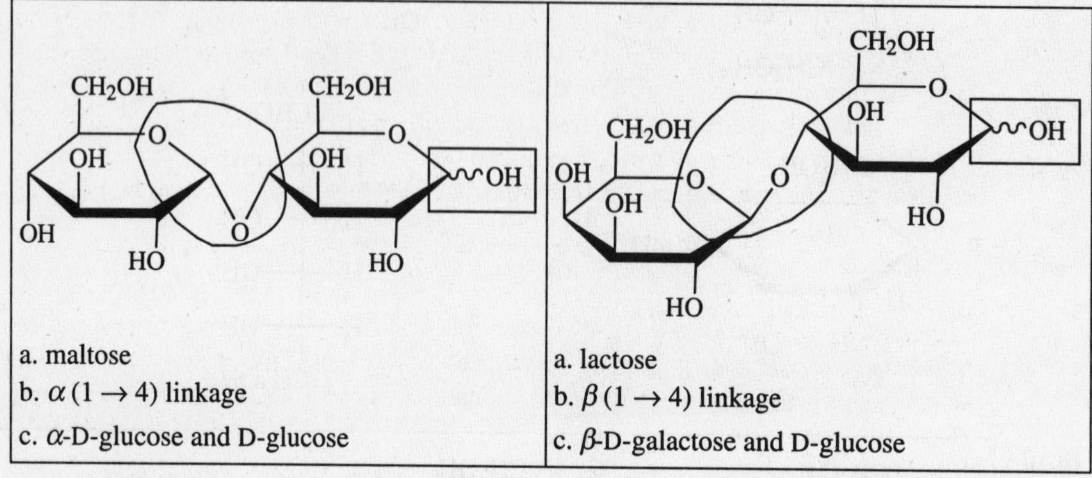

a. maltose	a. lactose
b. α (1 → 4) linkage	b. β (1 → 4) linkage
c. α-D-glucose and D-glucose	c. β-D-galactose and D-glucose

18.13 Both disaccharides are reducing sugars.

18.14

Polysaccharide	Linkage	Size (amu)	Branching	Function
glycogen	α(1 → 4) α(1 → 6)	3,000,000	very highly branched	storage form of glucose in animals
amylose	α(1 → 4)	50,000	straight-chain	storage form of glucose in plants (15-20%)
amylopectin	α(1 → 4)	300,000	highly branched	storage form of glucose in plants (80-85%)
cellulose	β(1 → 4)	900,000	straight-chain	structural component of cell walls in plants

Answers to Self-Test

The numbers in parentheses refer to sections in your textbook.
1. T (18.1) **2**. T (18.2) **3**. F; two to ten (18.3) **4**. F; an achiral molecule (18.4) **5**. F; no (18.4)
6. F; vertical lines (18.6) **7**. F; are (18.5) **8**. F; enantiomers (18.9) **9**. T (18.7)
10. F; (+) or (–) (18.6, 18.7) **11**. F; often different (18.7) **12**. F; a ketohexose (18.8)
13. T (18.14) **14**. F; H_2 (18.12) **15**. T (18.13) **16**. F; β(1 → 4) linkage (18.14) **17**. T (18.14)
18. F; heteropolysaccharide (18.15) **19**. b (18.2) **20**. c (18.4) **21**. d (18.9) **22**. a (18.9)
23. d (18.9) **24**. b (18.13, 18.14) **25**. c (18.13) **26**. a (18.14) **27**. b (18.4) **28**. a (18.13)

Lipids

Chapter 19

Chapter Overview

Lipids are compounds grouped according to their common solubility in nonpolar solvents and insolubility in water, but there are some structural features that also help to identify them. They have a variety of functions in the body, acting as energy-storage compounds and chemical messengers and providing structure to cell membranes.

In this chapter you will define fats and oils and explain how they differ in molecular structure. You will identify various groups of lipids according to their structures and components, and you will learn some of the functions that they perform in biological systems.

Practice Exercises

19.1 **Lipids** (Sec. 19.1) as a group do not have a common structural feature. However, there are subgroups of lipids that are united by certain characteristics. **Fatty acids** (Sec. 19.2) are naturally occurring carboxylic acids that contain long, unbranched hydrocarbon chains 12 to 26 carbon atoms in length.

Three types of fatty acids are **saturated** (no carbon-carbon double bonds), **monounsaturated** (one carbon-carbon double bond), and **polyunsaturated** (two or more carbon-carbon double bonds) (Sec. 19.2). This "type" designation may be abbreviated as SFA, MUFA, and PUFA. A shorthand notation for the structure of fatty acids uses the ratio of the number of carbon atoms to the number of double bonds.

In the omega classification system, the first double bond is identified by its number on the carbon chain, counted from the methyl end of the chain. Omega-3 and omega-6 are the most common "omega" families.

Double-bond positioning may be specified by using a delta with the carbon numbers of the double bonds superscripted. For this notation, the double-bond position is counted from the carboxyl end of the chain.

Complete the table below using information from Table 19.1 in your textbook. The first fatty acid has been filled in as an example.

Name of fatty acid	Carbon atoms and double bonds	Type designation	"Omega" designation	"Delta" designation
linolenic acid	18:3	PUFA	omega-3	$\Delta^{9,12,15}$
palmitoleic acid				
arachidonic acid				
linoleic acid				
oleic acid				
stearic acid				

19.2 **Triacylglycerols** (Sec. 19.4) are produced by the esterification of three fatty acid molecules with a glycerol molecule.

Draw structural formulas for the triacylglycerols that have the following fatty acid residues: a. three molecules of oleic acid; b. two molecules of stearic acid and one of oleic acid; c. three molecules of stearic acid.

a.	b.	c.

19.3 **Fats** (Sec. 19.4) are triacylglycerols with a high percentage of saturated fatty acids; **oils** (Sec. 19.4) are triacylglycerols with a high percentage of unsaturated fatty acids. Tell whether each of the compounds in Practice Exercise 19.2 would be a fat or an oil, and explain.

Compound	Fat or oil?	Explanation
a.		
b.		
c.		

19.4 Triacylglycerols undergo several characteristic reactions. In this exercise you will work with four of these types of reactions and the three triacylglycerol molecules in Practice Exercise 19.2.

1. Hydrolysis is the reverse of esterification. Complete hydrolysis of all three ester linkages in triacylglycerols produces glycerol and three fatty acid molecules. Write the names of the hydrolysis products of the three triacylglycerols from Practice Exercise 19.2.
2. Hydrolysis under alkaline conditions is called saponification. Write the products of the saponification with NaOH of the three triacyclglycerols above.
3. Hydrogenation involves adding H_2 molecules to the double bonds of the fatty acid residues. In the fourth column, tell how many molecules of hydrogen (H_2) would be used for the complete hydrogenation of each of the three triacylglycerol molecules.
4. Oxidation of triacylglycerols breaks the carbon-carbon double bonds and produces low-molecular-weight aldehydes and carboxylic acids that have unpleasant odors. Fats oxidized in this way are said to be rancid. In the table below, write "yes" in the fifth column if the triacylglycerol is likely to become rancid, "no" if it is not.

	Hydrolysis products	Saponification products	Moles of H_2	Oxidation (rancidity)
a.				
b.				
c.				

19.5 Triacylglycerols are the most common of the lipids that undergo hydrolysis. Other important lipids are the **glycerophospholipids** (Sec. 19.6), **sphingophospholipids** (Sec. 19.7), and the **glycolipids** (Section 19.8). These types of lipids have similar basic structures, but the component "building blocks" vary.

In the block diagrams below, fill in the name of each of the building blocks of each type of lipid named. Then follow the directions below the diagrams.

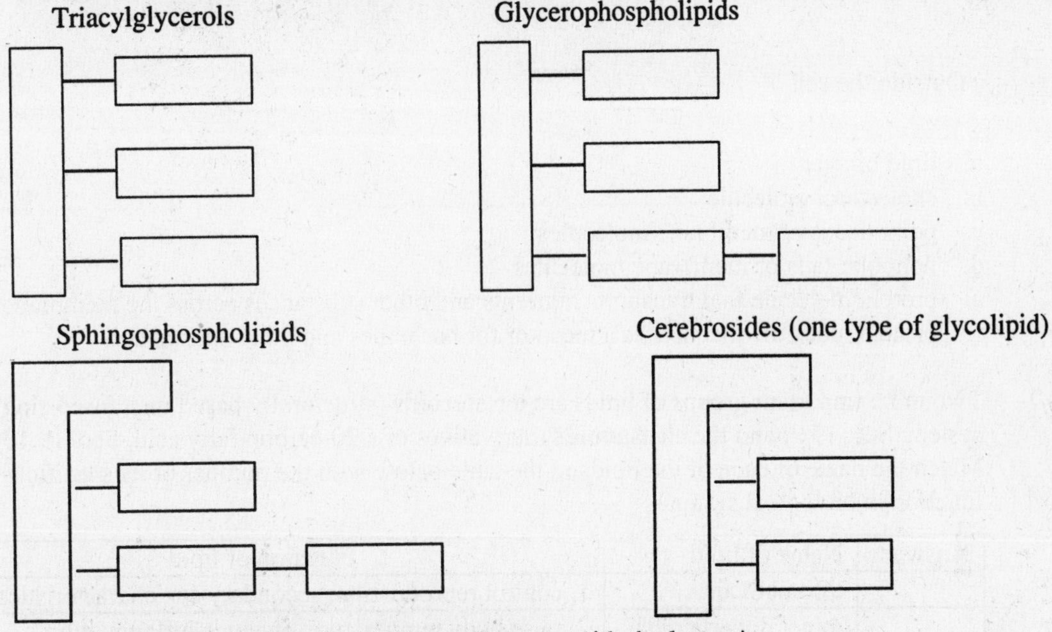

a. Label all ester linkages in the block diagrams with the letter A.
b. Label all amide linkages with the letter B.
c. Label all glycosidic linkages with the letter C.
d. Give the general similarities between triacylglycerols and glycerophospholipids.

____

Give the general differences between these two types of lipids. ______________

____

e. What do all of the lipids in the diagrams above have in common?

____

19.6 The aqueous material within a living cell is separated from its surrounding aqueous environment by a **cell membrane** (Sec. 19.10). Most of the mass of the cell membrane is lipid material, in the form of a **lipid bilayer** (Sec. 19.10). Label the diagram of the lipid bilayer shown in the diagram that follows, using the letters of the components listed below the diagram.

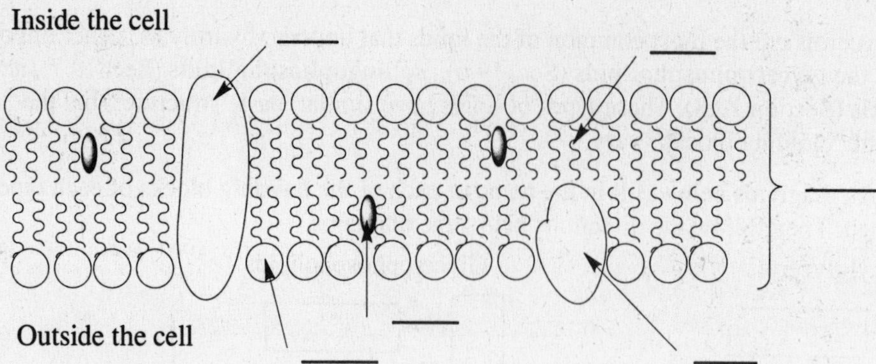

a. lipid bilayer
b. cholesterol molecule
c. polar heads of membrane molecules
d. nonpolar tails of membrane molecules
e. protein molecule that transports nutrients and other substances across the membrane
f. protein molecule that acts as a receptor for hormones and other substances

19.7 Two more important groups of lipids are the **steroids** (structurally based on a fused-ring system, Sec. 19.9) and the **eicosanoids** (derivatives of a 20-carbon fatty acid, Sec. 19.13). Match the name of each of the lipids in the table below with the number of its specific function in biological systems.

Answer	Name of lipid	Function of lipid
	thromboxanes	1. control reproduction, secondary sex characteristics
	mineralocorticoids	2. raise body temperature, enhance inflammation
	steroid hormones	3. starting material for synthesis of steroid hormones
	glucocorticoids	4. control glucose metabolism, reduce inflammation
	cholesterol	5. emulsifying agents produced by gall bladder
	bile acids	6. found in white blood cells, hypersensitivity response
	leukotrienes	7. promote formation of blood clots
	prostaglandins	8. control the balance of Na^+ and K^+ ions in cells

19.8 After each name in the table above, write A if the compound is a steroid or B if the compound is an eicosanoid.

19.9 As you have seen in this chapter, one of the principal ways to classify lipids is according to general function. Use the following letters to classify the lipids in the list below: A. energy-storage lipids, B. membrane lipids, C. emulsification lipids, D. messenger lipids, E. protective-coating lipids.

triacylglycerols _______
biological waxes _______
prostaglandin _______
cholic acid _______
adrenocorticoids _______
sphingomyelins _______

lecithins _______
androgens _______
thromboxanes _______
cerebrosides _______
cholesterol _______
steroid hormones _____

Self-Test

True-false: Indicate whether the following statements are true or false. If the statement is false, give the word or phrase that may be substituted for the underlined portion to make the statement true.

1. Fats contain a high proportion of unsaturated fatty acid chains.
2. Fats and oils become rancid when the double bonds on triacylglycerol side chains are oxidized.
3. The two essential fatty acids are linoleic acid and arachidonic acid.
4. Rancid fats contain low-molecular-mass acids.
5. Sphingophospholipids are complex lipids found in the brain and nerves.
6. In a cell membrane, the nonpolar tails of the phospholipids are on the outside surfaces of the lipid bilayer.
7. The presence of cholesterol molecules in the lipid bilayer of a cell membrane makes the bilayer more flexible.
8. Essential fatty acids cannot be synthesized within the human body from other substances.
9. Most naturally occurring triacylglycerols are formed with three identical fatty acid molecules.
10. Partial hydrogenation of vegetable oils produces a product of higher melting point than the original oil.
11. Glycerophopholipids are important components of cell membranes.
12. Bile acids play an important role in the human body in the process of digestion.
13. Cholesterol cannot be synthesized within the human body.
14. A fatty acid containing a *cis* double bond has a bent carbon chain.
15. Most fatty acids found in the human body have an odd number of carbon atoms.
16. The lipid molecules in the lipid bilayer of a cell membrane are held to one another by covalent bonds.
17. Waxes are esters of long-chain fatty acids and aromatic alcohols.
18. Some synthetic derivatives of adrenocorticoid hormones are used to control inflammatory diseases.
19. The eicosanoids are hormonelike substances that are transported in the bloodstream.
20. Eicosanoids are derivatives of polyunsaturated fatty acids that have 20 carbon atoms.
21. Facilitated transport involves movement of a substance across a membrane against a concentration gradient with the expenditure of cellular energy.

Multiple choice:

22. A triacylglycerol is prepared by combining glycerol and:

a. long-chain alcohols   b. fatty acids   c. unsaturated hydrocarbons
d. saturated hydrocarbons   e. none of these

23. Which of the following fatty acids is saturated?

a. palmitic acid   b. oleic acid   c. linoleic acid
d. arachidonic acid   e. none of these

24. An example of a glycerophospholipid is:

a. prostaglandin b. progesterone c. a lecithin
d. glycerol tristearate e. none of these

25. An example of a steroid hormone is:

a. progesterone b. cholesterol c. stearic acid
d. a cerebroside e. none of these

26. Cholesterol is the starting material within the human body for the synthesis of:

a. vitamin A b. vitamin B1 c. vitamin C
d. vitamin D e. vitamin E

27. The most abundant steroid in the human body is:

a. estradiol b. testosterone c. progesterone
d. cholesterol e. estrogen

Answers to Practice Exercises

19.1

Name of fatty acid	Carbon atoms and double bonds	Type designation	"Omega" designation	"Delta" designation
linolenic acid	18:3	PUFA	omega-3	$\Delta^{9,12,15}$
palmitoleic acid	16:1	MUFA	omega-6	Δ^{9}
arachidonic acid	20:4	PUFA	omega-6	$\Delta^{5,8,11,14}$
linoleic acid	18:2	PUFA	omega-6	$\Delta^{9,12}$
oleic acid	18:1	MUFA	omega-9	Δ^{9}
stearic acid	18:0	SFA	–	–

19.2 a.

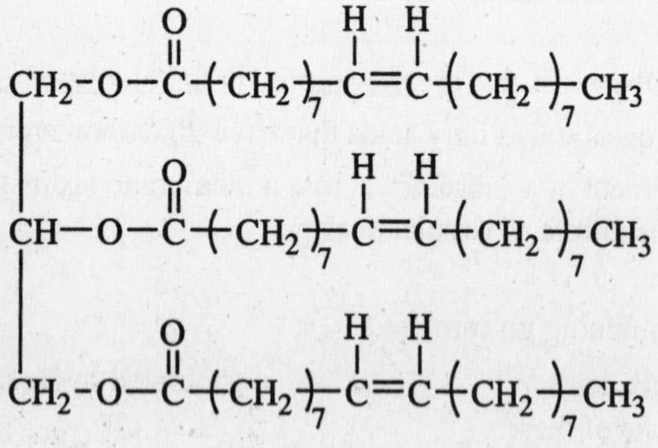

$$\begin{array}{l} CH_2{-}O{-}\overset{O}{\overset{\|}{C}}{-}(CH_2)_7{-}\overset{H}{\overset{|}{C}}{=}\overset{H}{\overset{|}{C}}{-}(CH_2)_7{-}CH_3 \\ | \\ CH{-}O{-}\overset{O}{\overset{\|}{C}}{-}(CH_2)_7{-}\overset{H}{\overset{|}{C}}{=}\overset{H}{\overset{|}{C}}{-}(CH_2)_7{-}CH_3 \\ | \\ CH_2{-}O{-}\overset{O}{\overset{\|}{C}}{-}(CH_2)_7{-}\overset{H}{\overset{|}{C}}{=}\overset{H}{\overset{|}{C}}{-}(CH_2)_7{-}CH_3 \end{array}$$

b.

$$CH_2{-}O{-}\overset{O}{\overset{\|}{C}}{-}\!(CH_2)_{16}{-}CH_3$$
$$CH{-}O{-}\overset{O}{\overset{\|}{C}}{-}\!(CH_2)_{16}{-}CH_3$$
$$CH_2{-}O{-}\overset{O}{\overset{\|}{C}}{-}\!(CH_2)_7{-}\overset{H}{\overset{|}{C}}{=}\overset{H}{\overset{|}{C}}{-}\!(CH_2)_7{-}CH_3$$

c.

$$CH_2{-}O{-}\overset{O}{\overset{\|}{C}}{-}\!(CH_2)_{16}{-}CH_3$$
$$CH{-}O{-}\overset{O}{\overset{\|}{C}}{-}\!(CH_2)_{16}{-}CH_3$$
$$CH_2{-}O{-}\overset{O}{\overset{\|}{C}}{-}\!(CH_2)_{16}{-}CH_3$$

19.3

	Fat or oil?	Explanation
a.	oil	All three fatty acid residues are unsaturated, so the molecules are bent and do not pack together closely, resulting in low melting point.
b.	fat	Only one fatty acid is unsaturated and two are saturated, so the molecules are less distorted than in the previous example and pack together more closely.
c.	fat	All three fatty acids are saturated and linear, resulting in close packing and high melting point.

19.4

	Hydrolysis products (per molecule)	Saponification products (per molecule)	Moles of hydrogen	Oxidation (rancidity)
a.	glycerol, 3 molecules of oleic acid	glycerol, 3 formula units of sodium oleate	3	yes
b.	glycerol, 1 molecule of oleic acid, 2 molecules of stearic acid	glycerol, 1 formula unit of sodium oleate, 2 units of sodium stearate	1	yes
c.	glycerol, 3 molecules of stearic acid	glycerol, 3 formula units of sodium stearate	0	no

19.5 Triacylglycerols

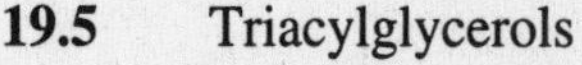

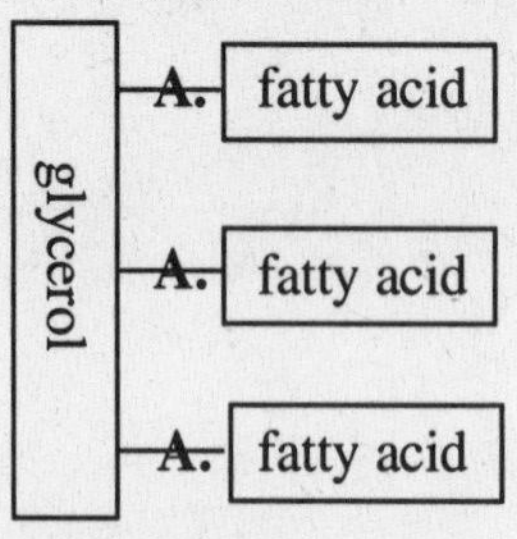

Glycerophospholipids

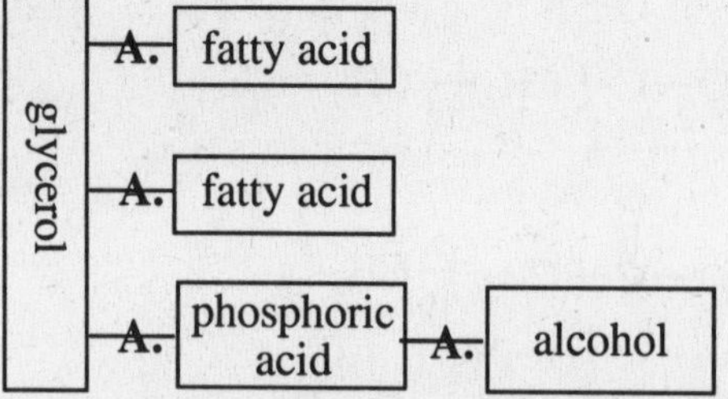

Sphingophopholipids

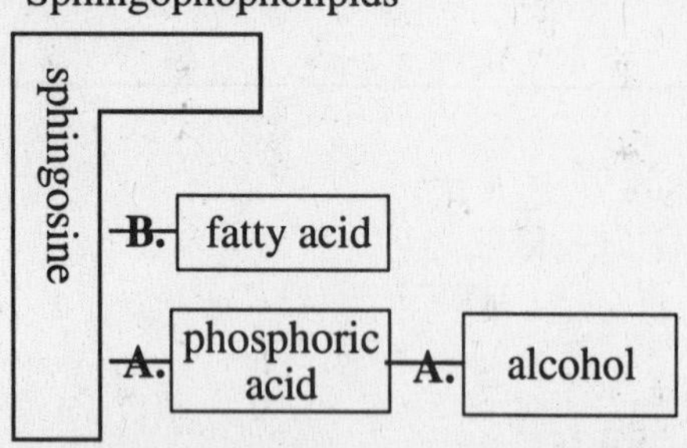

Cerebrosides (one type of glycolipid)

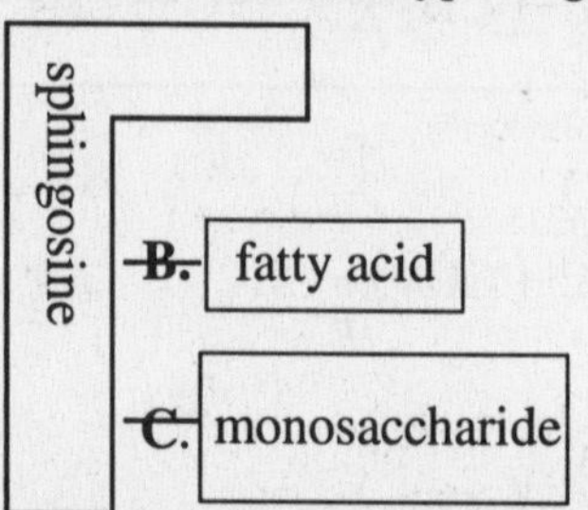

A. ≡ ester linkage **B.** ≡ amide linkage **C.** ≡ glycosidic linkage

d. Similarities: polar heads, nonpolar tails (fatty acid chains)

Differences: glycerophospholipids contain phosphoric acid and alcohol (in addition to the glycerol and fatty acids.)

e. They all have in common: polar heads, nonpolar tails, fatty acids, ester linkages.

19.6

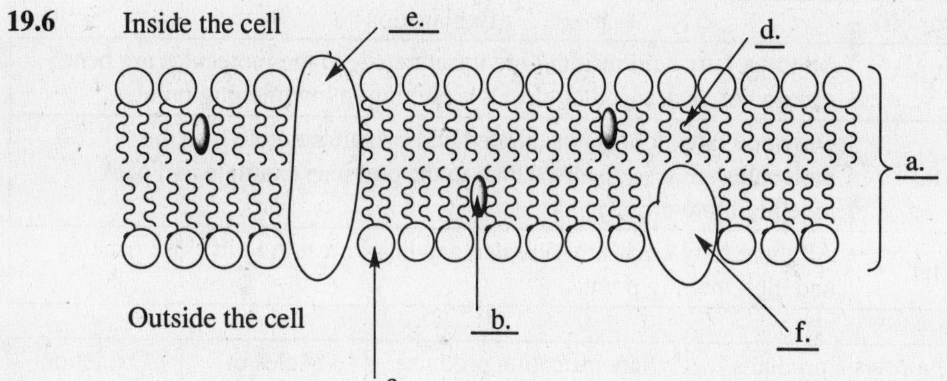

19.7

Answer	Name of lipid	Function of lipid
7.	thromboxanes	1. control reproduction, secondary sex characteristics
8.	mineralocorticoids	2. raise body temperature, enhance inflammation
1.	steroid hormones	3. starting material for synthesis of steroid hormones
4.	glucocorticoids	4. control glucose metabolism, reduce inflammation
3.	cholesterol	5. emulsifying agents produced by gall bladder
5.	bile salts	6. found in white blood cells, hypersensitivity response
6.	leukotrienes	7. promote formation of blood clots
2.	prostaglandins	8. control the balance of Na^+ and K^+ ions in cells

19.8 thromboxanes – B; mineralocorticoids – A; steroid hormones – A; glucocorticoids – A; cholesterol – A; bile salts – A; leukotrienes – B; prostaglandins – B

19.9

triacylglycerols A.	lecithins B.
biological waxes E.	androgens D.
prostaglandin D.	thromboxanes D.
cholic acid C.	cerebrosides B.
adrenocorticoids D.	cholesterol B.
sphingomyelins B.	steroid hormones D.

Answers to Self-Test

The numbers in parentheses refer to sections in your textbook.
1. F; saturated fatty acid (19.4) **2**. T (19.4) **3**. F; linoleic and linolenic (19.2) **4**. T (19.5) **5**. T (19.7) **6**. F; polar heads (19.10) **7**. F; more rigid (19.9) **8**. T (19.2) **9**. F; a mixture of (19.4) **10**. T (1.5) **11**. T (19.10) **12**. T (19.11) **13**. F; can (19.9) **14**. T (19.2) **15**. F; an even number (19.2) **16**. F; dipole-dipole interactions (19.10) **17**. F; long-chain alcohols (19.14) **18**. T (19.12) **19**. T (19.13) **20**. T (19.13) **21**. F; active transport (19.10) **22**. b (19.4) **23**. a (19.2) **24**. c (19.7) **25**. a (19.12) **26**. d (19.9) **27**. d (19.9)

Proteins Chapter 20

Chapter Overview

The functions of proteins in living systems are highly varied; they catalyze reactions, form structures, and transport substances. These functions depend on the properties of the small units that make up proteins (the amino acids) and on how these amino acids are joined together.

In this chapter you will learn to identify the basic structure of an amino acid and characterize some amino acid side chains. You will draw Fischer projections for amino acids and explain why an amino acid exists as a zwitterion. You will define the peptide bond and write abbreviated names for some peptides. You will compare the primary, secondary, tertiary, and quaternary structures of proteins.

Practice Exercises

20.1 A **protein** (Sec. 20.1) is a polymer in which the monomer units are amino acids. An **α-amino acid** (Sec. 20.2) contains an amino group and a carboxyl group, both attached to the same atom, the ***α*-carbon atom** (Sec. 20.2). Amino acids are classified as **nonpolar, polar neutral, polar basic,** or **polar acidic** (Sec. 20.2), depending on the nature of the side chain present.

Complete the following table for classification of different amino acids.

Name	Abbreviation	Structure	Classification
leucine			
aspartic acid			
lysine			
serine			
histidine			
asparagine			

20.2 The α-carbon atom of an α-amino acid is a chiral center (except in glycine); naturally occurring amino acids are L-isomers. The Fischer projection for an amino acid is drawn by putting the –COOH group at the top of the projection and the $–NH_2$ group to the left for the L-isomer.

a. Draw the Fischer projection for L-aspartic acid.
b. Draw the enantiomer of L-aspartic acid using a Fischer projection.

$HO-\overset{O}{\overset{\|}{C}}-CH_2-\underset{NH_2}{\underset{\|}{CH}}-\overset{O}{\overset{\|}{C}}-OH$ aspartic acid	a. Fischer projection of L-aspartic acid	b. enantiomer of L-aspartic acid

20.3 Because an amino acid has both an acidic group (the carboxyl group) and a basic group (the amino group), it exists as a **zwitterion** (Sec. 20.4), a molecule that has a positive charge on one atom and a negative charge on another atom. Draw the structural formula for neutral leucine and the zwitterionic structure of leucine.

Structural formula of leucine	Zwitterionic structure of leucine

20.4 In solution, three different amino acid forms can exist in equilibrium. The amount of each form present depends on the pH of the solution.
Draw the structure of leucine at each of the following pH values.

pH = 1	pH = 7	pH = 12

20.5 Referring to the structures in Practice Exercise 20.4, predict the direction (if any) of migration toward the positive or negative electrode for each structure. The pH at which no migration occurs is called the **isoelectric point** (Sec. 20.4). Write "isoelectric" if no migration occurs.

pH = 1 ____________________

pH = 7 ____________________

pH = 12 ____________________

20.6 The **peptide bond** (Sec. 20.6) is an amide linkage joining amino acids to form **peptides** (Sec. 20.6). The peptide bond forms between the carboxyl group of one amino acid and the amino group of another.

a. Draw the structural formulas of the two different dipeptides that could form from the amino acids aspartic acid (Asp) and lysine (Lys).
b. Circle the peptide bond in each structure.
c. Label the N-terminal end and the C-terminal end.

20.7 Peptides that contain the same amino acids but in different order are different molecules with different properties.

a. Using abbreviations for the amino acids, draw all possible tripeptides that could be formed from one molecule each of the following amino acids: Pro, Trp, and Lys.

b. Draw the abbreviated peptide formulas for all possible tripeptides that could be formed from two molecules of Pro and one molecule of Lys.

20.8 **Proteins** (Sec. 20.8) and polypeptides can undergo hydrolysis of their peptide bonds. Sometimes only a few of the bonds are hydrolyzed, yielding a mixture of smaller peptides. How many different di- and tripeptides could be present in a solution of partially hydrolyzed Arg-Arg-Cys-Lys?

20.9 A protein (Sec. 20.8) is a polypeptide in which at least 50 amino acid residues are present. The sequence of the amino acids in a protein is called its **primary structure** (Sec. 20.9). The shape of a protein molecule is the result of forces acting between various parts of the peptide chain. These forces determine the **secondary, tertiary, and quaternary structure of the protein** (Sec. 20.10, 20.11, and 20.12).

Complete the following table by writing the type of protein structure (primary, secondary, tertiary, or quaternary) most readily associated with each of the terms given.

Structural term	Type of protein structure
amino acid sequence	
beta-pleated sheet	
disulfide bonds	
alpha helix	
globular protein (Sec. 20.14)	
hemoglobin (tetramer)	
collagen (triple helix)	
fibrous protein (Sec. 20.14)	
hydrophobic interactions	

20.10 Determine the primary structure of a heptapeptide containing six different amino acids, if the following smaller peptides are among the partial-hydrolysis products:

Tyr-Trp, Glu-Asp-Tyr, Trp-Asp-Val, Val-Pro, Tyr-Trp-Asp. The C-terminal end is proline.

Structure of heptapeptide ______________________________

20.11 **Glycoproteins** (Sec. 20.17) are **conjugated proteins** (Sec. 20.12) containing carbohydrates as well as amino acids. They are important in several ways in biological systems. Match each term below with the statement that describes its role in a process involving glycoproteins.

Answer	Term	Statement
	collagen	1. glycoprotein produced as a protective response
	antigens	2. foreign substances that invade the body
	antibodies	3. aggregates caused by carbohydrate cross linking
	immunoglobulin	4. premilk substance containing immunoglobulins
	colostrum	5. triple helix of chains of amino acids
	collagen fibrils	6. molecules that counteract specific antigens

Self-Test

True-false: Indicate whether the following statements are true or false. If the statement is false, give the word or phrase that may be substituted for the underlined portion to make the statement true.

1. Naturally occurring amino acids are generally in the L form.
2. A peptide bond is the bond formed between the amino group of an amino acid and the carboxyl group of the same amino acid.
3. Because amino acids have both an acidic group and a basic group, they are able to undergo internal acid-base reactions.
4. When the pH of an amino acid solution is lowered, the amino acid zwitterion forms more of the positively charged species.
5. A peptide bond differs slightly from an amide bond.
6. A tripeptide has a COO^- group at each end of the molecule.
7. Vasopressin is a peptide in the human body that regulates uterine contractions and lactation.
8. The function of a protein is controlled by the protein's primary structure.
9. The alpha helix structure is a part of the primary structure of some proteins.
10. Denaturation of a protein involves changes in the protein's primary structure.
11. The quaternary structure of a protein involves associations among separate polypeptide chains.
12. An antibody is a foreign substance that invades the body.
13. The peptides in a beta-pleated sheet are held in place by hydrogen bonds.

Multiple choice:

14. An example of a polar basic amino acid is:

    a. lysine b. serine c. tryptophan
    d. leucine e. none of these

15. A tripeptide is formed from two alanine molecules and one glycine molecule. The maximum number of different tripeptides that could be formed from this combination is:

    a. two b. three c. four
    d. five e. none of these

16. The peptide that regulates the excretion of water by the kidneys is:

    a. glutathione b. insulin c. vasopressin
    d. lysine e. none of these

17. The partial hydrolysis of the pentapeptide Val-Ala-Ala-Gly-Ser could yield which of the following peptides? (The amino acid on the left is the N-terminal amino acid.)

    a. Val-Ala-Gly b. Ser-Gly-Ala c. Ala-Gly-Ser
    d. Ala-Ala-Ser e. none of these

18. Which of the following amino acids would have a net charge of zero at a pH of 7?

    a. arginine b. phenylalanine c. aspartic acid
    d. lysine e. none of these

19. Which of the following attractive interactions does *not* affect the formation of a protein's tertiary structure?

    a. disulfide bonds b. salt bridges c. hydrogen bonds
    d. peptide bonds e. none of these

20. Proteins may be denatured by:

    a. heat b. acid
    c. ethanol d. a, b, and c
    e. a and b only

21. Which of the following proteins is *not* a conjugated protein?

    a. hemoglobin b. insulin c. collagen
    d. immunoglobulin e. lipoprotein

22. Which of the following is a characteristic of globular proteins?

    a. stringlike molecules b. water-insoluble
    c. roughly spherical d. structural function in body
    e. none of these

Answers to Practice Exercises

20.1

Name	Abbreviation	Structure	Classification
leucine	Leu	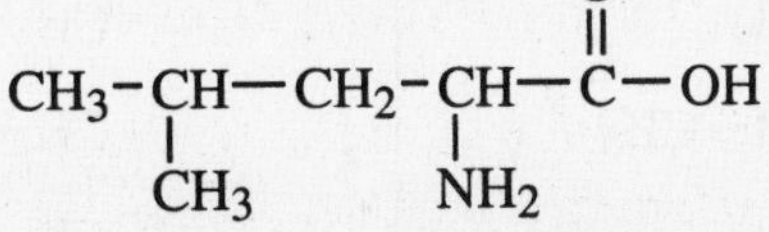 $CH_3-CH(CH_3)-CH_2-CH(NH_2)-C(=O)-OH$	nonpolar
aspartic acid	Asp	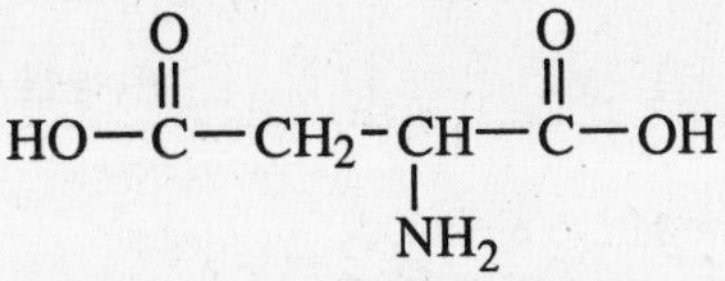 $HO-C(=O)-CH_2-CH(NH_2)-C(=O)-OH$	polar acidic
lysine	Lys	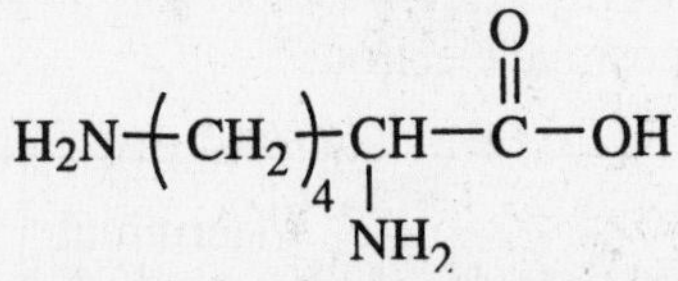 $H_2N-(CH_2)_4-CH(NH_2)-C(=O)-OH$	polar basic
serine	Ser	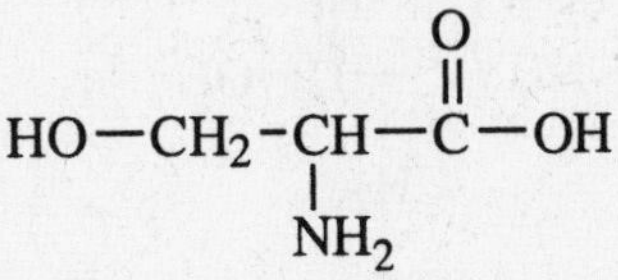 $HO-CH_2-CH(NH_2)-C(=O)-OH$	polar neutral
histidine	His	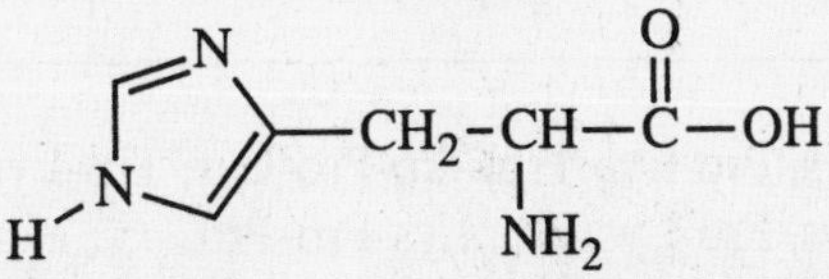 (imidazole ring, N–H)$-CH_2-CH(NH_2)-C(=O)-OH$	polar basic
asparagine	Asn	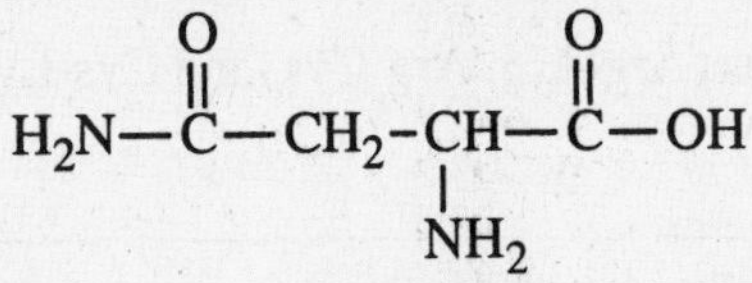 $H_2N-C(=O)-CH_2-CH(NH_2)-C(=O)-OH$	polar neutral

20.2

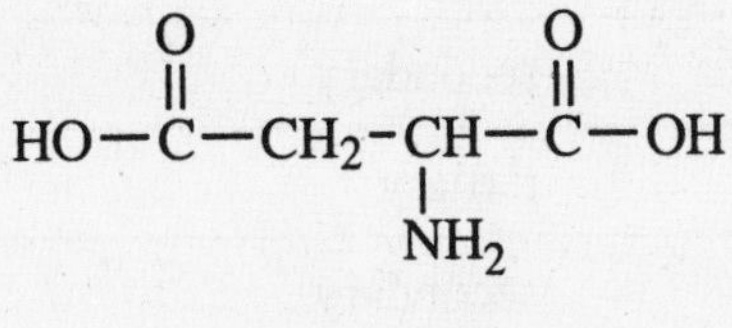 $HO-C(=O)-CH_2-CH(NH_2)-C(=O)-OH$ aspartic acid	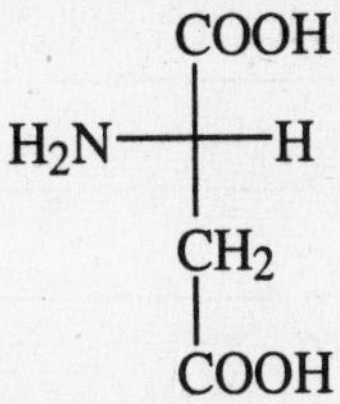 COOH H_2N—C—H CH_2 COOH a. Fischer projection of L- aspartic acid	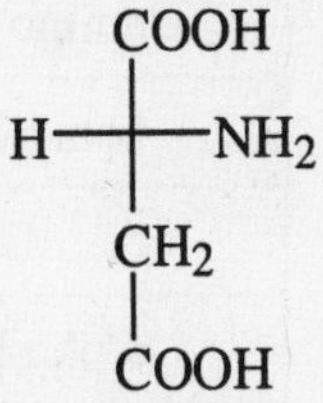 COOH H—C—NH_2 CH_2 COOH b. enantiomer of L-aspartic acid

20.3

Structural formula of leucine	Zwitterionic structure of leucine
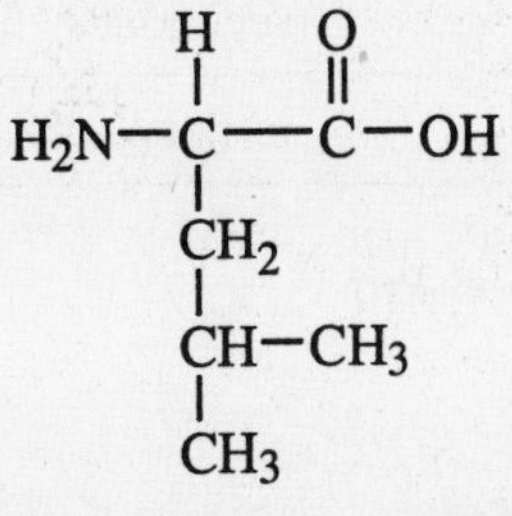 $H_2N-CH(CH_2-CH(CH_3)-CH_3)-C(=O)-OH$	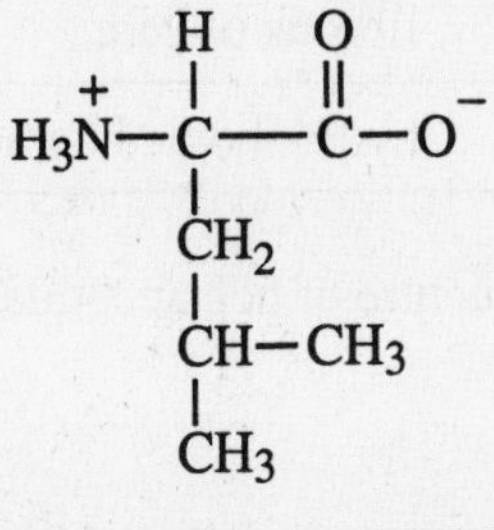 $H_3\overset{+}{N}-CH(CH_2-CH(CH_3)-CH_3)-C(=O)-O^-$

20.4

pH = 1	pH = 7	pH = 12
$\mathrm{H_3\overset{+}{N}-C(H)(CH_2-CH(CH_3)-CH_3)-C(=O)-OH}$	$\mathrm{H_3\overset{+}{N}-C(H)(CH_2-CH(CH_3)-CH_3)-C(=O)-O^-}$	$\mathrm{H_2N-C(H)(CH_2-CH(CH_3)-CH_3)-C(=O)-O^-}$

20.5 pH = 1: toward negative electrode
pH = 7: isoelectric
pH = 12: toward positive electrode

20.6

N-terminal end $\mathrm{H_2N-CH(CH(CH_3)-H_3C)-C(=O)-NH-C(H)(CH_2-OH)-C(=O)-OH}$ C-terminal end	N-terminal end $\mathrm{H_2N-CH(CH_2-OH)-C(=O)-NH-C(H)(CH(CH_3)-CH_3)-C(=O)-OH}$ C-terminal end

20.7 a. Pro-Trp-Lys, Pro-Lys-Trp, Trp-Pro-Lys, Trp-Lys-Pro, Lys-Trp-Pro, Lys-Pro-Trp
b. Pro-Pro-Lys, Pro-Lys-Pro, Lys-Pro-Pro

20.8 Three dipeptides (Arg-Arg, Arg-Cys, and Cys-Lys) and two tripeptides (Arg-Arg-Cys and Arg-Cys-Lys)

20.9

Structural Term	Classification of protein structure
amino acid sequence	primary
beta-pleated sheet	secondary
disulfide bonds	tertiary
alpha-helix	secondary
globular proteins	tertiary
hemoglobin (tetramer)	quaternary
collagen (triple helix)	secondary
fibrous protein	secondary
hydrophobic interactions	tertiary, quaternary

20.10 Structure of heptapeptide: Glu-Asp-Tyr-Trp-Asp-Val-Pro

20.11

Answer	Term	Statement
5.	collagen	1. glycoprotein produced as a protective response
2.	antigens	2. foreign substances that invade the body
6.	antibodies	3. aggregates caused by carbohydrate cross linking
1.	immunoglobulin	4. premilk substance containing immunoglobulins
4.	colostrum	5. triple helix of chains of amino acids
3.	collagen fibrils	6. molecules that counteract specific antigens

Answers to Self-Test

The numbers in parentheses refer to sections in your textbook.
1. T (20.3) **2**. F; another (20.5) **3**. T (20.4) **4**. T (20.4) **5**. F; is the same as (20.5) **6**. F; at one end and an amino group at the other end (20.5) **7**. F; oxytocin (20.5) **8**. F; primary, secondary, tertiary, and quaternary (20.6) **9**. F; secondary (20.8) **10**. F; secondary, tertiary, and quaternary (20.14) **11**. T (20.10) **12**. F; antigen (20.15) **13**. T (20.8) **14**. a (20.2) **15**. b (20.5) **16**. c (20.5) **17**. c (20.13) **18**. b (20.5) **19**. d (20.9) **20**. d (20.14) **21**. b (20.12) **22**. c (20.11)

Enzymes and Vitamins — Chapter 21

Chapter Overview

Enzymes provide the necessary "boost" for most of the chemical reactions of biological systems. Without the help of enzymes, most reactions would proceed so slowly that they would be ineffective. Vitamins are essential nutrients that often play a part in the activity of enzymes.

In this chapter you will learn the general characteristics of enzymes and how to predict the function of an enzyme from its name. You will describe the enzyme-substrate complex in terms of the lock-and-key model and the induced-fit model. You will learn how enzyme action can be inhibited and how it is controlled in biological systems, and you will identify the roles of vitamins in enzyme activity.

Practice Exercises

21.1 **Enzymes** (Sec. 21.1) are commonly named by taking the name of the **substrate** (Sec. 21.2), the compound undergoing change, and adding the ending *-ase*. Sometimes the enzyme name gives the type of reaction being catalyzed.

In the table below, predict the function for a given enzyme name or suggest a name for the enzyme that catalyzes the given type of reaction.

Name of enzyme	Type of reaction
α-amylase	
dehydrogenase	
	removal of a carboxyl group from a substrate
	transfer of an amino group from one molecule to another
peptidase	
oxidase	
	hydrolysis of ester linkages in lipids

21.2 Enzymes can be divided into two classes: **simple enzymes** and **conjugated enzymes** (Sec. 21.3). Other common terms used in describing enzymes are defined in Section 21.3 of your textbook. Match each of the following terms with the correct description.

Answer	Term	Description
	coenzyme	1. nonprotein portion of a conjugated enzyme
	cofactor	2. obtained from dietary minerals
	inorganic ion cofactor	3. protein portion of a conjugated enzyme
	apoenzyme	4. small organic molecule that serves as a cofactor

21.3 Before an enzyme-catalyzed reaction takes place, an **enzyme-substrate complex** (Sec. 21.4) is formed: the substrate binds to the **active site** (Sec. 21.4) of the enzyme.

a. What are two models that account for the specific way an enzyme selects a substrate?

b. What is the main difference between these two models?

21.4 **Enzyme activity** (Sec. 21.6) is a measure of the rate at which an enzyme converts substrate to products.

a. Name four factors that affect enzyme activity.

b. Describe the effects on rate of reaction of an increase in each of the four factors named above.

21.5 An **enzyme inhibitor** (Sec. 21.7) is a substance that slows or stops the normal catalytic function of an enzyme by binding to it. Enzyme activity can also be changed by regulators produced within a cell. After reading the sections on enzyme inhibition and regulation (Sections 21.7–21.10) in your textbook, match the following terms and their descriptions.

Answer	Term	Description
	penicillin	1. binds to a site other than the active site
	competitive inhibitor	2. contains both active and regulator sites
	zymogen	3. competes with substrate for active site
	sulfa drug	4. forms a covalent bond at the active site
	Cipro	5. inhibits cell wall formation in bacteria
	irreversible inhibitor	6. inactive precursor of an enzyme
	noncompetitive inhibitor	7. inhibitor of folic acid production
	allosteric enzyme	8. disrupts tertiary structure of DNA gyrase

21.6 **Vitamins** (Sec. 21.12) are organic chemicals that must be obtained in the diet and that are essential in trace amounts for proper functioning of the human body.

Match each of the following vitamins with its function in the human body.

Answer	Vitamin name	Function in human body
	vitamin A	1. prevents oxidation of polyunsaturated fatty acids
	vitamin C	2. helps form compounds that regulate blood clotting
	vitamin D	3. group of vitamins that are enzyme cofactors
	vitamin B	4. essential for formation of collagen
	vitamin E	5. promotes normal bone mineralization
	vitamin K	6. keeps skin and mucous membranes healthy

21.7 Vitamins are divided into two groups: water-soluble and fat-soluble. After each vitamin name in Practice Exercise 21.6, put a W if the vitamin is water-soluble or an F if it is fat-soluble.

21.8 Vitamins have a wide variety of structures, but there are a few common features.

a. What is the common structural feature that all of the fat-soluble vitamins share?

b. What type of system is found in all B-vitamin structures except that of pantothenic acid?

Self-Test

True-false: Indicate whether the following statements are true or false. If the statement is false, give the word or phrase that may be substituted for the underlined portion to make the statement true.

1. In an enzyme-catalyzed reaction, the compound that undergoes a chemical change is called a <u>substrate</u>.
2. Enzyme names are usually based on the <u>structure</u> of the enzyme.
3. The protein portion of a conjugated enzyme is called the <u>coenzyme</u>.
4. The <u>active site</u> of an enzyme is the small part of an enzyme where catalysis takes place.
5. According to the lock-and-key model of enzyme action, the active site of the enzyme is <u>flexible</u> in shape.
6. A carboxypeptidase is an enzyme that is specific for <u>one group of compounds</u>.
7. A substance that binds to an enzyme's active site and is not released is called <u>a noncompetitive</u> enzyme inhibitor.
8. Enzymes undergo all the reactions of proteins <u>except</u> denaturation.
9. A cofactor is a <u>protein part</u> of an enzyme necessary for the enzyme's function.

10. Large doses of <u>water-soluble vitamins</u> may be toxic, because they can be retained in the body in excess of need.
11. <u>Vitamin K</u> is necessary for the formation of prothrombin used in blood clotting.
12. Some vitamins act as <u>cofactors</u> in conjugated enzymes.
13. Tissue plasminogen activator is used in the <u>diagnosis</u> of heart attacks.

Multiple choice:

14. Enzymes assist chemical reactions by:

    a. increasing the rate of the reactions
    b. increasing the temperature of the reactions
    c. being consumed during the reactions
    d. all of these
    e. none of these

15. The enzyme that catalyzes the reaction of an alcohol to form an aldehyde would be:

    a. an oxidase
    b. a decarboxylase
    c. a dehydratase
    d. a reductase
    e. none of these

16. A competitive inhibitor of an enzymatic reaction:

    a. distorts the shape of the enzyme molecule
    b. attaches to the substrate
    c. blocks an enzyme's active site
    d. weakens the enzyme-substrate complex
    e. none of these

17. When the enzyme-substrate complex forms, the actual bond breaking and/or bond formation take place:

    a. at the active site
    b. at the regulator bonding site
    c. on the zymogen
    d. on the positive regulator
    e. none of these

18. α-Amylase does not catalyze the hydrolysis of β-glycosidic bonds because the enzyme α-amylase is:

    a. difficult to activate
    b. specific
    c. a proenzyme
    d. inhibited
    e. none of these

19. The antibiotic penicillin acts by:

    a. activating zymogens for bacterial enzymes
    b. cleaving peptide bonds in bacterial proteins
    c. catalyzing the hydrolysis of bacterial cell walls
    d. inhibiting bacterial enzymes that help form cell walls
    e. none of these

20. The vitamin that protects polyunsaturated fatty acids from oxidation is:

    a. vitamin B
    b. vitamin A
    c. vitamin D
    d. vitamin E
    e. none of these

Answers to Practice Exercises

21.1

Name of enzyme	Type of reaction
α-amylase	hydrolysis of α-linkage in starch molecules
dehydrogenase	removal of hydrogen from a substrate
decarboxylase	removal of a carboxyl group from a substrate
transaminase	transfer of an amino group from one molecule to another
peptidase	hydrolysis of peptide linkages
oxidase	oxidation of a substrate
lipase	hydrolysis of ester linkages in lipids

21.2

Answer	Term	Description
4.	coenzyme	1. nonprotein portion of a conjugated enzyme
1.	cofactor	2. obtained from dietary minerals
2.	inorganic ion cofactor	3. protein portion of a conjugated enzyme
3.	apoenzyme	4. small organic molecule that serves as a cofactor

21.3 a. The two models are the lock-and-key model and the induced-fit model.

b. The main difference between the two models is that according to the lock-and-key model, the active site of the enzyme is fixed and rigid, but in the induced-fit model, the active site can change its shape slightly to accommodate the shape of the substrate.

21.4 a. Four factors that affect the rate of enzyme activity are temperature, pH, substrate concentration, and enzyme concentration.

b. Increase in temperature: Temperature increases the rate of a reaction, but if the temperature is high enough to denature the protein enzyme, the rate will decrease.
Increase in pH: Each enzyme has an optimum pH for action; if the pH increases above this point, the reaction will slow down.
Increase in substrate concentration: The rate of reaction increases until maximum enzyme capacity is reached; after this there is no further increase.
Increase in enzyme concentration: Reaction rate increases.

21.5

Answer	Term	Description
5.	penicillin	1. binds to a site other than the active site
3.	competitive inhibitor	2. contains both active and regulator sites
6.	zymogen	3. competes with substrate for active site
7.	sulfa drug	4. forms a covalent bond at the active site
8.	Cipro	5. inhibits cell wall formation in bacteria
4.	irreversible inhibitor	6. inactive precursor of an enzyme
1.	noncompetitive inhibitor	7. inhibitor of folic acid production
2.	allosteric enzyme	8. disrupts tertiary structure of DNA gyrase

21.6 and 21.7

Answer	Vitamin name	Function in human body
6.	Vitamin A – F	1. prevents oxidation of unsaturated fatty acids in membranes
4.	Vitamin C – W	2. helps form compounds that regulate blood clotting
5.	Vitamin D – F	3. group of vitamins that are components of coenzymes
3.	Vitamin B – W	4. cosubstrate in the formation of collagen
1.	Vitamin E – F	5. promotes normal bone mineralization
2.	Vitamin K – F	6. keeps skin and mucous membranes healthy

21.8 a. They have a terpenelike structure made up of five-carbon isoprene units.
b. All except pantothenic acid involve a heterocyclic nitrogen ring system.

Answers to Self-Test

The numbers in parentheses refer to sections in your textbook.
1. T (21.2) **2**. F; function (21.2) **3**. F; apoenzyme (21.3) **4**. T (21.4)
5. F; fixed and rigid (21.4) **6**. T (21.5) **7**. F; an irreversible (21.7)
8. F; including (21.1) **9**. F; nonprotein part (21.3) **10**. F; fat-soluble vitamins (21.12)
11. T (21.14) **12**. T (21.12) **13**. F; treatment (21.11) **14**. a (21.1) **15**. a (21.2)
16. c (21.7) **17**. a (21.4) **18**. b (21.5) **19**. d (21.10) **20**. d (21.14)

Nucleic Acids — Chapter 22

Chapter Overview

Nucleic acids are the molecules of heredity. Every inherited trait of every living organism is coded in these huge molecules. The complexity of their structure has been unraveled in fairly recent times, and this new knowledge of the transmission of genetic information has led to the exciting field of recombinant DNA technology.

In this chapter you will name and identify the structures of nucleotides and nucleic acids. You will write shorthand forms for nucleotide sequences in segments of DNA and RNA. You will identify the amino acid sequence coded by a given segment of DNA and will describe the processes of replication, transcription, and translation leading to protein synthesis. You will learn the basic ideas of recombinant DNA technology and gene therapy.

Practice Exercises

22.1 **Nucleotides** (Sec. 22.2) are the structural units from which the polymeric **nucleic acids** (Sec. 22.1) are formed. A nucleotide is composed of a pentose sugar bonded to both a phosphate group and a nitrogen-containing heterocyclic base. The identities of the sugars and bases differ in ribonucleic acid (RNA) and deoxyribonucleic acid (DNA).

Complete the table below identifying the pentoses and bases found in the nucleotides that make up RNA and DNA.

Nucleotide components	RNA	DNA
pentose		
purine bases		
pyrimidine bases		

22.2 The names of eight nucleotides, the monomer units making up RNA and DNA, are listed in Table 22.1 in your textbook. Both the names and the abbreviations of these names are commonly used. Complete the table below to acquaint yourself with the way nucleotides are named.

Name of nucleotide	Abbreviation	Base	Sugar
deoxyadenosine 5′-monophosphate			
	dTMP		
		guanine	ribose
	CMP		

22.3 The formation of nucleotides from three constituent molecules involves condensation, with the formation of a water molecule. The nucleotide can undergo hydrolysis (addition of water) to yield the three molecules from which it was formed. Write the names for the products of the hydrolysis of each nucleotide below.

a. dCMP

b. UMP

22.4 Nucleotide monomers can be linked to each other through sugar–phosphate bonds. Draw the structural formula for the dinucleotide that forms between dAMP and dTMP, so that dAMP is the 5′ end and to the left in your drawing, and dTMP is the 3′ end and to the right in your drawing.

22.5 The **primary structure** (Sec. 22.3) of a nucleic acid consists of the order in which the nucleotides are linked together. Both RNA and DNA have an alternating sugar–phosphate backbone with the nitrogen-containing bases as side-chain components.

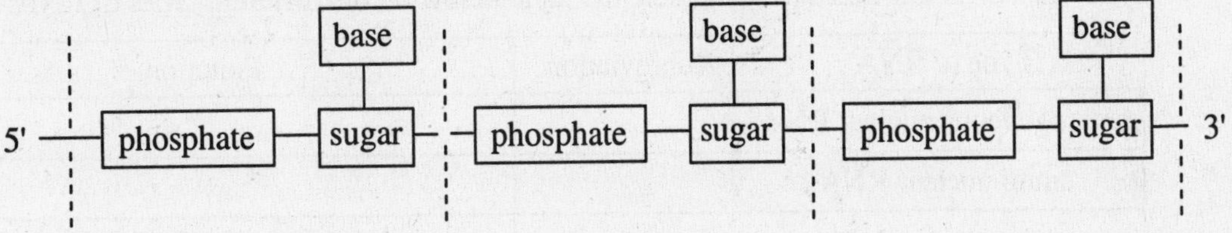

The end of the nucleotide chain that has a free phospate group attached to the 5′ carbon is called the 5′ end, and the end with a free hydroxyl group attached to the 3′ carbon atom is the 3′ end. The strand is read from the 5′ end to the 3′ end.

Draw the structural formula for the trinucleotide that forms among dAMP, dTMP, and dCMP, so that dAMP is the 5′ end and to the left in your drawing, and dTMP is the 3′ end and to the right in your drawing. Use a structural block diagram similar to the one above, but replace "base" with the specific name of the base, and replace "sugar" with the name of the sugar.

22.6 The structure of DNA is that of a double helix, in which two strands of DNA are coiled around each other in a spiral. The two strands are held together by hydrogen bonds between two pairs of **complementary bases** (Sec. 22.4), A–T and G–C. The relative amounts of these base pairs are constant for a given life form.

If, in a DNA molecule, the percentage of the base adenine is 20% of the total bases present, what would be the percentages of the bases thymine, cytosine, and guanine?

22.7 In the DNA double helix, the two complementary strands run in opposite directions, one in the 5′ to 3′ direction and the other 3′ to 5′. Complete the following segment of a DNA double helix. Write symbols for the missing bases. Indicate the correct number of hydrogen bonds between the bases in each pair.

```
5′ T – C –    – C –   – G –   – A  3′
   ||  |||
3′ A – G – T –    – T –   – A –    5′
```

22.8 During **DNA replication** (Sec. 22.5), the DNA molecule makes an exact duplicate of itself. The two strands unwind, and free nucleotides line up along each strand, with complementary base pairs attracted to one another by hydrogen bonding. Polymerization of the new strand takes place. The daughter strand (3′ to 5′) is in the direction opposite to that of the parent strand (5′ to 3′).

Write the sequence of bases for the replication of the DNA segment below:

5′ T–A–A–G–C–G–T–G–G 3′

22.9 There are five types of RNA molecules involved in the process of protein synthesis. Each type has a different function. Complete the table below on the different types of RNA.

Type of RNA	Abbreviation	Function
heterogeneous nuclear RNA		
small nuclear RNA		
messenger RNA		
ribosomal RNA		
transfer RNA		

22.10 During **transcription** (Sec. 22.6), one strand of a DNA molecule acts as the template for the formation of a molecule of hnRNA. The nucleotides that line up next to the DNA strand have ribose as a sugar; the same bases are present except that uracil is substituted for thymine.

DNA contains certain segments, called **introns** (Sec. 22.8), that do not convey genetic information, as well as **exons** (Sec. 22.8), which do carry the genetic code. Both exons and introns are transcribed to hnRNA. In post-transcription processing of the hnRNA molecule, introns are deleted, leaving the exons to be joined together as mRNA.

In the DNA template strand below, sections A, C, and E are exons, and B and D are introns. Write the hnRNA segment transcribed from this DNA strand. Below the hnRNA strand write the structure of the mRNA strand with the introns deleted.

A B C D E

5′ G–C–C–T–G–T–A–C–T–T–C–G–A–T–T–G–G–A 3′ DNA

hnRNA

mRNA

22.11 During **translation** (Sec. 22.11) mRNA directs the synthesis of proteins by carrying the genetic code from the nucleus to a **ribosome** (Sec. 22.11), where the code is translated into the correct series of amino acids in the protein. A **codon** (Sec. 22.9) is a sequence of three nucleotides in an mRNA molecule that codes for a specific amino acid.

a. Complete the tables below with correct amino acid names and codons. The information can be obtained from Table 22.2 in your textbook.

Codon	Amino acid
UCA	
	asparagine
GAC	

Codon	Amino acid
	methionine
GAU	
	tryptophan

b. Are there any synonyms among the codons in the table in part a?

c. Why is ATC not listed in Table 22.2 as one of the codon sequences?

22.12 a. Write a base sequence for mRNA that codes for the tripeptide Gly-Pro-Leu.

b. Will there be only one answer? Explain.

22.13 An **anticodon** (Sec. 22.10) is a three-nucleotide sequence on tRNA that complements the mRNA sequence for the amino acid that bonds to that tRNA. Complete the table below for codons, their anticodons, and the amino acids they specify.

Codon	CAU		
Anticodon		GAG	
Amino acid			Trp

22.14 The mRNA segment below is the one that was determined in Practice Exercise 22.10. Write the amino acid sequence of a tetrapeptide coded by the mRNA base sequence below.

5′ U–C–C–C–G–A–A–G–U–G–G–C 3′ mRNA

tetrapeptide

22.15 The two main processes of protein synthesis are transcription, in which DNA directs the synthesis of RNA molecules, and translation, in which RNA directs the synthesis of proteins. These two processes consist of various steps, which are reviewed in the table below. Tell what happens in each step and which molecules are involved.

Step	Process	Molecules involved
1. Formation of hnRNA		
2. Removal of introns		
3. mRNA to cytoplasm		
4. Activation of tRNA		
5. Initiation		
6. Elongation		
7. Termination		

22.16 **Mutations** (Sec. 22.12) are changes in the base sequence. This change in genetic information can cause a change in the amino acid sequence in protein synthesis. Consider the following segment of mRNA:

5′ G–C–C–U–A–C–A–A–U–G–C–G 3′

a. What is the amino acid sequence formed by **translation** (Sec. 22.6)?

b. What amino acid sequence would result if adenine were substituted for the first uracil?

Self-Test

True-false: Indicate whether the following statements are true or false. If the statement is false, give the word or phrase that may be substituted for the underlined portion to make the statement true.

1. The sugar unit found in DNA molecules is <u>ribose</u>.
2. In DNA the amount of adenine is equal to the amount of <u>guanine</u>.
3. In DNA strands, a phosphate ester bridge connects hydroxyl groups on the 3′ and 5′ positions of the <u>sugar</u> units.
4. DNA molecules are <u>the same</u> for individuals of the same species.
5. <u>Transfer RNA</u> molecules carry the genetic code from DNA to the ribosomes.
6. A codon is a series of <u>three</u> adjacent bases that carry the code for a specific amino acid.
7. A <u>gene</u> is an individual DNA molecule bound to a group of proteins.
8. The two strands of a DNA molecule are connected to each other by <u>hydrogen bonds</u> between the base units.

9. The process by which a DNA molecule forms an exact duplicate of itself is called transcription.
10. Heterogeneous nuclear RNA is edited under the direction of snRNA and joined together to form messenger RNA.
11. Two different codons that specify the same amino acid are called synonyms.
12. Different species of organisms usually have the same genetic code for an amino acid.
13. A single mRNA molecule can serve as a codon sequence for the synthesis of one protein molecule at a time.
14. Mutagens are agents that cause a change in the structure of a DNA molecule.
15. Viruses are tiny disease-causing agents composed of a protein coat and a glycogen core.
16. Viruses can reproduce in water with dissolved organic nutrients.
17. A virus that contains RNA rather than DNA is called a retrovirus.

Multiple choice:

18. The codon 5′ UGC 3′ would have as its anticodon:

a. 5′ GCA 3′ b. 5′ UAG 3′ c. 5′ GAC 3′
d. 5′ AUC 3′ e. none of these

19. Sections of DNA that carry noncoding base sequences are called:

a. introns b. exons c. codons
d. anticodons e. none of these

20. Fifteen nucleotide units in a DNA molecule can contain the code for no more than:

a. 3 amino acids b. 5 amino acids c. 10 amino acids
d. 15 amino acids e. none of these

21. Which of the following types of molecules does *not* carry information for protein synthesis?

a. DNA b. ribosomal RNA c. messenger RNA
d. transfer RNA e. none of these

22. A sequence of three nucleotides in an mRNA molecule is a(n):

a. exon b. intron c. codon
d. anticodon e. none of these

23. The intermediary molecules that deliver amino acids to the ribosomes are:

a. snRNA b. tRNA c. rRNA
d. mRNA e. none of these

24. The codon that initiates protein synthesis when it occurs as the first codon in an amino acid sequence is:

a. GTA b. UGA c. GAC
d. AUG e. none of these

25. The process of inserting recombinant DNA into a host cell is:

a. translation b. transformation c. transcription
d. translocation e. none of these

26. Cells that have descended from a single cell and have identical DNA are called:

a. mutagens b. mutations c. clones
d. plasmids e. none of these

Answers to Practice Exercises

22.1

Nucleotide components	RNA	DNA
pentose	ribose	deoxyribose
purine bases	adenine, guanine	adenine, guanine
pyrimidine bases	uracil, cytosine	cytosine, thymine

22.2

Name of nucleotide	Abbreviation	Base	Sugar
deoxyadenosine 5′-monophosphate	dAMP	adenine	deoxyribose
deoxythymidine 5′-monophosphate	dTMP	thymine	deoxyribose
guanosine 5′-monophosphate	GMP	guanine	ribose
cytidine 5′-monophosphate	CMP	cytosine	ribose

22.3 a. dCMP – cytosine, phosphate, deoxyribose

b. UMP – uracil, phosphate, ribose

22.4

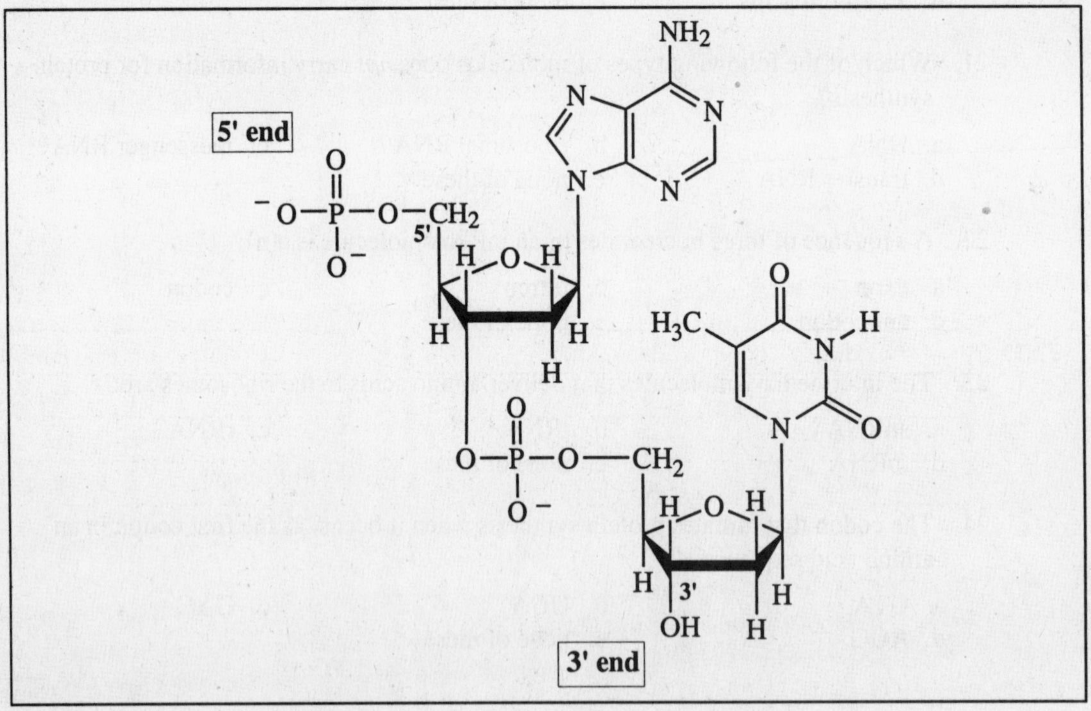

5' end

3' end

22.5

	adenine		cytosine		thymine
5' — phosphate	deoxyribose	phosphate	deoxyribose	phosphate	deoxyribose — 3'

22.6 We know that %A = %T and %G = %C, because these bases are paired in DNA.
If %A = 20%, then %T = 20%. %A + %T = 40%,
so %G + %C = 60% and %G = %C = 30%.

22.7
```
5′ T – C – A – C – A – G – T – A 3′
   ||  |||  ||  |||  ||  |||  ||  ||
3′ A – G – T – G – T – C – A – T 5′
```

22.8 3′ A–T–T–C–G–C–A–C–C 5′ or 5′ C–C–A–C–G–C–T–T–A 3′

22.9

Type of RNA	Abbreviation	Function
heterogeneous nuclear RNA	hnRNA	material from which mRNA is made
small nuclear RNA	snRNA	governs the conversion of hnRNA to rRNA
messenger RNA	mRNA	carries genetic information from the nucleus to the ribosomes
ribosomal RNA	rRNA	site for protein synthesis
transfer RNA	tRNA	delivers individual amino acids to ribosomes for protein synthesis

22.10
```
     A      B        C         D      E
5′ G–C–C–T–G–T–A–C–T–T–C–G–A–T–T–G–G–A 3′   DNA
3′ C–G–G–A–C–A–U–G–A–A–G–C–U–A–A–C–C–U 5′   hnRNA
3′ C–G–G–U–G–A–A–G–C–C–C–U 5′               mRNA
```

22.11

Codon	Amino acid
UCA	serine
AAU and AAC	asparagine
GAC	aspartic acid

Codon	Amino acid
AUG	methionine
GAU	aspartic acid
UGG	tryptophan

b. Yes; GAU and GAC both code for aspartic acid, and AAU and AAC both code for asparagine.

c. ATC is not listed as a codon because RNA does not contain thymine.

22.12 a. 5′ G–G–U–C–C–C–C–U–U 3′ is one possible answer.

b. No, because there is more than one codon for most amino acids.

22.13

Codon	CAU	CUC	UGG
Anticodon	GUA	GAG	ACC
Amino acid	His	Leu	Trp

22.14 5′ U–C–C–C–G–A–A–G–U–G–G–C 3′ mRNA
Ser–Arg–Ser–Gly tetrapeptide

22.15

Step	Process	Molecules involved*
1. Formation of hnRNA	DNA unwinds, acts as template for hnRNA formation	DNA, hnRNA, nucleotides
2. Removal of introns	hnRNA strand cut, introns removed, mRNA bonds form	hnRNA, snRNA, mRNA
3. mRNA to cytoplasm	mRNA moves out of the nucleus into the cytoplasm	mRNA
4. Activation of tRNA	tRNA attaches to an amino acid and becomes energized	tRNA, amino acid, ATP
5. Initiation	mRNA attaches to ribosome, tRNA with amino acid moves to first codon	mRNA, tRNA with attached amino acid, rRNA
6. Elongation	more tRNA molecules move to next codons, polypeptide chain transfers to each new tRNA	mRNA, tRNA with attached amino acid, rRNA
7. Termination	stop codon appears on mRNA, peptide chain (protein) is cleaved from tRNA	mRNA, tRNA, protein

*Enzymes are also involved in each step of the process.

22.16 a. Ala–Tyr–Asn–Ala
b. Ala–Asn–Asn–Ala Asparagine would replace tyrosine.

Answers to Self-Test

The numbers in parentheses refer to sections in your textbook.
1. F; deoxyribose (22.2) **2**. F; thymine (22.4) **3**. T (22.3) **4**. F; different (22.3)
5. F; Messenger RNA (22.7) **6**. T (22.9) **7**. F; chromosome (22.5, 22.8)
8. T (22.4) **9**. F; replication (22.5) **10**. T (22.7) **11**. T (22.9) **12**. T (22.9)
13. F; many protein molecules at a time (22.11) **14**. T (22.12)
15. F; DNA or RNA core (22.13) **16**. F; only in cells of living organisms (22.13) **17**. T (22.13)
18. a (22.11) **19**. a (22.8) **20**. b (22.9) **21**. b (22.7) **22**. c (22.9) **23**. b (22.7)
24. d (22.9) **25**. b (22.14) **26**. c (22.14)

Chapter Overview

The most important job of the body's cells is the production of energy to be utilized in carrying out all the complex processes known as life. The production and use of energy by living organisms involve an important intermediate called ATP.

In this chapter you will study the formation of acetyl CoA from the products of the digestion of food, as well as the further oxidation of acetyl CoA during the individual steps of the citric acid cycle. You will study the function and processes of the electron transport chain and the important role of ATP in energy transfer and release.

Practice Exercises

23.1 **Metabolism** (Sec. 23.1) is the sum total of all the chemical reactions of a living organism. It may be divided into **catabolism** (in which molecules are broken down) and **anabolism** (in which molecules are put together, Sec. 23.1) Metabolic reactions take place in various locations within the cell. A eukaryotic cell, in which DNA is inside a membrane-enclosed nucleus, is shown below. Write the letters from the diagram next to the appropriate descriptions below the diagram.

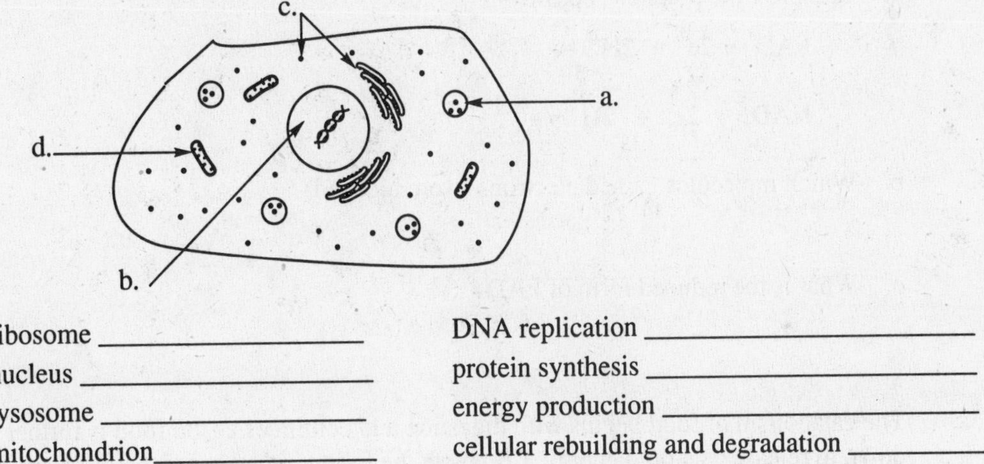

ribosome ______	DNA replication ______
nucleus ______	protein synthesis ______
lysosome ______	energy production ______
mitochondrion ______	cellular rebuilding and degradation ______

23.2 During the catabolic reactions that convert food to energy, several compounds function as key intermediates. To show how these compounds are structurally related to one another, complete the block diagrams below. Use information from Section 23.3 of your textbook.

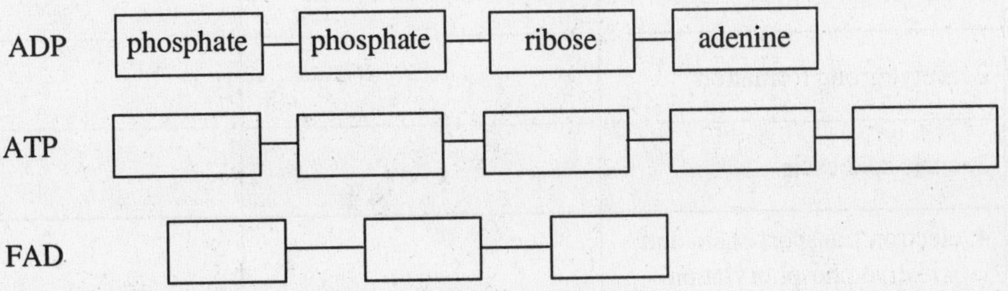

NAD$^+$ [ ]—[ ]—[ ]

Coenzyme A [ ]—[ ]—[ ]

23.3 High-energy compounds are important in the processes of metabolism because they contain one or more very reactive bonds that release a large amount of free energy when they are broken by hydrolysis.

a. Complete the equations below showing the hydrolysis of two of these compounds.

$$ATP + H_2O \rightarrow \ ? \ + P_i$$

$$? + H_2O \rightarrow AMP + P_i$$

b. In the equations for part a, which molecules contained high-energy bonds that were hydrolyzed?

23.4 Two important coenzymes, FAD, NAD$^+$, transport hydrogen atoms and electron pairs.

a. Complete the reactions below.

$$FAD + 2e^- + 2H^+ \rightarrow$$

$$NAD^+ + 2e^- + 2H^+ \rightarrow$$

b. Which molecules gained electrons (were reduced)?

c. What is the reduced form of FAD?

23.5 The catabolism of food begins with digestion and continues as the food is further broken down to release its stored energy. Complete the following table for the different stages of biochemical energy production by the catabolism of food.

Stage of catabolism	Where process occurs	Products
1. digestion		
2. acetyl group formation		
3. citric acid cycle		
4. electron transport chain and oxidative phosphorylation		

23.6 The **citric acid cycle** (Sec. 23.6) is the series of reactions in which the acetyl group of acetyl CoA is oxidized to CO_2, and $FADH_2$ and NADH are produced. Complete the following table summarizing the steps of the citric acid cycle. Use the discussion and equations in Section 23.6 of your textbook.

Step	Type of reaction	Final product(s)	Energy transfer intermediates
1.	condensation		none
2.		isocitrate	
3.			NADH
4.			
5.			
6.			
7.			
8.			

23.7 Four of the steps summarized in Practice Exercise 23.6 involve oxidation. For each of these steps, give the name and/or symbol for the substance that was oxidized and for the one that was reduced.

Step	Substance oxidized	Substance reduced
3.		
4.	α-ketoglutarate	
6.		FAD (to $FADH_2$)
8.		

23.8 The **electron transport chain** (Sec. 23.7) is a series of reactions in which electrons and hydrogen ions from NADH and $FADH_2$ are passed to intermediate carriers and ultimately react with molecular oxygen to produce water.

Use the summary of the electron transport chain given in Section 23.7 in your textbook to complete the following table.

Step	Substance oxidized	Substance reduced
1.		
2.		
3.		
4.		
5.		
6.		
7.		
8.		
final step		

23.9 **Oxidative phophorylation** (Sec. 23.8) is the process by which ATP is synthesized from ADP using energy released in the electron transport chain. Each NADH from the citric acid cycle produces 2.5 ATP molecules, each $FADH_2$ produces 1.5 ATP molecules, and each GTP produces 1 ATP molecule. Using Practice Exercise 23.3, summarize the number of ATP molecules produced in one turn of the citric acid cycle (Steps 1 through 8).

Step	Energy-rich compound formed	Number of ATP's produced
3.		
4,		
5.		
6.		
8.		
	Total:	

Self-Test

True-false: Indicate whether the following statements are true or false. If the statement is false, give the word or phrase that may be substituted for the underlined portion to make the statement true.

1. Catabolic reactions usually <u>release</u> energy.
2. Bacterial cells do not contain a nucleus and so are classified as <u>eukaryotic</u>.
3. <u>Ribosomes</u> are organelles that have a central role in the production of energy.
4. The active portion of the FAD molecule is the <u>flavin</u> subunit.
5. <u>NAD^+</u> is the oxidized form of nicotine adenine dinucleotide.
6. The citric acid cycle occurs in the <u>cytoplasm</u> of cells.

7. In the first step of the citric acid cycle, oxaloacetate reacts with <u>glucose</u>.
8. The electron transport chain receives electrons and hydrogen ions from <u>NAD^+ and FAD</u>.
9. The electrons that pass through the electron transport chain <u>gain energy</u> in each transfer along the chain.
10. Cytochromes contain <u>iron</u> atoms that are reversibly oxidized and reduced.
11. ATP is produced from ADP using energy <u>released</u> in the electron transport chain.

Multiple choice:

12. Which of the following is *not* an organelle?

a. mitochondrion b. hemoglobin c. ribosome
d. lysosome e. none of these

13. Which of these molecules is *not* part of the electron transport chain?

a. coenzyme Q b. acetyl CoA c. cytochrome b
d. ATP e. none of these

14. Energy used in cells is obtained directly from:

a. oxidation of NADH b. activation of acetyl CoA c. oxidation of $FADH_2$
d. hydrolysis of ATP e. none of these

15. The final acceptor of electrons in the electron transport chain is:

a. NAD^+ b. water c. oxygen
d. FAD e. none of these

Answers to Practice Exercises

23.1

ribosome	c.	DNA replication	b.
nucleus	b.	protein synthesis	c.
lysosome	a.	energy production	d.
mitochondrion	d.	cellular rebuilding and degradation	a.

23.2

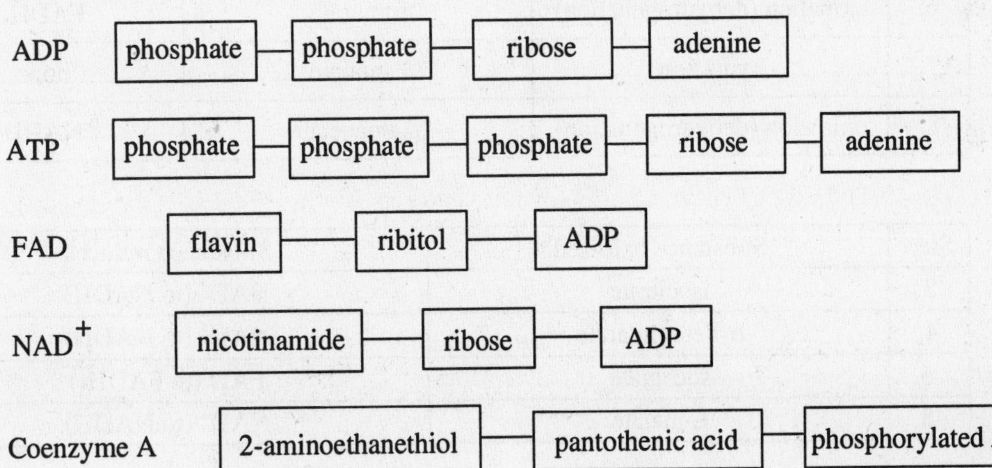

23.3 a. $ATP + H_2O \rightarrow ADP + P_i$

$ADP + H_2O \rightarrow AMP + P_i$

b. ATP and ADP

23.4 a. $FAD + 2e^- + 2H^+ \rightarrow FADH_2$

$NAD^+ + 2e^- + 2H^+ \rightarrow NADH + H^+$

b. FAD and NAD^+

c. $FADH_2$

23.5

Stage of catabolism	Where process occurs	Major products
1. digestion	mouth, stomach, small intestine	glucose, fatty acids and glycerol, amino acids
2. acetyl group formation	cytoplasm of cell and inside mitochondria	acetyl CoA
3. citric acid cycle	inside mitochondria	CO_2, water, NADH, and $FADH_2$
4. electron transport chain and oxidative phosphorylation	inside mitochondria	water and ATP

23.6

Step	Type of reaction	Final product(s)	Energy transfer intermediates
1.	condensation	citrate, free coenzyme A	none
2.	isomerization	isocitrate	none
3.	oxidation, decarboxylation	α-ketoglutarate, CO_2	NADH
4.	oxidation, decarboxylation	succinyl CoA, CO_2	NADH
5.	phosphorylation	succinate, free coenzyme A	GTP
6.	oxidation (dehydrogenation)	fumarate	$FADH_2$
7.	hydration	L-malate	none
8.	oxidation (dehydrogenation)	oxaloacetate	NADH

23.7

Step	Substance oxidized	Substance reduced
3.	isocitrate	NAD^+ (to NADH)
4.	α-ketoglutarate	NAD^+ (to NADH)
6.	succinate	FAD (to $FADH_2$)
8.	L-malate	NAD^+ (to NADH)

23.8

Step	Substance oxidized	Substance reduced
1.	NADH	FMN
2.	$FMNH_2$	Fe(III)
3.	Fe(II) (or $FADH_2$)	CoQ
4.	$CoQH_2$	Fe(III) cytochrome b
5.	Fe(II) cytochrome b	Fe(III) cytochrome c_1
6.	Fe(II) cytochrome c_1	Fe(III) cytochrome c
7.	Fe(II) cytochrome c	Fe(III) cytochrome a
8.	Fe(II) cytochrome a	Fe(III) cytochrome a_3
final step	Fe(II) cytochrome a_3	O_2

23.9

Step	Energy-rich compound formed	Number of ATPs produced
3.	NADH	2.5
4.	NADH	2.5
5.	GTP	1
6.	$FADH_2$	1.5
8.	NADH	2.5
	Total:	10

Answers to Self-Test

The numbers in parentheses refer to sections in your textbook.
1. T (23.1) **2.** F; prokaryotic (23.2) **3.** F; mitochondria (23.2) **4.** T (23.3) **5.** T (23.3) **6.** F; mitochondria (23.5) **7.** F; acetyl CoA (23.6) **8.** F; NADH and $FADH_2$ (23.6) **9.** F; lose energy (23.7) **10.** T (23.7) **11.** T (23.8) **12.** b (23.2) **13.** b (23.7) **14.** d (23.9) **15.** c (23.7)

Carbohydrate Metabolism — Chapter 24

Chapter Overview

The complete oxidation of glucose supplies the energy needed by cells to carry out their many vital functions. This oxidation takes place in a number of steps that conserve the energy contained in the chemical bonds of glucose and transfer it efficiently.

In this chapter you will study the reactions of glycolysis to produce pyruvate, the pathways of pyruvate under aerobic and anaerobic conditions, and the pentose phosphate pathway. You will study the ways in which glycogen is synthesized in the body and broken down and the hormones that control these processes.

Practice Exercises

24.1 **Glycolysis** (Sec. 24.2) is the metabolic pathway by which glucose is converted into two molecules of pyruvate. Using the steps of glycolysis discussed in Section 24.2 of your textbook, give the number(s) of the reaction steps for each of the following:

a. Where are ATPs produced? ____________________

b. Where are ATPs used? ____________________

c. Where is NAD^+ reduced? ____________________

d. Where is the carbon chain split? ____________________

e. Where is a ketone isomerized to an aldehyde? ____________________

f. Where are phosphate groups added to sugar molecules? ____________________

g. Where are phosphate groups removed from sugar molecules? ____________________

h. Where is water lost? ____________________

24.2 The pyruvate produced by glycolysis reacts further in several different ways according to the conditions and the type of organism. Complete the following table on the fates of pyruvate produced by glycolysis.

Conditions	Name of process	Name of product	Number of NADH used or produced
1. aerobic			
2. anaerobic (humans)			
3. anaerobic (yeasts)			

24.3 The very different metabolic processes of **glycogenesis** (Sec. 24.5), **glycogenolysis** (Sec. 24.4), and **gluconeogenesis** (Sec. 24.6) have like-sounding names. Fill in the following table showing the differences among these processes.

Name of process	What the process accomplishes	Where the process takes place	High-energy phosphate molecules used
glycogenesis			
glycogenolysis			
gluconeogenesis			

24.4 Another pathway by which glucose may be degraded is the **pentose phosphate pathway** (Sec. 24.8).

a. What are the two main functions of the pentose phosphate pathway?

1.

2.

b. In the table below, compare the coenzyme NADPH to NADH.

Coenzyme	Oxidized form	Function in human body
NADH		
NADPH		

24.5 Three hormones affect carbohydrate metabolism. Complete the table below comparing these hormones.

Hormone	Source	Effect
insulin		
glucagon		
epinephrine		

24.6 Review the following terms, introduced in this chapter, by matching each with its correct description.

Answer	Term	Description
	fermentation	1. lactate → pyruvate → glucose
	glycogenesis	2. stimulates: glycogen (liver) → glucose
	gluconeogenesis	3. glucose → glycogen
	glycolysis	4. stimulates: glycogen (muscle) → glucose-6-phosphate
	insulin	5. glucose → ethanol
	epinephrine	6. glycogen → glucose or glucose-6-phosphate
	glycogenolysis	7. stimulates: glucose (blood) → glucose (cells)
	glucagon	8. glucose → pyruvate

24.7 You have studied the many reactions involved in the complete oxidation of one glucose molecule. Below is a summary of the reactions in each of the four stages of this oxidation. Because of the complexity of the processes, the "equations" are unbalanced. To get an overview of the process, work with them as you would with ordinary balanced equations.

Add the equations together by crossing out molecules that occur on both the reactant side of one equation and the product side of another. Write the molecules that remain as the final net equation for the entire process. Balance each type of atom in the final equation. Sum up the ATPs produced.

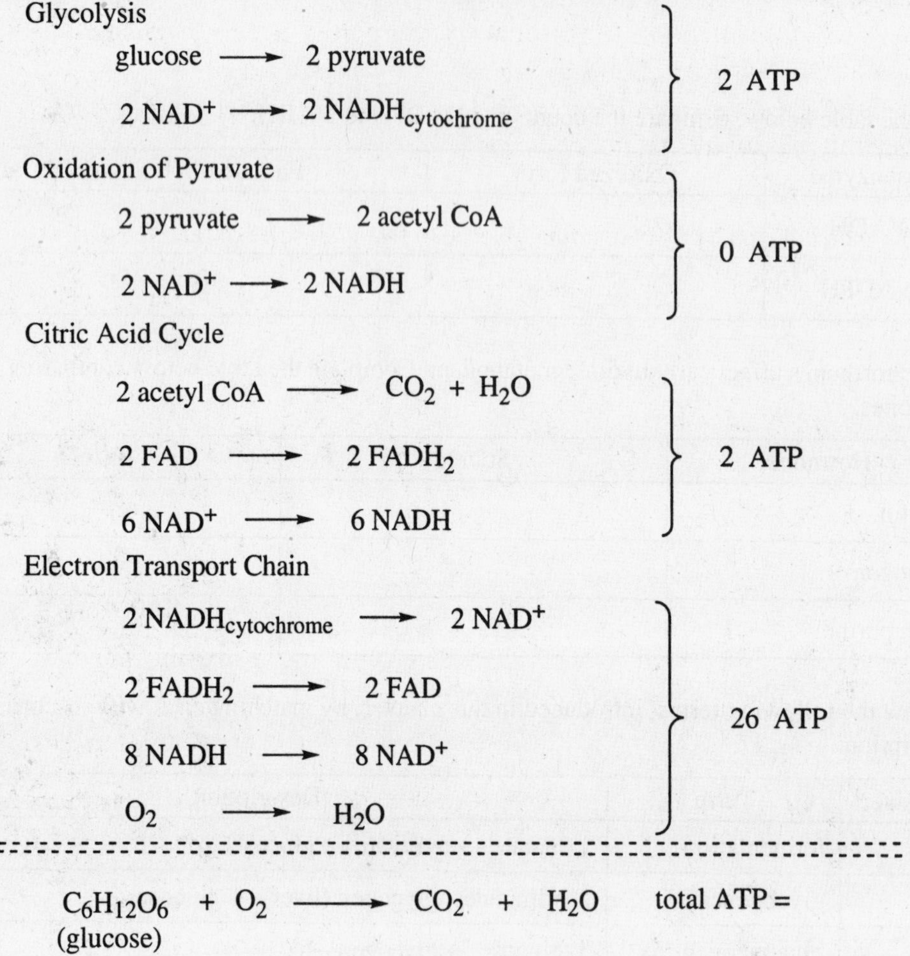

Self-Test

True-false: Indicate whether the following statements are true or false. If the statement is false, give the word or phrase that may be substituted for the underlined portion to make the statement true.

1. The main site for carbohydrate digestion is <u>the small intestine</u>.
2. Glycolysis takes place in the <u>cytoplasm</u> of cells.
3. The process by which glucose is degraded to ethanol is called <u>fermentation</u>.
4. The enzymes maltase, sucrase, and lactase convert <u>starch to disaccharides</u>.
5. Substrate-level phosphorylation is the direct transfer of <u>phosphate ions in solution</u> to ADP molecules to produce ATP.
6. Fructose and galactose are converted, <u>in the liver</u>, to intermediates that enter the glycolysis pathway.
7. In human metabolism, under anaerobic conditions, pyruvate is reduced to <u>acetyl CoA</u>.

8. The complete oxidation of glucose in skeletal muscle and nerve cells yields <u>30</u> ATP molecules per glucose molecule.
9. When glycogen stored in muscle and liver tissues is depleted by strenuous exercise, glucose molecules can be synthesized by the process of <u>gluconeogenesis</u>.
10. <u>Glucagon</u> speeds up the rate of glycogenolysis in the muscle cells.
11. The breakdown of glycogen in order to maintain normal glucose levels in the bloodstream is called <u>glycolysis</u>.
12. The pentose phosphate pathway metabolizes glucose to produce <u>ribose and other sugars needed for biosynthesis</u>.

Multiple choice:

13. The metabolic pathway converting glucose to two molecules of pyruvate is:

    a. glycogenolysis b. glycogenesis c. glycolysis
    d. gluconeogenesis e. none of these

14. The hormone that lowers blood glucose levels is:

    a. glucagon b. insulin c. epinephrine
    d. norepinephrine e. none of these

15. Carbohydrate digestion produces:

    a. glucose b. galactose c. fructose
    d. all of these e. none of these

16. The lactate produced during strenuous exercise is converted to pyruvate in the:

    a. kidneys b. liver c. muscles
    d. small intestine e. none of these

17. One of the control mechanisms of glycolysis is feedback inhibition of hexokinase by:

    a. glucose b. glucose 6-phosphate c. glucose 1-phosphate
    d. fructose 6-phosphate e. none of these

18. In order for the electrons from NADH produced during glycolysis to enter the electron transport chain, they must:

    a. react with acetyl CoA and then enter the citric acid cycle
    b. reduce FAD, which then passes through the mitochondrial membrane
    c. be shuttled by an intermediate through the mitochondrial membrane to FAD
    d. pass energy directly to ATP in the cytoplasm
    e. none of these

Answers to Practice Exercises

24.1 a. Steps 7 and 10; b. Steps 1 and 3; c. Step 6; d. Step 4
e. Step 5; f. Steps 1, 3, and 6; g. Steps 7 and 10; h. Step 9

24.2

Conditions	Name of process	Product	Number of NADH
1. aerobic	oxidation	acetyl	1 produced
2. anaerobic (humans)	reduction	lactate	1 used
3. anaerobic (yeasts)	fermentation	ethanol	1 used

24.3

Name of process	What the process accomplishes	Where the process takes place	Phosphate molecules used
glycogenesis	synthesis of glycogen from glucose	muscle and liver tissue	2 ATP
glycogenolysis	breakdown of glycogen to glucose	muscle and brain (glucose-6-phosphate) and liver (free glucose)	none
gluconeogenesis	glucose synthesis from noncarbohydrates	liver	4 ATP, 2 GTP

24.4 a. 1. synthesis of NADPH
2. production of ribose-5-phosphate for synthesis of nucleic acids and coenzymes

b.

Coenzyme	Oxidized form	Function in human body
NADH	NAD^+	promotes common metabolic pathways of energy transfer
NADPH	$NADP^+$	promotes biosynthetic reactions of lipids and nucleic acids

24.5

Hormone	Source	Effect
insulin	beta cells of pancreas	increases uptake and utilization of glucose by cells; lowers blood glucose
glucagon	alpha cells of pancreas	increases blood glucose by speeding up glycogenolysis in liver
epinephrine	adrenal glands	stimulates glycogenolysis in muscle cells; gives quick energy to muscle cells

24.6

Answer	Term	Description
5.	fermentation	1. lactate → pyruvate → glucose
3.	glycogenesi	2. stimulates: glycogen (liver) → glucose
1.	glyconeogenesis	3. glucose → glycogen
8.	glycolysis	4. stimulates: glycogen (muscle) → glucose-6-phosphate
7.	insulin	5. glucose → ethanol
4.	epinephrine	6. glycogen → glucose or glucose-6-phosphate
6.	glycogenolysis	7. stimulates: glucose (blood) → glucose (cells)
2.	glucagon	8. glucose → pyruvate

24.7

Glycolysis

glucose ⟶ 2 pyruvate

$2\ NAD^+$ ⟶ $2\ NADH_{cytochrome}$

} 2 ATP

Oxidation of Pyruvate

2 pyruvate ⟶ 2 acetyl CoA

$2\ NAD^+$ ⟶ 2 NADH

} 0 ATP

Citric Acid Cycle

2 acetyl CoA ⟶ $CO_2 + H_2O$

2 FAD ⟶ $2\ FADH_2$

$6\ NAD^+$ ⟶ 6 NADH

} 2 ATP

Electron Transport Chain

$2\ NADH_{cytochrome}$ ⟶ $2\ NAD^+$

$2\ FADH_2$ ⟶ 2 FAD

8 NADH ⟶ $8\ NAD^+$

O_2 ⟶ H_2O

} 26 ATP

$C_6H_{12}O_6 + 6\ O_2 \longrightarrow 6\ CO_2 + 6\ H_2O$ total ATP = 30 ATP
(glucose)

Answers to Self-Test

The numbers in parentheses refer to sections in your textbook.
1. T (24.1) **2.** T (24.2) **3.** T (24.3) **4.** F; disaccharides to monosaccharides (24.1) **5.** F; high-energy phosphate groups from substrate molecules (24.2) **6.** T (24.2) **7.** F; lactate (24.3) **8.** T (24.4) **9.** T (24.6) **10.** F; epinephrine (24.9) **11.** F; glycogenolysis (24.5) **12.** T (24.8) **13.** c (24.2) **14.** b (24.9) **15.** d (24.1) **16.** b (24.3) **17.** b (24.2) **18.** c (24.4)

Lipid Metabolism — Chapter 25

Chapter Overview

Lipids are the most efficient energy-storage compounds of the body. They are also important materials in membranes. This chapter discusses the biosynthesis and storage of lipids and their degradation to produce energy.

In this chapter you will study the processes of digestion of triacyglycerols and their absorption into the bloodstream, their storage, and their degradation by means of the fatty acid spiral. You will define ketone bodies and learn the conditions under which they are produced. You will compare fatty acid synthesis to fatty acid oxidation, and you will study the biosynthesis of cholesterol.

Practice Exercises

25.1 Most dietary lipids are triacylglycerols (TAGs). Complete the following diagram showing the stages of digestion and absorption of TAGs. In each box, write the processes that take place at that location.

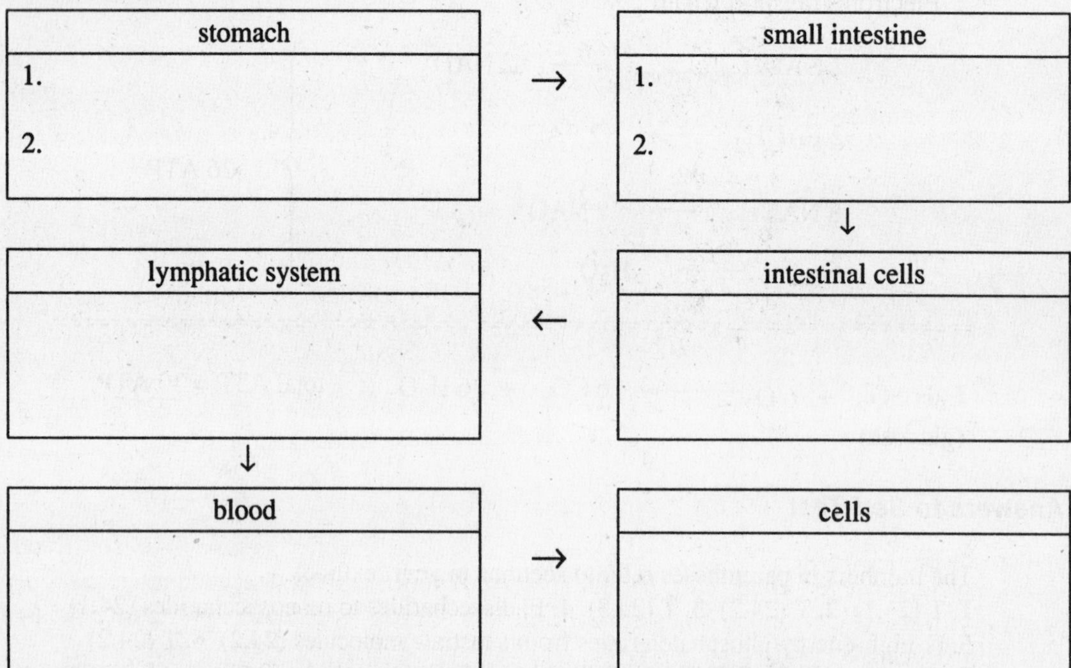

25.2 TAG molecules can be stored in adipose tissue until needed for energy production.

a. What is the process of **triacylglycerol mobilization** (Sec. 25.2)?

b. What happens to the products of triacylglycerol mobilization?

25.3 The **fatty acid spiral** (Sec. 25.4) is a four-step process in which two carbon atoms are cleaved from a fatty acid for each turn of the spiral.

a. Summarize the type of reaction in each of the following steps of the fatty acid spiral by completing the table below. (Use the discussion of the fatty acid spiral in Section 25.4 of your textbook.)

Step	Type of reaction	Coenzyme oxidizing agent
1. alkane → alkene		
2. alkene → secondary alcohol		
3. secondary alcohol → ketone		
4. ketone → shortened fatty acid		

b. Give the numbers of the reaction steps for these parts of the process.

1. Where is FAD reduced? ____________________
2. Where is NAD^+ reduced? ____________________
3. Where is CoA-SH added? ____________________
4. Where is water added to the carbon chain? ____________________
5. Where is acetyl CoA removed from the carbon chain? ____________________

25.4 a. Calculate the total number of acetyl CoA molecules, NADH molecules, and $FADH_2$ molecules produced by the aerobic catabolism of capric acid, a saturated 10-carbon fatty acid.

b. Determine the total number of molecules of ATP produced for energy use by the total oxidation of capric acid to carbon dioxide and water.

25.5 Acetyl CoA from the fatty acid spiral is usually processed further through the citric acid cycle; however, under some conditions there is not enough oxaloacetate produced to react with acetyl CoA in the first step of the citric acid cycle, and **ketone bodies** (Sec. 25.6) are produced.

a. Name the three ketone bodies produced in the human body.

b. Under what conditions do ketone bodies form?

c. What is ketosis?

d. Which ketone bodies could produce a lowered blood pH (acidosis)?

25.6 **Lipogenesis** (Sec. 25.7) is the synthesis of fatty acids from acetyl CoA. It is not simply the reverse of the fatty acid spiral for degradation of fatty acids. Complete the table below comparing these two processes.

Comparisons	Degradation of fatty acids	Lipogenesis
reaction site in cell		
carbons lost/gained per turn		
intermediate carriers		
types of enzymes		
coenzymes for energy transfer		

25.7 Four reactions occurring in a cyclic pattern constitute the chain elongation process of lipogenesis. Complete the table below summarizing the types of reactions that take place in each turn of the cycle.

Step	Type of reaction	Coenzyme reducing agent
1.		
2.		
3.		
4.		

25.8 Answer the following questions about the biosynthesis of capric acid, a 10-carbon saturated fatty acid, from acetyl CoA molecules.

a. How many rounds of the fatty acid biosynthesis pathway are needed? ______________

b. How many molecules of malonyl ACP must be formed? ______________

c. How many high-energy ATP bonds are consumed? ______________

d. How many NADPH molecules are needed? ______________

25.9 Biosynthesis of cholesterol takes place in the liver. There are at least 27 steps in the process, but these can be considered to occur in biosynthetic stages. Fill in the missing parts of the diagram below summarizing the stages of cholesterol biosynthesis.

1. 3 acetyl CoA + 3ATP →	1 isoprene unit
2. →	1 squalene molecule
3. →	
4. lanosterol →	

25.10 Acetyl CoA is a key intermediate for many metabolic processes. Fill in the diagram below showing the products or reactants in these processes.

[ ] (degradation) → acetyl CoA → (biosynthesis) [ ]

↓ (energy production)

oxidation in citric acid cycle

Self-Test

True-false: Indicate whether the following statements are true or false. If the statement is false, give the word or phrase that may be substituted for the underlined portion to make the statement true.

1. Bile is released into the small intestine where it acts as <u>a hydrolytic enzyme</u> for lipids.
2. A meal high in triacylglycerols will cause the concentration of <u>fatty acid micelles</u> in the blood and lymph to peak in 4 to 6 hours.
3. Adipocytes are triacylglycerol storage cells found mainly in <u>liver tissue</u>.
4. Chylomicrons consist of lipoproteins combined with <u>fatty acids</u> in the intestinal cells.
5. Glycerol is metabolized to <u>dihydroxyacetone phosphate</u> before entering glycolysis.
6. A stearic acid molecule produces <u>more than 4 times</u> as much energy as a glucose molecule.
7. <u>Isoprene</u> is an important intermediate in the synthesis of cholesterol.
8. Acetyl CoA molecules can be further oxidized in the <u>fatty acid spiral</u>.
9. Cholesterol synthesis takes place in the <u>liver</u>.
10. When glucose is not available to the body, acetyl CoA molecules are used to manufacture <u>pyruvate</u>.
11. The liver uses <u>fatty acids</u> as the preferred fuel.
12. The last turn of the fatty acid spiral produces <u>two</u> acetyl CoA molecules.

Multiple choice:

13. Triacylglycerol mobilization is the process in which triacylglycerols are:

    a. oxidized b. hydrolyzed c. synthesized
    d. hydrogenated e. none of these

14. Acetyl CoA may *not* be used to:

    a. synthesize ketone bodies b. synthesize fatty acids c. synthesize glucose
    d. synthesize cholesterol e. none of these

15. Diabetic ketosis results in:

    a. ketonemia b. ketonuria c. metabolic acidosis
    d. all of these e. none of these

16. One round of the fatty acid spiral produces:

    a. one NADH and one FAD b. two NADH and one FAD
    c. one NADH and two $FADH_2$ d. one NADH and one $FADH_2$
    e. none of these

17. Which of the following does *not* aid in the digestion of lipids?

    a. pancreatic lipases b. cholecytokinin
    c. salivary enzymes d. churning action of stomach
    e. all of the above are used

18. A fatty acid micelle produced by hydrolysis of lipids contains:

    a. triacylglycerols b. diacylglycerols c. monoacylglycerols
    d. lipoproteins e. none of these

19. A 20-carbon saturated fatty acid will produce 10 acetyl CoA molecules from the fatty acid spiral after:

    a. 1 turn b. 9 turns c. 10 turns d. 20 turns e. none of these

20. The total number of ATPs produced in the body by complete oxidation of an 18-carbon saturated fatty acid is:

    a. 120 b. 118 c. 152 d. 98 e. none of these

Answers to Practice Exercises

25.1

stomach
1. churning action, TAGs broken into small globules 2. beginning of hydrolysis of TAGS with gastric lipase

→

small intestine
1. emulsification by bile, hydrolysis of TAGs to form glycerol, monoacylglycerols, and fatty acids 2. formation of fatty acid micelles, which are absorbed into the intestinal cells

↓

intestinal cells
TAGs are reassembled and combined with water-soluble protein in chylomicrons

←

lymphatic system
chylomicrons enter lymphatic vessels and are transported to the thoracic duct, where fluid enters a vein

↓

blood
TAGs are hydrolyzed to fatty acids and glycerol, which are absorbed by the cells

→

cells
fatty acids and glycerol are broken down for energy or stored as TAGs

25.2 a. Triacylglycerol mobilization is the hydrolysis of stored TAGs and the release of the fatty acids and glycerol produced into the bloodstream.

b. The fatty acids and glycerol are oxidized to produce energy, the fatty acids through the fatty acid spiral and the glycerol in a two-step process that produces dihydroxyacetone phosphate.

25.3 a.

Step	Type of reaction	Oxidizing agent
1. alkane → alkene	oxidation (dehydrogenation)	FAD
2. alkene → secondary alcohol	hydration	none
3. secondary alcohol → ketone	oxidation (dehydrogenation)	NAD^+
4. ketone → shortened fatty acid	chain cleavage	none

b. 1. Step 1; 2. Step 3; 3. Step 4; 4. Step 2; 5. Step 4

25.4 a. In four times through the fatty acid spiral, a 10-carbon fatty acid produces: 5 acetyl CoA molecules, 4 NADH, and 4 $FADH_2$.

b. Citric acid cycle: 5 acetyl CoA x 10 ATP/ 1 acetyl CoA = 50 ATP
Electron transport chain: 4 NADH x 2 ATP/ 1 NADH = 8 ATP
4 $FADH_2$ x 1.5 ATP/ 1 $FADH_2$ = 6 ATP
Activation of fatty acid spiral: –2 ATP
Total: 62 ATP

25.5 a. Acetoacetate, β-hydroxybutyrate, acetone
b. Ketone bodies form from acetyl CoA when there is insufficient oxaloacetate being formed from pyruvate. This would occur when dietary intakes are high in fat and low in carbohydrates, when the body cannot adequately process glucose (as in diabetes), and during prolonged fasting or starvation.
c. Ketosis is the accumulation of ketone bodies in the blood and urine.
d. Acetoacetate and β-hydroxybutyrate are acids and could produce a lowered blood pH.

25.6

Comparisons	Fatty acid degradation	Lipogenesis
reaction site in cell	mitochondrial matrix	cytoplasm
carbons lost/gained per turn	two carbons lost	two carbons gained
intermediate carriers	coenzyme A	ACP (acyl carrier protein)
types of enzymes	independent enzymes	fatty acid synthetase complex
coenzymes -- energy transfer	FAD and NAD^+	NADPH

25.7

Step	Type of reaction	Coenzyme reducing agent
1.	condensation	none
2.	hydrogenation	NADPH
3.	dehydration	none
4.	hydrogenation	NADPH

25.8 a. 4 rounds; b. 4 malonyl ACP molecules; c. 4 ATP bonds; d. 8 NADPH molecules

25.9

1. 3 acetyl CoA + 3ATP	→	1 isoprene unit
2. 6 isoprene units	→	squalene
3. squalene	→	lanosterol (4-ring system)
4. lanosterol	→	cholesterol

25.10 [fatty acids, glucose, glycerol] (degradation) → acetyl CoA → (biosynthesis) [fatty acids, ketone bodies, cholesterol]

Answers to Self-Test

The numbers in parentheses refer to sections in your textbook.
1. F; an emulsifier (25.1) **2.** F; chylomicrons (25.1) **3.** F; adipose tissue (25.2) **4.** F; triacylglycerols (25.1) **5.** T (25.3) **6.** T (25.5) **7.** T (25.8) **8.** F; citric acid cycle (25.4) **9.** T (25.8) **10.** F; ketone bodies (25.6) **11.** T (25.5) **12.** T (25.4) **13.** b (25.2) **14.** c (25.9) **15.** d (25.6) **16.** d (25.4) **17.** c (25.1) **18.** c (25.1) **19.** b (25.4) **20.** b (25.5)

Protein Metabolism — Chapter 26

Chapter Overview

Like other organic compounds, proteins and amino acids are metabolized by the body. Although their primary uses are in biosynthesis, they can also be used as sources of energy for the body.

In this chapter you will study the digestion of proteins and the absorption of amino acids, the breakdown of amino acids in the body, and the entry of the products of catabolism into the metabolic pathways of carbohydrates and fatty acids. You will study the biosynthesis of the nonessential amino acids and learn how the urea cycle is used to rid the body of toxic ammonium ion. You will learn how the body uses the degradation products of hemoglobin.

Practice Exercises

26.1 The digestion of proteins produces amino acids, which enter the bloodstream for distribution throughout the body. This **amino acid pool** (Sec. 26.2) is utilized in four different ways.

a. Complete the diagram below for the digestion of proteins and absorption of amino acids.

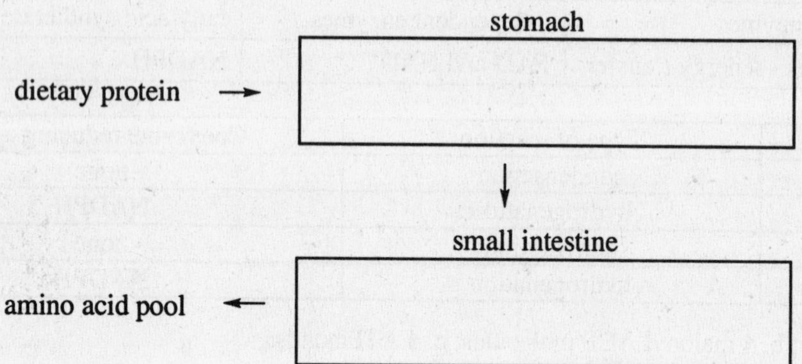

b. What are the four ways in which the amino acids from the amino acid pool are utilized in the human body?

1.
2.
3.
4.

26.2 The degradation of amino acids takes place in two stages, **transamination** (Sec. 26.3) and **oxidative deamination** (Sec. 26.3). Complete the equations below by drawing structural formulas for the missing products and reactants. (Equation a is a transamination reaction; equation b is an oxidative deamination reaction.)

a.

$$CH_3-\underset{}{\overset{\overset{+}{NH_3}|}{CH}}-COO^- + {}^-OOC-CH_2-\overset{\overset{O}{\|}}{C}-COO^- \xrightarrow{\text{trans-aminase}} CH_3-\overset{\overset{O}{\|}}{C}-COO^- + \ ?$$

b. $$^{-}OOC-CH_2-\underset{\displaystyle |}{CH}(\overset{+}{N}H_3)-COO^- + \ ? \longrightarrow \overset{+}{N}H_4 + \ ? \ + NADH + H^+$$

26.3 The ammonium ion produced by oxidative deamination is toxic; it is either used by the human body in biosynthesis reactions or removed from the body as urea in the **urea cycle** (Sec. 26.4). Using the steps of the urea cycle shown in your textbook, answer the following questions:

a. Where is urea removed from the cycle? ____________________

b. Where is carbamoyl phosphate added? ____________________

c. Where is water added? ____________________

d. Where is fumarate produced? ____________________

e. Where is ATP used? ____________________

f. How much ATP is consumed in the production of one urea molecule? ____________

26.4 The carbon skeletons produced by transamination or oxidative deamination of amino acids undergo a sequence of degradations. The seven degradation products of these series of reactions are further metabolized. The original amino acids are classified in terms of the pathways taken by their degradation products.

Classification of original amino acids	Products of degradation sequences	Further metabolic reactions
glucogenic (Sec. 26.5)		
ketogenic (Sec. 26.5)		
glucogenic and ketogenic		

26.5 The essential amino acids must be consumed by humans, but the 11 nonessential amino acids can be synthesized in the human body from intermediates of glycolysis and the citric acid cycle. Complete the equations below for the three amino acids that can be biosynthesized by transamination of the appropriate α-keto acid. Name the amino acids produced.

a. $$^{-}OOC-CH_2-\overset{\displaystyle O}{\overset{\|}{C}}-COO^- \xrightarrow{\text{transaminase}}$$

b. $$CH_3-\overset{\displaystyle O}{\overset{\|}{C}}-COO^- \xrightarrow{\text{transaminase}}$$

c. $$^{-}OOC-CH_2-CH_2-\overset{\displaystyle O}{\overset{\|}{C}}-COO^- \xrightarrow{\text{transaminase}}$$

26.6 Hemoglobin is the conjugated protein responsible for the oxygen-carrying ability of red blood cells. Hemoglobin catabolism takes place in the spleen and liver.

Complete the diagram below showing the degradation products of hemoglobin.

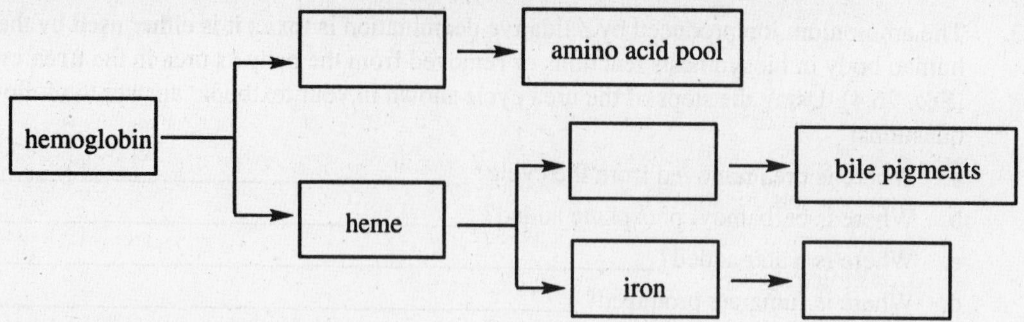

26.7 Complete the table below summarizing some of the metabolism of the three groups of nutrients you have been studying.

Class of nutrient	Basic structural unit	Storage compounds	Degradation pathway
carbohydrates			
lipids			
proteins			

Self-Test

True-false: Indicate whether the following statements are true or false. If the statement is false, give the word or phrase that may be substituted for the underlined portion to make the statement true.

1. Protein digestion begins in the <u>stomach</u>.
2. Proteins are <u>denatured</u> in the stomach by hydrochloric acid.
3. The pH of the pancreatic juice that aids in protein digestion is between <u>1.5 and 2.0</u>.
4. <u>Transamination</u> of an amino acid releases the amino group as an ammonium ion.
5. Transaminases require the presence of the coenzyme <u>pyridoxal phosphate</u>.
6. Oxidative deamination is the conversion of an amino acid into a keto acid with the release of <u>urea</u>.
7. <u>Glucogenic</u> amino acids are converted to acetyl CoA.
8. <u>All</u> of the nonessential amino acids can be synthesized in the body.
9. <u>Two</u> of the essential amino acids can be synthesized in the body.
10. A <u>positive</u> nitrogen balance means that more nitrogen is excreted than is taken in.
11. Phenylketonuria (PKU) is a lack of the enzyme needed to <u>synthesize</u> phenylalanine.
12. During periods of negative nitrogen balance, the body uses amino acids obtained from <u>the amino acid pool</u>.
13. Only two amino acids are purely ketogenic: <u>leucine and lysine</u>.

Multiple choice:

14. The urea cycle is important because it:
    a. removes ammonium ion from the body
    b. regenerates α-ketoglutarate by oxidative deamination
    c. results in transamination of amino acids
    d. all of these
    e. none of these

15. Before entering the urea cycle, ammonia is converted to carbamoyl phosphate. This requires how much ATP?
    a. 1 molecule b. 2 molecules c. 3 molecules
    d. 4 molecules e. none of these

16. Which of the following amino acids can be synthesized by the human body?
    a. valine b. lysine c. alanine
    d. tryptophan e. phenylalanine

17. The degradation products of the globin portion of hemoglobin contribute to the formation of:
    a. bile pigments b. ferritin molecules c. tetrapyrrole
    d. the amino acid pool e. none of these

18. The net effect of the transamination of amino acids is to produce the single amino acid:
    a. glutamate b. aspartate c. pyridoxine
    d. oxaloacetate e. none of these

19. The amino acids whose carbon skeletons are degraded to citric acid cycle intermediates are classified as:
    a. essential amino acids b. nonessential amino acids
    c. ketogenic amino acids d. glucogenic amino acids
    e. both glucogenic and ketogenic

20. Jaundice can occur because of:
    a. a negative nitrogen balance
    b. a low rate of protein turnover
    c. a low rate of heme degradation
    d. a low rate of bilirubin excretion by the liver
    e. none of these

Answers to Practice Exercises

26.1 a.

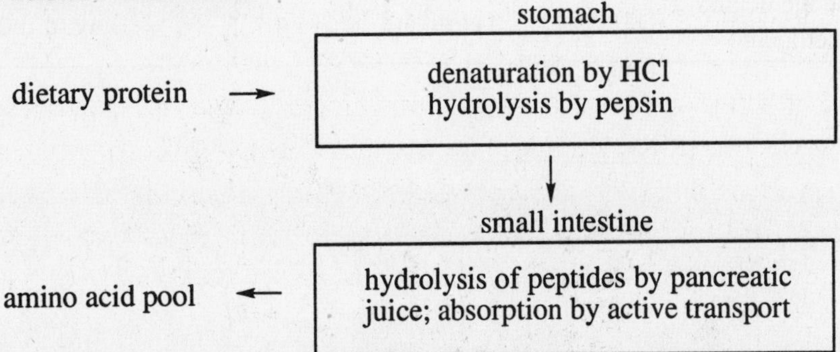

b. Four ways in which the amino acids from the amino acid pool are utilized:
1. energy production
2. biosynthesis of functional proteins
3. biosynthesis of nonessential amino acids
4. biosynthesis of nonprotein nitrogen-containing compounds

26.2 a.

$$CH_3{-}\underset{\overset{|}{\overset{+}{N}H_3}}{CH}{-}COO^- + {}^-OOC{-}CH_2{-}\overset{\overset{O}{\|}}{C}{-}COO^- \xrightarrow{\text{trans-aminase}} CH_3{-}\overset{\overset{O}{\|}}{C}{-}COO^- + {}^-OOC{-}CH_2{-}\overset{\overset{+}{N}H_3}{\overset{|}{C}H}{-}COO^-$$

b.

$${}^-OOC{-}CH_2{-}\overset{\overset{+}{N}H_3}{\overset{|}{C}H}{-}COO^- + NAD^+ + H_2O \longrightarrow \overset{+}{N}H_4 + {}^-OOC{-}CH_2{-}\overset{\overset{O}{\|}}{C}{-}COO^- + NADH + H^+$$

26.3
a. Step 4
b. Step 1
c. Step 4
d. Step 3
e. in the formation of carbamoyl phosphate (2) and Step 2 (2 high-energy bonds)
f. The equivalent of 4 ATP molecules (2 ATP → 2 ADP and 1 ATP → 1 AMP)

26.4

Classification of original amino acids	Products of degradation sequences	Further metabolic reactions
glucogenic	pyruvate and citric acid cycle intermediates	gluconeogenesis, ATP production
ketogenic	acetyl CoA, acetoacetyl CoA	production of ketone bodies or fatty acids, ATP production
glucogenic and ketogenic	pyruvate	any of the above

26.5 a. $^{-}OOC-CH_2-\overset{\overset{O}{\|}}{C}-COO^{-} \xrightarrow{\text{transaminase}} {}^{-}OOC-CH_2-\overset{\overset{\overset{+}{NH_3}}{|}}{CH}-COO^{-}$ aspartate

b. $CH_3-\overset{\overset{O}{\|}}{C}-COO^{-} \xrightarrow{\text{transaminase}} CH_3-\overset{\overset{\overset{+}{NH_3}}{|}}{CH}-COO^{-}$ alanine

c. $^{-}OOC-CH_2-CH_2-\overset{\overset{O}{\|}}{C}-COO^{-} \xrightarrow{\text{transaminase}} {}^{-}OOC-CH_2-CH_2-\overset{\overset{\overset{+}{NH_3}}{|}}{CH}-COO^{-}$ glutamate

26.6

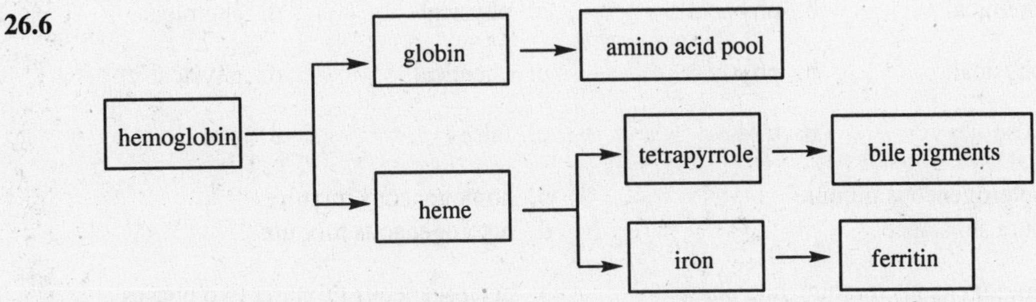

26.7

Nutrient class	Basic structural unit	Storage compounds	Degradation pathway
carbohydrates	monosaccharides	glycogen	glycolysis, citric acid cycle
lipids	fatty acids, glycerol	triacylglycerols	fatty acid spiral, citric acid cycle
proteins	amino acids	functional proteins (amino acids are not "stored")	transamination, oxidative deamination, citric acid cycle

Answers to Self-Test

The numbers in parentheses refer to sections in your textbook.
1. T (26.1) **2**. T (26.1) **3**. F; 7 and 8 (26.1) **4**. F; oxidative deamination (26.3)
5. T (26.3) **6**. F; ammonium ion (26.3) **7**. F; ketogenic (26.5) **8**. T (26.6)
9. F; none (26.6) **10**. F; negative (26.2) **11**. F; oxidize (26.6)
12. F; degradation of functional proteins (26.2) **13**. T (26.5) **14**. a (26.4)
15. b (26.4) **16**. c (26.6) **17**. d (26.7) **18**. a (26.3) **19**. d (26.3) **20**. d (26.7)

Solutions to Selected Problems

1.1 a. matter b. matter c. energy d. energy e. matter f. matter

1.3 a. indefinite shape versus definite shape b. indefinite volume versus definite volume

1.5 a. no b. no c. yes d. yes

1.7 a. physical b. chemical c. chemical d. physical

1.9 a. chemical b. physical c. chemical d. physical

1.11 a. chemical b. physical c. physical d. chemical

1.13 a. physical b. physical c. chemical d. physical

1.15 a. false b. true c. false d. true

1.17 a. heterogeneous mixture b. homogeneous mixture
c. pure substance d. heterogeneous mixture

1.19 a. homogeneous mixture, one phase b. heterogeneous mixture, two phases
c. heterogeneous mixture, three phases d. heterogeneous mixture, three phases

1.21 a. compound b. compound
c. classification not possible d. classification not possible

1.23 a. A, classification not possible; B, classification not possible; C, compound
b. D, compound; E, classification not possible; F, classification not possible; G, classification not possible

1.25 a. true b. false c. false d. false

1.27 a. false b. true c. false d. true

1.29 a. silver b. gold c. calcium d. sodium e. phosphorus f. sulfur

1.31 a. Sn b. Cu c. Al d. B e. Ba f. Ar

1.33 a. no b. yes c. yes d. no

1.35 a. true
b. false; triatomic molecules must contain at least one kind of atom
c. true
d. false; both homoatomic and heteroatomic molecules may contain three or more atoms

1.37 a. heteroatomic, diatomic, compound b. heteroatomic, triatomic, compound
c. homoatomic, diatomic, element d. heteroatomic, triatomic, compound

1.39 a.  b.  c. 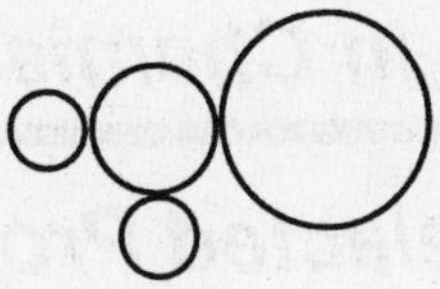 d.

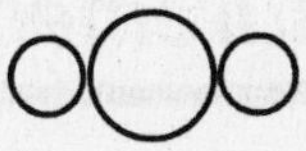

1.41 a. QX b. QZX c. X_2 d. X_2Q

1.43 a. compound b. compound c. element d. compound
e. compound f. compound g. element h. element

1.45 a. $C_8H_{10}N_4O_2$ b. $C_{12}H_{22}O_{11}$ c. HCN d. H_2SO_4

1.47 a. 3 elements, 2 H atoms, 1 C atom, 3 O atoms
b. 4 elements, 1 N atom, 4 H atoms, 1 Cl atom, 4 O atoms
c. 3 elements, 1 Ca atom, 1 S atom, 4 O atoms
d. 2 elements, 4 C atoms, 10 H atoms

1.49 a. solid is pulverized, granules are heated
b. discoloration, bursts into flame and burns

1.51 a. homogeneous mixture b. heterogeneous mixture
c. compound d. compound

1.53 a. B-Ar-Ba-Ra b. Eu-Ge-Ne c. He-At-H-Er d. Al-La-N

1.55 a. same, both 4 b. more, 6 and 5 c. same, both 5 d. fewer, 13 and 15

1.57 a. 2 (N_2, NH_3) b. 4 (N, H, C, Cl)
c. 110; 5(2 + 6 + 4 + 5 + 5) d. 56; 4(4 + 3 + 4 + 3)

Solutions to Selected Problems

2.1 a. kilo b. milli c. micro d. deci

2.3 a. centimeter b. kiloliter c. microliter d. nanogram

2.5 a. nanogram, milligram, centigram b. kilometer, megameter, gigameter c. picoliter, microliter, deciliter d. microgram, milligram, kilogram

2.7 a. exact b. exact c. inexact d. exact

2.9 a. inexact b. exact c. exact d. inexact

2.11 a. 0.1°C b. 0.01 mL c. 1 mL d. 0.1 mm

2.13 a. 4 b. 2 c. 4 d. 3 e. 5 f. 4

2.15 a. same b. different c. same d. same

2.17 a. last zero b. the 2 c. last 1 d. the 4 e. last zero f. last zero

2.19 a. ±0.001 b. ±0.0001 c. ±0.00001 d. ±100 e. ±0.001 f. ±0.00001

2.21 a. 0.351 b. 653,900 c. 22.556 d. 0.2777

2.23 a. 2 b. 2 c. 2 d. 2

2.25 a. 0.0080 b. 0.0143 c. 14 d. 0.182 e. 1.1 f. 5.72

2.27 a. 162 b. 9.3 c. 1261 d. 20.0

2.29 a. 1.207×10^2 b. 3.4×10^{-3} c. 2.3100×10^2 d. 2.3×10^4 e. 2.00×10^{-1} f. 1.011×10^{-1}

2.31 a. 1.0×10^{-3} b. 1.0×10^3 c. 6.3×10^4 d. 6.3×10^{-4}

2.33 a. 2 b. 3 c. 3 d. 4

2.35 a. 5.50×10^{12} b. 4.14×10^{-2} c. 1.5×10^4 d. 2.0×10^{-7} e. 1.5×10^{11} f. 1.2×10^6

2.37 a. $\dfrac{1\text{ kg}}{10^3\text{ g}}$ and $\dfrac{10^3\text{ g}}{1\text{ kg}}$ b. $\dfrac{1\text{ nm}}{10^{-9}\text{ m}}$ and $\dfrac{10^{-9}\text{ m}}{1\text{ nm}}$

c. $\dfrac{1\text{ mL}}{10^{-3}\text{ L}}$ and $\dfrac{10^{-3}\text{ L}}{1\text{ mL}}$ d. $\dfrac{1.00\text{ lb}}{454\text{ g}}$ and $\dfrac{454\text{ g}}{1.00\text{ lb}}$

e. $\dfrac{1.00\text{ km}}{0.621\text{ mi}}$ and $\dfrac{0.621\text{ mi}}{1.00\text{ km}}$ f. $\dfrac{1.00\text{ L}}{0.265\text{ gal}}$ and $\dfrac{0.265\text{ gal}}{1.00\text{ L}}$

2.39 a. $1.6 \times 10^3 \text{ dm} \times \left(\frac{10^{-1} \text{ m}}{1 \text{ dm}}\right) = 1.6 \times 10^2 \text{ m}$ b. $24 \text{ nm} \times \left(\frac{10^{-9} \text{ m}}{1 \text{ nm}}\right) = 2.4 \times 10^{-8} \text{ m}$

c. $0.003 \text{ km} \times \left(\frac{10^3 \text{ m}}{1 \text{ km}}\right) = 3 \text{ m}$ d. $3.0 \times 10^8 \text{ mm} \times \left(\frac{10^{-3} \text{ m}}{1 \text{ mm}}\right) = 3.0 \times 10^5 \text{ m}$

2.41 $2500 \text{ mL} \times \left(\frac{10^{-3} \text{ L}}{1 \text{ mL}}\right) = 2.5 \text{ L}$

2.43 $1550 \text{ g} \times \left(\frac{1.00 \text{ lb}}{454 \text{ g}}\right) = 3.41 \text{ lb}$

2.45 $25 \text{ mL} \times \left(\frac{10^{-3} \text{ L}}{1 \text{ mL}}\right) \times \left(\frac{0.265 \text{ gal}}{1.00 \text{ L}}\right) = 0.0066 \text{ gal}$

2.47 $83.2 \text{ kg} \times \left(\frac{2.20 \text{ lb}}{1.00 \text{ kg}}\right) = 183 \text{ lb}$

$1.92 \text{ m} \times \left(\frac{39.4 \text{ in.}}{1.00 \text{ m}}\right) \times \left(\frac{1 \text{ ft}}{12 \text{ in.}}\right) = 6.30 \text{ ft}$ (6 ft 4 in.)

2.49 $\frac{524.5 \text{ g}}{38.72 \text{ cm}^3} = 13.55 \frac{\text{g}}{\text{cm}^3}$

2.51 $20.0 \text{ g} \times \left(\frac{1 \text{ mL}}{0.791 \text{ g}}\right) = 25.3 \text{ mL}$

2.53 $236 \text{ mL} \times \left(\frac{1.03 \text{ g}}{1 \text{ mL}}\right) = 243 \text{ g}$

2.55 $\frac{5}{9}\left(525^\circ - 32^\circ\right) = 274^\circ\text{C}$

2.57 $\frac{9}{5}\left(-38.9^\circ\right) + 32.0^\circ = -38.0^\circ\text{F}$

2.59 $\frac{9}{5}\left(-10^\circ\right) + 32^\circ = 14^\circ\text{F}$; -10°C is higher

2.61 $0.63 \frac{\text{cal}}{\text{g}^\circ\text{C}} \times \left(\frac{4.184 \text{ J}}{1 \text{ cal}}\right) = 2.6 \frac{\text{J}}{\text{g}^\circ\text{C}}$

2.63 $\dfrac{18.6\text{ cal}}{12.0\text{ g x }10.0°\text{C}} = 0.155\ \dfrac{\text{cal}}{\text{g°C}}$

2.65 a. $0.057\ \dfrac{\text{cal}}{\text{g°C}} \text{ x } (42.0\text{ g}) \text{ x } (20.0°\text{C}) = 48\text{ cal}$

b. $1.00\ \dfrac{\text{cal}}{\text{g°C}} \text{ x } (42.0\text{ g}) \text{ x } (20.0°\text{C}) = 8.40 \text{ x } 10^2\text{ cal}$

c. $0.21\ \dfrac{\text{cal}}{\text{g°C}} \text{ x } (42.0\text{ g}) \text{ x } (20.0°\text{C}) = 180\text{ cal}$

2.67 The first 12 is an exact number (no uncertainty), and the second 12 is an inexact number (contains uncertainty).

2.69 a. $3.00 \text{ x } 10^{-3}$ b. $9.4 \text{ x } 10^5$ c. $2.35 \text{ x } 10^1$ d. $4.50000 \text{ x } 10^8$

2.71 a. 4 significant figures b. 4 significant figures
c. 3 significant figures d. exact

2.73 a. $75.0\text{ g x } \dfrac{1\text{ mL}}{0.56\text{ g}} = 1.3 \text{ x } 10^2\text{ mL}$

b. $75.0\text{ g x } \dfrac{1\text{ cm}^3}{0.93\text{ g}} \text{ x } \dfrac{1\text{ mL}}{1\text{ cm}^3} = 81\text{ mL}$

c. $75.0\text{ g x } \dfrac{1\text{ L}}{0.759\text{ g}} \text{ x } \dfrac{1\text{ mL}}{10^{-3}\text{ L}} = 9.88 \text{ x } 10^4\text{ mL}$

d. $75.0\text{ g x } \dfrac{1\text{ mL}}{13.6\text{ g}} = 5.51\text{ mL}$

2.75 a. $\dfrac{4.5\text{ mg}}{1\text{ mL}} \text{ x } \dfrac{1\text{ mL}}{10^{-3}\text{ L}} = 4.5 \text{ x } 10^3\text{ mg / L}$

b. $\dfrac{4.5\text{ mg}}{1\text{ mL}} \text{ x } \dfrac{10^{-3}\text{ g}}{1\text{ mg}} \text{ x } \dfrac{1\text{ pg}}{10^{-12}\text{ g}} = 4.5 \text{ x } 10^9\text{ pg / mL}$

c. $\dfrac{4.5\text{ mg}}{1\text{ mL}} \text{ x } \dfrac{10^{-3}\text{ g}}{1\text{ mg}} \text{ x } \dfrac{1\text{ mL}}{10^{-3}\text{ L}} = 4.5\text{ g / L}$

d. $\dfrac{4.5\text{ mg}}{1\text{ mL}} \text{ x } \dfrac{10^{-3}\text{ g}}{1\text{ mg}} \text{ x } \dfrac{1\text{ kg}}{10^3\text{ g}} \text{ x } \dfrac{1\text{ mL}}{1\text{ cm}^3} \text{ x } \dfrac{1\text{ cm}^3}{(10^{-2}\text{ m})^3} = 4.5\text{ kg / m}^3$

Solutions to Selected Problems

3.1 a. electron b. neutron c. proton d. proton

3.3 a. false b. false c. false d. true

3.5 a. atomic number = 2, mass number = 4
b. atomic number = 4, mass number = 9
c. atomic number = 5, mass number = 9
d. atomic number = 28, mass number = 58

3.7 a. 8 protons, 8 neutrons, 8 electrons
b. 8 protons, 10 neutrons, 8 electrons
c. 20 protons, 24 neutrons, 20 electrons
d. 100 protons, 157 neutrons, 100 electrons

3.9 a. atomic number
b. both atomic number and mass number
c. mass number
d. both atomic number and mass number

3.11 a. S, Cl, Ar, K b. Ar, K, Cl, S c. S, Cl, Ar, K d. S, Cl, K, Ar

3.13 a. 24, 29, 24, 53, 77
b. 101, 155, 101, 256, 357
c. 30, 37, 30, 67, 97
d. 20, 20, 20, 40, 60

3.15 $^{90}_{40}Zr$, $^{91}_{40}Zr$, $^{92}_{40}Zr$, $^{94}_{40}Zr$, $^{96}_{40}Zr$

3.17 a. false b. false c. true d. true

3.19 a. not the same b. same c. not the same d. same

3.21 8.91 x 12.0 amu = 106.92 amu = 107 amu (3 significant figures)

3.23 a. 0.0742 x 6.01 amu = 0.446 amu
0.9258 x 7.02 amu = 6.50 amu
6.946 amu = 6.95 amu (answer is limited to the hundredths place)

b. 0.7899 x 23.99 amu = 18.95 amu
0.1000 x 24.99 amu = 2.499 amu
0.1101 x 25.98 amu = 2.860 amu
24.309 amu = 24.31 amu (answer is limited to the hundredths place)

3.25 The exact number 12 applies only to ^{12}C. The number 12.011 is an average obtained by considering *all* naturally occurring isotopes of C.

3.27 a. Ca b. Mo c. Li d. Sn

3.29 a. 6 b. 28.09 amu c. 39 d. 9.01 amu

3.31 a. K, Rb b. P, As c. F, I d. Na, Cs

3.33 a. group b. periodic law c. periodic law d. group

3.35 a. alkali metal b. alkali metal c. noble gas d. alkaline earth metal
e. halogen f. noble gas g. alkaline earth metal h. halogen

3.37 a. no b. no c. yes d. yes

3.39 a. S b. P c. I d. Cl

3.41 a. metal b. nonmetal c. metal d. nonmetal

3.43 a. orbital b. orbital c. shell d. shell

3.45 a. true b. true c. false d. true

3.47 a. 2 b. 2 c. 6 d. 18

3.49 a. $1s^22s^22p^2$ b. $1s^22s^22p^63s^1$ c. $1s^22s^22p^63s^23p^4$ d. $1s^22s^22p^63s^23p^6$

3.51 a. oxygen b. neon c. aluminum d. calcium

3.53 a. $1s^22s^22p^63s^23p^5$ b. $1s^22s^22p^63s^23p^64s^23d^{10}4p^65s^24d^7$
c. $1s^22s^22p^63s^23p^64s^2$ d. $1s^22s^22p^63s^23p^64s^23d^1$

3.55 a. [↓↑] [↓↑] [↑][↑][ ]

b. [↓↑] [↓↑] [↓↑][↓↑][↓↑] [↓↑]

c. [↓↑] [↓↑] [↓↑][↓↑][↓↑] [↓↑] [↑][↑][↑]

d. [↓↑] [↓↑] [↓↑][↓↑][↓↑] [↓↑] [↓↑][↓↑][↓↑] [↓↑] [↓↑][↓↑][↑][↑][↑]

3.57 a. 3 b. 0 c. 1 d. 5

3.59 a. no b. yes c. no d. yes

3.61 a. p^1 b. d^3 c. s^2 d. p^6

3.63 a. representative b. noble gas c. transition d. inner transition

3.65 a. noble gas b. representative c. transition d. representative

3.67 a. true b. false c. false d. false

3.69 a. $^{44}_{20}Ca$ b. $^{211}_{86}Rn$ c. $^{110}_{47}Ag$ d. $^{9}_{4}Be$

3.71 a. same b. different c. different d. same

3.73 9[(12 x 6) + (22 x 1) + (11 x 8)] = 1638 electrons

3.75 the same

3.77 a. $_{8}O$ b. $_{10}Ne$ c. $_{30}Zn$ d $_{12}Mg$

Chemical Bonding: The Ionic Bond Model — Chapter 4

Solutions to Selected Problems

4.1 a. 2 b. 2 c. 3 d. 4

4.3 a. group IA, 1 valence electron b. group VIIIA, 8 valence electrons c. group IIA, 2 valence electrons d. group VIIA, 7 valence electrons

4.5 a. $1s^22s^22p^2$ b. $1s^22s^22p^5$ c. $1s^22s^22p^63s^2$ d. $1s^22s^22p^63s^23p^3$

4.7 a. ·Mg· b. K· c. ·P̈· d. :K̈r:

4.9 a. Li b. F c. Be d. N

4.11 a. O^{2-} b. Mg^{2+} c. F^- d. Al^{3+}

4.13 a. Ca^{2+} b. O^{2-} c. Na^+ d. Al^{3+}

4.15 a. 15p, 18e b. 7p, 10e c. 12p, 10e d. 3p, 2e

4.17 a. 2+ b. 3– c. 1+ d. 1–

4.19 a. 2 lost b. 1 gained c. 2 lost d. 2 gained

4.21 a. neon b. argon c. argon d. argon

4.23 a. group IIA b. group VIA c. group VA d. group IA

4.25 a $1s^22s^22p^63s^23p^1$ b. $1s^22s^22p^6$

4.27 a. 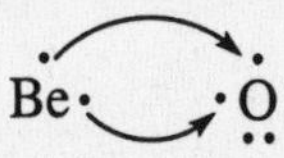 Be: ⟶ ·Ö: b. 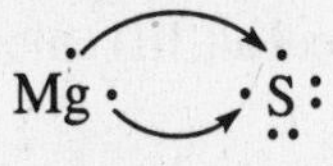 Mg: ⟶ ·S̈:

c. 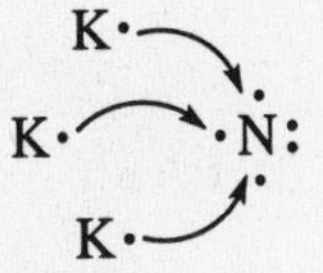 K· K· K· ⟶ ·N: d. 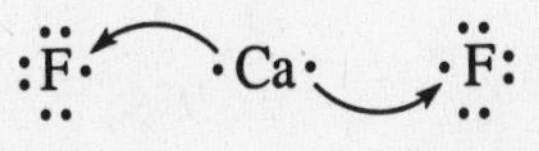 :F̈· ⟵ ·Ca· ⟶ ·F̈:

4.29 a. $BaCl_2$ b. $BaBr_2$ c. Ba_3N_2 d. BaO

4.31 a. MgF_2 b. BeF_2 c. LiF d. AlF_3

4.33 a. Na_2S b. CaI_2 c. Li_3N d. $AlBr_3$

4.35 a. ionic b. ionic c. not ionic d. ionic

4.37 a. ionic b. not ionic c. ionic d. not ionic

4.39 a. potassium iodide b. beryllium oxide c. aluminum fluoride d. sodium phosphide

4.41 a. +1 b. +2 c. +4 d. +2

4.43 a. iron(II) oxide b. gold(III) oxide c. copper(II) sulfide d. cobalt(II) bromide

4.45 a. gold(I) chloride b. potassium chloride c. silver chloride d. copper(II) chloride

4.47 a. KBr b. Ag_2O c. BeF_2 d. Ba_3P_2

4.49 a. CoS b. Co_2S_3 c. SnI_4 d. Pb_3N_2

4.51 a. SO_4^{2-} b. ClO_3^- c. OH^- d. CN^-

4.53 a. PO_4^{3-} and HPO_4^{2-} b. NO_3^- and NO_2^- c. H_3O^+ and OH^- d. CrO_4^{2-} and $Cr_2O_7^{2-}$

4.55 a. $NaClO_4$ b. $Fe(OH)_3$ c. $Ba(NO_3)_2$ d. $Al_2(CO_3)_3$

4.57 a. magnesium carbonate b.. zinc sulfate c. beryllium nitrate d. silver phosphate

4.59 a. iron(II) hydroxide b. copper(II) carbonate c. gold(I) cyanide d. manganese(II) phosphate

4.61 a. $KHCO_3$ b. $Au_2(SO_4)_3$ c. $AgNO_3$ d. $Cu_3(PO_4)_2$

4.63 a. Na^+ b. F^- c. S^{2-} d. Ca^{2+}

4.65 a. S b. Mg c. P d. Al

4.67 a. K^+, Cl^- b. Ca^{2+}, S^{2-} c. Be^{2+}, two F^- d. two Al^{3+}, three S^{2-}

4.69 a. tin(IV) chloride, tin(II) chloride b. iron(II) sulfide, iron(III) sulfide c. copper(I) nitride, copper(II) nitride d. nickel(II) iodide, nickel(III) iodide

4.71 a. copper(I) nitrate, copper(II) nitrate
b. lead(II) phosphate, lead(IV) phosphate
c. manganese(III) cyanide, manganese(II) cyanide
d. cobalt(II) chlorate, cobalt(III) chlorate

Chemical Bonding: The Covalent Bond Model

Chapter 5

Solutions to Selected Problems

5.1 a. :Br:Br: b. H:I: c. :I:Br: d. :Br:F:

5.3 a. 2 b. 2 c. 0 d. 6

5.5 a. one triple bond b. three single bonds
c. one double and two single bonds d. one double and four single bonds

5.7 a. :N≡N: b. H—C=C—H (each C bonded to H) c. H—C—H (C=O, :O:) d. H—O—O—H

5.9 a. NF_3 b. Cl_2O c. H_2S d. CH_4

5.11 a. N b. C c. N d. C

5.13 Oxygen forms three bonds instead of the normal two.

5.15 a. 2 (Cl) + 1 (O) = 2(7) + 1(6) = 20 electrons
b. 2 (H) + 1 (S) = 2(1) + 1(6) = 8 electrons
c. 1 (N) + 3 (H) = 1(5) + 3(1) = 8 electrons
d. 1 (S) + 3 (O) = 1(6) + 3(6) = 24 electrons

5.17 a. H:P:H (H below P) b. :Cl:P:Cl: (:Cl: below P) c. :Br:Si:Br: (:Br: above and below Si) d. :F:O:F:

5.19 a. :F:S:F: b. :I:C:I: (:I: above and below C) c. :Br:N:Br: (:Br: below N) d. H:Se:H

5.21 a. H:C::C::C:H (H below each terminal C) b. :F:N::N:F:

c. H:C:C:::N: (H above and below first C) d. H:C:C:::C:H (H above and below first C)

5.23 a. [:O:H]$^-$ b. [H:Be:H (H above and below Be)]$^{2-}$ c. [:Cl:Al:Cl: (:Cl: above and below Al)]$^-$ d. [:O:N:O: (:O below N)]$^-$

5.25 a. $Na^+[:C:::N:]^-$ b. $3[K^+]\left[\begin{array}{c} :\ddot{O}: \\ :\ddot{O}:\ddot{P}:\ddot{O}: \\ :\ddot{O}: \end{array}\right]^{3-}$

5.27 a. angular b. angular c. angular d. linear

5.29 a. trigonal pyramidal b. trigonal planar c. tetrahedral d. tetrahedral

5.31 a. trigonal pyramidal b. tetrahedral c. angular d. angular

5.33 a. trigonal planar about each carbon atom
b. tetrahedral about carbon atom and angular about oxygen atom

5.35 a. Na, Mg, Al, P b. I, Br, Cl, F c. Al, P, S, O d. Ca, Mg, C, O

5.37 a. Br, Cl, N, O, F b. K, Na, Ca, Li c. Cl, N, O, F d. 0.5

5.39 a. $\overset{\delta+}{B}—\overset{\delta-}{N}$ b. $\overset{\delta+}{Cl}—\overset{\delta-}{F}$ c. $\overset{\delta-}{N}—\overset{\delta+}{C}$ d. $\overset{\delta-}{F}—\overset{\delta+}{O}$

5.41 a. H–Br, H–Cl, H–O b. O–F, P–O, Al–O
c. Br–Br, H–Cl, B–N d. P–N, S–O, Br–F

5.43 a. polar covalent b. ionic c. nonpolar covalent d. polar covalent

5.45 a. nonpolar b. polar c. polar d. polar

5.47 a. nonpolar b. polar c. polar d. polar

5.49 a. polar b. polar c. nonpolar d. polar

5.51 a. sulfur tetrafluoride b. tetraphosphorus hexoxide
c. chlorine dioxide d. hydrogen sulfide

5.53 a. ICl b. N_2O c. NCl_3 d. HBr

5.55 a. H_2O_2 b. CH_4 c. NH_3 d. PH_3

5.57 a. 26 b. 24 c. 14 d. 8

5.59 a. not enough electron dots
b. not enough electron dots
c. improper placement of a correct number of electron dots
d. too many electron dots

5.61 a. can't classify; you do not know the geometry
b. nonpolar; polarity requires the presence of polar bonds
c. can't classify; you do not know the geometry
d. polar; a nonpolar bond and a polar bond cannot "cancel" each other

5.63 DA, DC, CA, CB

5.65 a. sodium chloride b. bromine monochloride
c. potassium sulfide d. dichlorine monoxide

Chemical Calculations: Formula Masses, Moles, and Chemical Equations

Chapter 6

Solutions to Selected Problems

6.1 a. [12(12.01) + 22(1.01) + 11(16.00)] amu = 342.34 amu
b. [7(12.01) + 16(1.01)] amu = 100.23 amu
c. [7(12.01) + 5(1.01) + 14.01 + 3(16.00) + 32.07] amu = 183.20 amu
d. [2(14.01) + 8(1.01) + 32.07 + 4(16.00)] amu = 132.17 amu

6.3 a. 6.02×10^{23} apples
b. 6.02×10^{23} elephants
c. 6.02×10^{23} Zn atoms
d. 6.02×10^{23} CO_2 molecules

6.5 a. $1.50 \text{ moles Fe} \times \left(\frac{6.02 \times 10^{23} \text{ atoms Fe}}{1 \text{ mole Fe}}\right) = 9.03 \times 10^{23} \text{ atoms Fe}$

b. $1.50 \text{ moles Ni} \times \left(\frac{6.02 \times 10^{23} \text{ atoms Ni}}{1 \text{ mole Ni}}\right) = 9.03 \times 10^{23} \text{ atoms Ni}$

c. $1.50 \text{ moles C} \times \left(\frac{6.02 \times 10^{23} \text{ atoms C}}{1 \text{ mole C}}\right) = 9.03 \times 10^{23} \text{ atoms C}$

d. $1.50 \text{ moles Ne} \times \left(\frac{6.02 \times 10^{23} \text{ atoms Ne}}{1 \text{ mole Ne}}\right) = 9.03 \times 10^{23} \text{ atoms Ne}$

6.7 a. 0.200 mole
b. Avogadro's number
c. 1.50 moles
d. 6.50×10^{23} atoms

6.9 a. [12.01 + 16.00] g = 28.0 g
b. [12.01 + 2(16.00)] g = 44.0 g
c. [22.99 + 35.45] g = 58.4 g
d. [12(12.01) + 22(1.01) + 11(16.00)] g = 342 g

6.11 a. $0.034 \text{ mole Au} \times \left(\frac{196.97 \text{ g Au}}{1 \text{ mole Au}}\right) = 6.7 \text{ g Au}$

b. $0.034 \text{ mole Ag} \times \left(\frac{107.87 \text{ g Ag}}{1 \text{ mole Ag}}\right) = 3.7 \text{ g Ag}$

c. $3.00 \text{ moles O} \times \left(\frac{16.00 \text{ g O}}{1 \text{ mole O}}\right) = 48.0 \text{ g O}$

d. $3.00 \text{ moles } O_2 \times \left(\frac{32.00 \text{ g } O_2}{1 \text{ mole } O_2}\right) = 96.0 \text{ g } O_2$

6.13 a. $5.00 \text{ g CO} \times \left(\frac{1 \text{ mole CO}}{28.01 \text{ g CO}}\right) = 0.179 \text{ mole CO}$

b. $5.00 \text{ g } CO_2 \times \left(\frac{1 \text{ mole } CO_2}{44.01 \text{ g } CO_2}\right) = 0.114 \text{ mole } CO_2$

c. $5.00 \text{ g } B_4H_{10} \times \left(\frac{1 \text{ mole } B_4H_{10}}{53.34 \text{ g } B_4H_{10}}\right) = 0.0937 \text{ mole } B_4H_{10}$

d. $5.00 \text{ g U} \times \left(\frac{1 \text{ mole U}}{238 \text{ g U}}\right) = 0.0210 \text{ mole U}$

6.15 a. $\frac{2 \text{ moles H}}{1 \text{ mole } H_2SO_4}, \frac{1 \text{ mole } H_2SO_4}{2 \text{ moles H}}, \frac{1 \text{ mole S}}{1 \text{ mole } H_2SO_4}, \frac{1 \text{ mole } H_2SO_4}{1 \text{ mole S}},$

$\frac{4 \text{ moles O}}{1 \text{ mole } H_2SO_4}, \frac{1 \text{ mole } H_2SO_4}{4 \text{ moles O}}$

b. $\frac{1 \text{ mole P}}{1 \text{ mole } POCl_3}, \frac{1 \text{ mole } POCl_3}{1 \text{ mole P}}, \frac{1 \text{ mole O}}{1 \text{ mole } POCl_3}, \frac{1 \text{ mole } POCl_3}{1 \text{ mole O}},$

$\frac{3 \text{ moles Cl}}{1 \text{ mole } POCl_3}, \frac{1 \text{ mole } POCl_3}{3 \text{ moles Cl}}$

6.17 a. $2.00 \text{ mole } SO_2 \times \left(\frac{1 \text{ mole S}}{1 \text{ mole } SO_2}\right) = 2.00 \text{ moles S}$

$2.00 \text{ mole } SO_2 \times \left(\frac{2 \text{ moles O}}{1 \text{ mole } SO_2}\right) = 4.00 \text{ moles O}$

b. $2.00 \text{ mole } SO_3 \times \left(\frac{1 \text{ mole S}}{1 \text{ mole } SO_3}\right) = 2.00 \text{ moles S}$

$2.00 \text{ mole } SO_3 \times \left(\frac{3 \text{ moles O}}{1 \text{ mole } SO_3}\right) = 6.00 \text{ moles O}$

c. $3.00 \text{ mole } NH_3 \times \left(\frac{1 \text{ mole N}}{1 \text{ mole } NH_3}\right) = 3.00 \text{ moles N}$

$3.00 \text{ mole } NH_3 \times \left(\frac{3 \text{ moles H}}{1 \text{ mole } NH_3}\right) = 9.00 \text{ moles H}$

d. $3.00 \text{ mole } N_2H_4 \times \left(\frac{2 \text{ mole N}}{1 \text{ mole } N_2H_4}\right) = 6.00 \text{ moles N}$

$3.00 \text{ mole } N_2H_4 \times \left(\frac{4 \text{ moles H}}{1 \text{ mole } N_2H_4}\right) = 12.0 \text{ moles H}$

6.19 a. $4.00 \text{ moles } SO_3 \times \left(\frac{4 \text{ moles atoms}}{1.00 \text{ mole } SO_3}\right) = 16.0 \text{ moles atoms}$

b. $2.00 \text{ moles } H_2SO_4 \times \left(\frac{7 \text{ moles atoms}}{1.00 \text{ mole } H_2SO_4}\right) = 14.0 \text{ moles atoms}$

c. $1.00 \text{ mole } C_{12}H_{22}O_{11} \times \left(\frac{45 \text{ moles atoms}}{1.00 \text{ mole } C_{12}H_{22}O_{11}}\right) = 45.0 \text{ moles atoms}$

d. $3.00 \text{ mole } Mg(OH)_2 \times \left(\frac{5 \text{ moles atoms}}{1.00 \text{ mole } Mg(OH)_2}\right) = 15.0 \text{ moles atoms}$

6.21 a. $10.0 \text{ g B} \times \left(\frac{1 \text{ mole B}}{10.81 \text{ g B}}\right) \times \left(\frac{6.02 \times 10^{23} \text{ atoms B}}{1 \text{ mole B}}\right) = 5.57 \times 10^{23} \text{ atoms B}$

b. $32.0 \text{ g Ca} \times \left(\frac{1 \text{ mole Ca}}{40.08 \text{ g Ca}}\right) \times \left(\frac{6.02 \times 10^{23} \text{ atoms Ca}}{1 \text{ mole Ca}}\right) = 4.81 \times 10^{23} \text{ atoms Ca}$

c. $2.0 \text{ g Ne} \times \left(\frac{1 \text{ mole Ne}}{20.18 \text{ g Ne}}\right) \times \left(\frac{6.02 \times 10^{23} \text{ atoms Ne}}{1 \text{ mole Ne}}\right) = 6.0 \times 10^{22} \text{ atoms Ne}$

d. $7.0 \text{ g N} \times \left(\frac{1 \text{ mole N}}{14.01 \text{ g N}}\right) \times \left(\frac{6.02 \times 10^{23} \text{ atoms N}}{1 \text{ mole N}}\right) = 3.0 \times 10^{23} \text{ atoms N}$

6.23 a. $6.02 \times 10^{23} \text{ atoms Cu} \times \left(\frac{1 \text{ mole Cu}}{6.02 \times 10^{23} \text{ atoms Cu}}\right) \times \left(\frac{63.55 \text{ g Cu}}{1 \text{ mole Cu}}\right) = 63.6 \text{ g Cu}$

b. $3.01 \times 10^{23} \text{ atoms Cu} \times \left(\frac{1 \text{ mole Cu}}{6.02 \times 10^{23} \text{ atoms Cu}}\right) \times \left(\frac{63.55 \text{ g Cu}}{1 \text{ mole Cu}}\right) = 31.8 \text{ g Cu}$

c. $557 \text{ atoms Cu} \times \left(\frac{1 \text{ mole Cu}}{6.02 \times 10^{23} \text{ atoms Cu}}\right) \times \left(\frac{63.55 \text{ g Cu}}{1 \text{ mole Cu}}\right) = 5.88 \times 10^{-20} \text{ g Cu}$

d. $1 \text{ atom Cu} \times \left(\frac{1 \text{ mole Cu}}{6.02 \times 10^{23} \text{ atoms Cu}}\right) \times \left(\frac{63.55 \text{ g Cu}}{1 \text{ mole Cu}}\right) = 1.06 \times 10^{-22} \text{ g Cu}$

6.25 a. $10.0 \text{ g He} \times \left(\frac{1 \text{ mole He}}{4.00 \text{ g He}}\right) = 2.50 \text{ moles He}$

b. $10.0 \text{ g } N_2O \times \left(\frac{1 \text{ mole } N_2O}{44.02 \text{ g } N_2O}\right) = 0.227 \text{ moles He}$

c. $4.0 \times 10^{10} \text{ atoms P} \times \left(\frac{1 \text{ mole P}}{6.02 \times 10^{23} \text{ atoms P}}\right) = 6.6 \times 10^{-14} \text{ mole P}$

d. $4.0 \times 10^{10} \text{ atoms Be} \times \left(\frac{1 \text{ mole Be}}{6.02 \times 10^{23} \text{ atoms Be}}\right) = 6.6 \times 10^{-14} \text{ mole Be}$

6.27 a. $10.0 \text{ g } H_2SO_4 \times \left(\frac{1 \text{ mole } H_2SO_4}{98.09 \text{ g } H_2SO_4}\right) \times \left(\frac{1 \text{ mole S}}{1 \text{ mole } H_2SO_4}\right) \times \left(\frac{6.02 \times 10^{23} \text{ atoms S}}{1 \text{ mole S}}\right)$

$= 6.14 \times 10^{22} \text{ atoms S}$

b. $20.0 \text{ g } SO_3 \times \left(\frac{1 \text{ mole } SO_3}{80.07 \text{ g } SO_3}\right) \times \left(\frac{1 \text{ mole S}}{1 \text{ mole } SO_3}\right) \times \left(\frac{6.02 \times 10^{23} \text{ atoms S}}{1 \text{ mole S}}\right)$

$= 1.50 \times 10^{23} \text{ atoms S}$

c. $30.0 \text{ g } Al_2S_3 \times \left(\frac{1 \text{ mole } Al_2S_3}{150.17 \text{ g } Al_2S_3}\right) \times \left(\frac{3 \text{ moles S}}{1 \text{ mole } Al_2S_3}\right) \times \left(\frac{6.02 \times 10^{23} \text{ atoms S}}{1 \text{ mole S}}\right)$

$= 3.61 \times 10^{23} \text{ atoms S}$

d. $2 \text{ moles } S_2O \times \left(\frac{2 \text{ moles S}}{1 \text{ mole } S_2O}\right) \times \left(\frac{6.02 \times 10^{23} \text{ atoms S}}{1 \text{ mole S}}\right) = 2.41 \times 10^{24} \text{ atoms S}$

6.29 a. $3.01 \times 10^{23} \text{ molecules } S_2O \times \left(\frac{1 \text{ mole } S_2O}{6.02 \times 10^{23} \text{ molecules } S_2O}\right) \times \left(\frac{2 \text{ moles S}}{1 \text{ mole } S_2O}\right) \times \left(\frac{32.07 \text{ g S}}{1 \text{ mole S}}\right)$

$= 32.1 \text{ g S}$

b. $3 \text{ molecules } S_4N_4 \times \left(\frac{1 \text{ mole } S_4N_4}{6.02 \times 10^{23} \text{ molecules } S_4N_4}\right) \times \left(\frac{4 \text{ moles S}}{1 \text{ mole } S_4N_4}\right) \times \left(\frac{32.07 \text{ g S}}{1 \text{ mole S}}\right)$

$= 6.39 \times 10^{-22} \text{ g S}$

c. $2.00 \text{ moles } SO_2 \times \left(\frac{1 \text{ mole S}}{1 \text{ mole } SO_2}\right) \times \left(\frac{32.07 \text{ g S}}{1 \text{ mole S}}\right) = 64.1 \text{ g S}$

d. $4.50 \text{ moles } S_8 \times \left(\frac{8 \text{ mole S}}{1 \text{ mole } S_8}\right) \times \left(\frac{32.07 \text{ g S}}{1 \text{ mole S}}\right) = 1150 \text{ g S}$

6.31 a. balanced b. balanced c. not balanced d. balanced

6.33 a. 4 N, 6 O b. 10 N, 12 H, 6 O
c. 1 P, 3 Cl, 6 H d. 2 Al, 3 O, 6 H, 6 Cl

6.35 a. $2Na + 2H_2O \rightarrow 2NaOH + H_2$ b. $2Na + ZnSO_4 \rightarrow Na_2SO_4 + Zn$
c. $2NaBr + Cl_2 \rightarrow 2NaCl + Br_2$ d. $2ZnS + 3O_2 \rightarrow 2ZnO + 2SO_2$

6.37 a. $CH_4 + 2O_2 \rightarrow CO_2 + 2H_2O$ b. $2C_6H_6 + 15O_2 \rightarrow 12CO_2 + 6H_2O$

c. $C_4H_8O_2 + 5O_2 \rightarrow 4CO_2 + 4H_2O$ d. $C_5H_{10}O + 7O_2 \rightarrow 5CO_2 + 5H_2O$

6.39 a. $3PbO + 2NH_3 \rightarrow 3Pb + N_2 + 3H_2O$

b. $2Fe(OH)_3 + 3H_2SO_4 \rightarrow Fe_2(SO_4)_3 + 6H_2O$

6.41 $\frac{2 \text{ moles } Ag_2CO_3}{4 \text{ moles Ag}}, \frac{2 \text{ moles } Ag_2CO_3}{2 \text{ moles } CO_2}, \frac{2 \text{ moles } Ag_2CO_3}{1 \text{ mole } O_2}, \frac{4 \text{ moles Ag}}{2 \text{ moles } CO_2}, \frac{4 \text{ moles Ag}}{1 \text{ mole } O_2},$

$\frac{2 \text{ moles } CO_2}{1 \text{ mole } O_2}$; the other six are reciprocals of these six factors

6.43 a. $2.00 \text{ moles } C_7H_{16} \text{ x} \left(\frac{7 \text{ moles } CO_2}{1 \text{ mole } C_7H_{16}}\right) = 14.0 \text{ moles } CO_2$

b. $2.00 \text{ moles HCl x} \left(\frac{1 \text{ mole } CO_2}{2.00 \text{ moles HCl}}\right) = 1.00 \text{ mole } CO_2$

c. $2.00 \text{ moles } Na_2SO_4 \text{ x} \left(\frac{2 \text{ moles } CO_2}{1.00 \text{ mole } Na_2SO_4}\right) = 4.00 \text{ moles } CO_2$

d. $2.00 \text{ moles } Fe_3O_4 \text{ x} \left(\frac{1 \text{ mole } CO_2}{1.00 \text{ mole } Fe_3O_4}\right) = 2.00 \text{ moles } CO_2$

6.45 a. $20.0 \text{ g } N_2 \text{ x} \left(\frac{1 \text{ mole } N_2}{28.02 \text{ g } N_2}\right) \text{ x} \left(\frac{4 \text{ moles } NH_3}{2 \text{ moles } N_2}\right) \text{ x} \left(\frac{17.04 \text{ g } NH_3}{1 \text{ mole } NH_3}\right) = 24.3 \text{ g } NH_3$

b. $20.0 \text{ g } N_2 \text{ x} \left(\frac{1 \text{ mole } N_2}{28.02 \text{ g } N_2}\right) \text{ x} \left(\frac{1 \text{ mole } (NH_4)_2Cr_2O_7}{1 \text{ mole } N_2}\right) \text{ x} \left(\frac{252.10 \text{ g } (NH_4)_2Cr_2O_7}{1 \text{ mole } (NH_4)_2Cr_2O_7}\right)$

$= 1.80 \times 10^2 \text{ g } (NH_4)_2Cr_2O_7$

c. $20.0 \text{ g } N_2 \text{ x} \left(\frac{1 \text{ mole } N_2}{28.02 \text{ g } N_2}\right) \text{ x} \left(\frac{1 \text{ mole } N_2H_4}{1 \text{ mole } N_2}\right) \text{ x} \left(\frac{32.06 \text{ g } N_2H_4}{1 \text{ mole } N_2H_4}\right) = 22.9 \text{ g } N_2H_4$

d. $20.0 \text{ g } N_2 \text{ x} \left(\frac{1 \text{ mole } N_2}{28.02 \text{ g } N_2}\right) \text{ x} \left(\frac{2 \text{ moles } NH_3}{1 \text{ mole } N_2}\right) \text{ x} \left(\frac{17.04 \text{ g } NH_3}{1 \text{ mole } NH_3}\right) = 24.3 \text{ g } NH_3$

6.47 $3.50 \text{ g } CO_2 \text{ x} \left(\frac{1 \text{ mole } CO_2}{44.01 \text{ g } CO_2}\right) \text{ x} \left(\frac{2 \text{ moles } O_2}{1 \text{ mole } CO_2}\right) \text{ x} \left(\frac{32.00 \text{ g } O_2}{1 \text{ mole } O_2}\right) = 5.09 \text{ g } O_2$

6.49 $25.0 \text{ g CO x} \left(\frac{1 \text{ mole CO}}{28.01 \text{ g CO}}\right) \text{ x} \left(\frac{1 \text{ mole } O_2}{2 \text{ moles CO}}\right) \text{ x} \left(\frac{32.00 \text{ g } O_2}{1 \text{ mole } O_2}\right) = 14.3 \text{ g } O_2$

6.51 $10.0 \text{ g } SO_2 \times \left(\frac{1 \text{ mole } SO_2}{64.07 \text{ g } SO_2}\right) \times \left(\frac{2 \text{ moles } H_2O}{1 \text{ mole } SO_2}\right) \times \left(\frac{18.02 \text{ g } H_2O}{1 \text{ mole } H_2O}\right) = 5.63 \text{ g } H_2O$

6.53
$$\begin{aligned}[3(12.01) + y(1.01) + 32.07] \text{ amu} &= 76.18 \text{ amu} \\ 1.01y &= 8.08 \\ y &= 8.00\end{aligned}$$

6.55 a. $1.000 \text{ g Si} \times \left(\frac{1 \text{ mole Si}}{28.09 \text{ g Si}}\right) \times \left(\frac{1 \text{ mole } SiH_4}{1 \text{ mole Si}}\right) = 0.03560 \text{ mole } SiH_4$

b. $1.000 \text{ g Si} \times \left(\frac{1 \text{ mole Si}}{28.09 \text{ g Si}}\right) \times \left(\frac{1 \text{ mole } SiO_2}{1 \text{ mole Si}}\right) \times \left(\frac{60.09 \text{ g } SiO_2}{1 \text{ mole } SiO_2}\right) = 2.139 \text{ g } SiO_2$

c. $1.000 \text{ g Si} \times \left(\frac{1 \text{ mole Si}}{28.09 \text{ g Si}}\right) \times \left(\frac{1 \text{ mole } (CH_3)_3SiCl}{1 \text{ mole Si}}\right) \times \left(\frac{6.022 \times 10^{23} \text{ molecules } (CH_3)_3SiCl}{1 \text{ mole } (CH_3)_3SiCl}\right)$

$= 2.144 \times 10^{22} \text{ molecules } (CH_3)_3SiCl$

d. $1.000 \text{ g Si} \times \left(\frac{1 \text{ mole Si}}{28.09 \text{ g Si}}\right) \times \left(\frac{6.022 \times 10^{23} \text{ atoms Si}}{1 \text{ mole Si}}\right) = 2.144 \times 10^{22} \text{ atoms Si}$

6.57 Because the oxygen is balanced with 22 atoms on each side of the equation, the compound butyne contains only C and H.

$$2C_xH_y + 11O_2 \rightarrow 8CO_2 + 6H_2O$$

Carbon balance:	$2x = 8$	$x = 4$
Hydrogen balance:	$2y = 6(2)$	$y = 6$

Butyne has the formula C_4H_6.

6.59 $125 \text{ g } Ag_2S \times \left(\frac{1 \text{ mole } Ag_2S}{248 \text{ g } Ag_2S}\right) \times \left(\frac{2 \text{ moles Ag}}{1 \text{ mole } Ag_2S}\right) \times \left(\frac{108 \text{ g Ag}}{1 \text{ mole Ag}}\right) = 109 \text{ g Ag}$

$125 \text{ g } Ag_2S \times \left(\frac{1 \text{ mole } Ag_2S}{248 \text{ g } Ag_2S}\right) \times \left(\frac{1 \text{ mole S}}{1 \text{ mole } Ag_2S}\right) \times \left(\frac{32.1 \text{ g S}}{1 \text{ mole S}}\right) = 16.2 \text{ g S}$

Solutions to Selected Problems

7.1 a. velocity increases with increasing temperature
b. potential energy
c. disruptive force magnitude increases with increasing temperature
d. gaseous state

7.3 a. Cohesive forces are strong enough that the space between particles changes very little with a temperature increase.
b. Particles of a gas are widely separated because disruptive forces are greater than cohesive forces.

7.5 a. $735 \text{ mm of Hg} \times \left(\frac{1 \text{ atm}}{760 \text{ mm Hg}}\right) = 0.967 \text{ atm}$

b. $0.530 \text{ atm} \times \left(\frac{760 \text{ mm Hg}}{1 \text{ atm}}\right) = 403 \text{ mm Hg}$

c. $0.530 \text{ atm} \times \left(\frac{760 \text{ torr}}{1 \text{ atm}}\right) = 403 \text{ torr}$

d. $12.0 \text{ psi} \times \left(\frac{1 \text{ atm}}{14.7 \text{ psi}}\right) = 0.816 \text{ atm}$

7.7 $P_2 = 3.0 \text{ atm} \times \left(\frac{6.0 \text{ L}}{2.5 \text{ L}}\right) = 7.2 \text{ atm}$

7.9 $V_2 = 3.00 \text{ L} \times \left(\frac{655 \text{ mm Hg}}{725 \text{ mm Hg}}\right) = 2.71 \text{ L}$

7.11 $V_2 = 2.73 \text{ L} \times \left(\frac{400 \text{ K}}{300 \text{ K}}\right) = 3.64 \text{ L}$

7.13 $T_2 = 298 \text{ K} \times \left(\frac{525 \text{ mL}}{375 \text{ mL}}\right) = 417 \text{ K}$

$417 \text{ K} - 273 = 144^\circ\text{C}$

7.15 a. $T_1 = T_2 \times \frac{P_1}{P_2} \times \frac{V_1}{V_2}$

b. $P_2 = P_1 \times \frac{V_1}{V_2} \times \frac{T_2}{T_1}$

c. $V_1 = V_2 \times \frac{P_2}{P_1} \times \frac{T_1}{T_2}$

7.17 a. $V_2 = 15.2\text{ L} \times \left(\frac{1.35\text{ atm}}{3.50\text{ atm}}\right) \times \left(\frac{308\text{ K}}{306\text{ K}}\right) = 5.90\text{ L}$

b. $P_2 = 1.35\text{ atm} \times \left(\frac{15.2\text{ L}}{10.0\text{ L}}\right) \times \left(\frac{315\text{ K}}{306\text{ K}}\right) = 2.11\text{ atm}$

c. $T_2 = 306\text{ K} \times \left(\frac{7.00\text{ atm}}{1.35\text{ atm}}\right) \times \left(\frac{0.973\text{ L}}{15.2\text{ L}}\right) = 102\text{ K}$

$102\text{ K} - 273 = -171^\circ\text{C}$

d. $V_2 = 15{,}200\text{ mL} \times \left(\frac{1.35\text{ atm}}{6.70\text{ atm}}\right) \times \left(\frac{370\text{ K}}{306\text{ K}}\right) = 3.70 \times 10^3\text{ mL}$

7.19 $T = \dfrac{(5.23\text{ atm})(5.23\text{ L})}{(5.23\text{ moles})\left(0.0821\frac{\text{atm L}}{\text{mole K}}\right)} = 63.7\text{ K}$

$63.7\text{ K} - 273 = -209^\circ\text{C}$

7.21 $V = \dfrac{(0.100\text{ mole})\left(0.0821\frac{\text{atm L}}{\text{mole K}}\right)(400\text{ K})}{(2.00\text{ atm})} = 1.21\text{ L}$

7.23 a. $V = \dfrac{(0.250\text{ mole})\left(0.0821\frac{\text{atm L}}{\text{mole K}}\right)(300\text{ K})}{(1.50\text{ atm})} = 4.11\text{ L}$

b. $P = \dfrac{(0.250\text{ mole})\left(0.0821\frac{\text{atm L}}{\text{mole K}}\right)(308\text{ K})}{(2.00\text{ L})} = 3.16\text{ atm}$

c. $T = \dfrac{(1.20\text{ atm})(3.00\text{ L})}{(0.250\text{ mole})\left(0.0821\frac{\text{atm L}}{\text{mole K}}\right)} = 175\text{ K}$

$175\text{ K} - 273 = -98^\circ\text{C}$

d. $V = \dfrac{(0.250\text{ mole})\left(0.0821\frac{\text{atm L}}{\text{mole K}}\right)(398\text{ K})}{(0.500\text{ atm})} = 16.3\text{ L} = 16{,}300\text{ mL}$

7.25 $(1.50 - 0.75 - 0.33)\text{ atm} = 0.42\text{ atm}$

7.27 $(623 - 125 - 175 - 225)\text{ mm Hg} = 98\text{ mm Hg}$

7.29 a. endothermic b. endothermic c. exothermic

7.31 a. no b. yes c. yes

7.33 a. boiling point b. vapor pressure c. boiling d. boiling point

7.35 a. Different intermolecular force strengths result in different vapor pressures.
b. Vapor pressure becomes equal to atmospheric pressure at a lower temperature.
c. Evaporation is a cooling process because of the loss of the most energetic molecules.
d. Low-heat and high-heat boiling water have the same temperature and the same heat content.

7.37 The molecule must be polar.

7.39 Boiling point increases as intermolecular force strength increases.

7.41 a. London b. hydrogen bonding
c. dipole-dipole d. London

7.43 a. no b. yes c. yes d. no

7.45 four

7.47 a. $P_2 = 1.25 \text{ atm} \times \left(\frac{575 \text{ mL}}{825 \text{ mL}}\right) = 0.871 \text{ atm}$

b. $T_2 = 398 \text{ K} \times \left(\frac{825 \text{ mL}}{575 \text{ mL}}\right) = 571 \text{ K} = 298°\text{C}$

c. $T_2 = 398 \text{ K} \times \left(\frac{50 \text{ atm}}{25 \text{ atm}}\right) \times \left(\frac{825 \text{ mL}}{575 \text{ mL}}\right) = 1142 \text{ K} = 869°\text{C}$

7.49 a. Boyle's law, Charles's law, and combined gas law
b. Charles's law
c. Boyle's law
d. Boyle's law

7.51 a. $P = \frac{(0.72 \text{ mole})\left(0.0821 \frac{\text{atm L}}{\text{mole K}}\right)(313 \text{ K})}{4.00 \text{ L}} = 4.6 \text{ atm}$

b. $P = \frac{(4.5 \text{ moles})\left(0.0821 \frac{\text{atm L}}{\text{mole K}}\right)(313 \text{ K})}{4.00 \text{ L}} = 29 \text{ atm}$

c. $P = \frac{\left(0.72 \text{ g} \times \frac{1 \text{ mole}}{32 \text{ g}}\right)\left(0.0821 \frac{\text{atm L}}{\text{mole K}}\right)(313 \text{ K})}{4.00 \text{ L}} = 0.14 \text{ atm}$

d. $$P = \frac{\left(4.5\text{ g x }\frac{1\text{ mole}}{32\text{ g}}\right)\left(0.0821\frac{\text{atm L}}{\text{mole K}}\right)(313\text{ K})}{4.00\text{ L}} = 0.90\text{ atm}$$

7.53

7.53 $$V = \frac{(1.00\text{ mole})\left(0.0821\frac{\text{atm L}}{\text{mole K}}\right)(296\text{ K})}{0.983\text{ atm}} = 24.7\text{ L}$$

7.55 He = 9.0 atm; Ne = (14.0 – 9.0) atm = 5.0 atm; Ar = (29.0 – 14.0) atm = 15.0 atm

7.57 a. PBr_3, lower vapor pressure
b. PI_3, higher vapor pressure
c. PI_3, higher vapor pressure

Solutions — Chapter 8

Solutions to Selected Problems

8.1 a. true b. true c. true d. false

8.3 a. solute: sodium chloride; solvent: water
b. solute: sucrose ; solvent: water
c. solute: water; solvent: ethyl alcohol
d. solute: ethyl alcohol; solvent: methyl alcohol

8.5 a. first solution b. first solution c. first solution d. second solution

8.7 a. saturated b. unsaturated c. unsaturated d. saturated

8.9 a. dilute b. concentrated c. dilute d. concentrated

8.11 a. hydrated ion b. hydrated ion c. oxygen atom d. hydrogen atom

8.13 a. decrease b. increase c. increase d. increase

8.15 a. slightly soluble b. very soluble c. slightly soluble d. slightly soluble

8.17 a. soluble with exceptions b. soluble
c. insoluble with exceptions d. soluble

8.19 a. all are soluble b. all are soluble
c. $CaBr_2$, $Ca(OH)_2$, $CaCl_2$ d. $NiSO_4$

8.21 a. $\frac{6.50\text{ g}}{91.5\text{ g}} \times 100 = 7.10\%\text{(m/m)}$

b. $\frac{2.31\text{ g}}{37.3\text{ g}} \times 100 = 6.19\%\text{(m/m)}$

c. $\frac{12.5\text{ g}}{138\text{ g}} \times 100 = 9.06\%\text{(m/m)}$

d. $\frac{0.0032\text{ g}}{1.2\text{ g}} \times 100 = 0.27\%\text{(m/m)}$

8.23 a. $275\text{ g H}_2\text{O} \times \left(\frac{1.30\text{ g glucose}}{98.70\text{ g H}_2\text{O}}\right) = 3.62\text{ g glucose}$

b. $275\text{ g H}_2\text{O} \times \left(\frac{5.00\text{ g glucose}}{95.00\text{ g H}_2\text{O}}\right) = 14.5\text{ g glucose}$

c. $275\text{ g H}_2\text{O} \times \left(\frac{20.0\text{ g glucose}}{80.0\text{ g H}_2\text{O}}\right) = 68.8\text{ g glucose}$

d. $275 \text{ g } H_2O \times \left(\frac{31.0 \text{ g glucose}}{69.0 \text{ g } H_2O}\right) = 124 \text{ g glucose}$

8.25 $32.00 \text{ g solution} \times \left(\frac{2.000 \text{ g } K_2SO_4}{100.0 \text{ g solution}}\right) = 0.6400 \text{ g } K_2SO_4$

8.27 $20.0 \text{ g NaOH} \times \left(\frac{93.25 \text{ g } H_2O}{6.75 \text{ g NaOH}}\right) = 276 \text{ g } H_2O$

8.29 a. $\frac{20.0 \text{ mL}}{475 \text{ mL}} \times 100 = 4.21\%(\text{v/v})$

b. $\frac{4.00 \text{ mL}}{87.0 \text{ mL}} \times 100 = 4.60\%(\text{v/v})$

8.31 $\frac{22 \text{ mL}}{125 \text{ mL}} \times 100 = 18\%(\text{v/v})$

8.33 a. $\frac{5.0 \text{ g}}{250 \text{ mL}} \times 100 = 2.0\%(\text{m/v})$

b. $\frac{85 \text{ g}}{580 \text{ mL}} \times 100 = 15\%(\text{m/v})$

8.35 $25.0 \text{ mL solution} \times \left(\frac{2.00 \text{ g } Na_2CO_3}{100.0 \text{ mL solution}}\right) = 0.500 \text{ g } Na_2CO_3$

8.37 $50.0 \text{ mL solution} \times \left(\frac{7.50 \text{ g NaCl}}{100.0 \text{ mL solution}}\right) = 3.75 \text{ g NaCl}$

8.39 a. $\frac{3.0 \text{ moles}}{0.50 \text{ L}} = 6.0 \text{ M}$

b. $12.5 \text{ g } C_{12}H_{22}O_{11} \times \left(\frac{1 \text{ mole } C_{12}H_{22}O_{11}}{342.34 \text{ g } C_{12}H_{22}O_{11}}\right) = 0.0365 \text{ mole } C_{12}H_{22}O_{11}$

$\frac{0.0365 \text{ mole}}{0.0800 \text{ L}} = 0.456 \text{ M}$

c. $25.0 \text{ g NaCl} \times \left(\frac{1 \text{ mole NaCl}}{58.44 \text{ g NaCl}}\right) = 0.428 \text{ mole NaCl}$

$\frac{0.428 \text{ mole}}{1.250 \text{ L}} = 0.342 \text{ M}$

d. $\dfrac{0.00125 \text{ mole}}{0.00250 \text{ L}} = 0.500 \text{ M}$

8.41 a. $2.50 \text{ L solution} \times \left(\dfrac{3.00 \text{ moles HCl}}{1.00 \text{ L solution}}\right) \times \left(\dfrac{36.46 \text{ g HCl}}{1 \text{ mole HCl}}\right) = 273 \text{ g HCl}$

b. $0.0100 \text{ L solution} \times \left(\dfrac{0.500 \text{ mole KCl}}{1.00 \text{ L solution}}\right) \times \left(\dfrac{74.55 \text{ g KCl}}{1 \text{ mole KCl}}\right) = 0.373 \text{ g KCl}$

c. $0.875 \text{ L solution} \times \left(\dfrac{1.83 \text{ moles NaNO}_3}{1.00 \text{ L solution}}\right) \times \left(\dfrac{85.00 \text{ g NaNO}_3}{1 \text{ mole NaNO}_3}\right) = 136 \text{ g NaNO}_3$

d. $0.075 \text{ L solution} \times \left(\dfrac{12.0 \text{ moles H}_2\text{SO}_4}{1.00 \text{ L solution}}\right) \times \left(\dfrac{98.09 \text{ g H}_2\text{SO}_4}{1 \text{ mole H}_2\text{SO}_4}\right) = 88 \text{ g H}_2\text{SO}_4$

8.43 a. $1.00 \text{ g NaCl} \times \left(\dfrac{1 \text{ mole NaCl}}{58.44 \text{ g NaCl}}\right) \times \left(\dfrac{1000 \text{ mL solution}}{0.200 \text{ mole NaCl}}\right) = 85.6 \text{ mL solution}$

b. $2.00 \text{ g C}_6\text{H}_{12}\text{O}_6 \times \left(\dfrac{1 \text{ mole C}_6\text{H}_{12}\text{O}_6}{180.18 \text{ g C}_6\text{H}_{12}\text{O}_6}\right) \times \left(\dfrac{1000 \text{ mL solution}}{4.20 \text{ moles C}_6\text{H}_{12}\text{O}_6}\right) = 2.64 \text{ mL solution}$

c. $3.67 \text{ moles AgNO}_3 \times \left(\dfrac{1000 \text{ mL solution}}{0.400 \text{ mole AgNO}_3}\right) = 9180 \text{ mL solution}$

d. $0.0021 \text{ moles C}_{12}\text{H}_{22}\text{O}_{11} \times \left(\dfrac{1000 \text{ mL solution}}{8.7 \text{ moles C}_{12}\text{H}_{22}\text{O}_{11}}\right) = 0.24 \text{ mL solution}$

8.45 a. $0.220 \text{ M} \times \left(\dfrac{25.0 \text{ mL}}{30.0 \text{ mL}}\right) = 0.183 \text{ M}$

b. $0.220 \text{ M} \times \left(\dfrac{25.0 \text{ mL}}{75.0 \text{ mL}}\right) = 0.0733 \text{ M}$

c. $0.220 \text{ M} \times \left(\dfrac{25.0 \text{ mL}}{457 \text{ mL}}\right) = 0.0120 \text{ M}$

d. $0.220 \text{ M} \times \left(\dfrac{25.0 \text{ mL}}{2000 \text{ mL}}\right) = 0.00275 \text{ M}$

8.47 a. $50.0 \text{ mL} \times \left(\frac{3.00 \text{ M}}{0.0100 \text{ M}}\right) = 1500 \text{ mL}$

$1500 \text{ mL} - 50.0 \text{ mL} = 1450 \text{ mL}$

b. $2.00 \text{ mL} \times \left(\frac{1.00 \text{ M}}{0.100 \text{ M}}\right) = 20.0 \text{ mL}$

$20.0 \text{ mL} - 2.00 \text{ mL} = 18.0 \text{ mL}$

c. $1450 \text{ mL} \times \left(\frac{6.00 \text{ M}}{0.100 \text{ M}}\right) = 87000 \text{ mL}$

$87000 \text{ mL} - 1450 \text{ mL} = 85600 \text{ mL}$

d. $75.0 \text{ mL} \times \left(\frac{0.110 \text{ M}}{0.100 \text{ M}}\right) = 82.5 \text{ mL}$

$82.5 \text{ mL} - 75.0 \text{ mL} = 7.5 \text{ mL}$

8.49 a. $5.0 \text{ M} \times \left(\frac{30.0 \text{ mL}}{50.0 \text{ mL}}\right) = 3.0 \text{ M}$

b. $5.0 \text{ M} \times \left(\frac{30.0 \text{ mL}}{50.0 \text{ mL}}\right) = 3.0 \text{ M}$

c. $7.5 \text{ M} \times \left(\frac{30.0 \text{ mL}}{50.0 \text{ mL}}\right) = 4.5 \text{ M}$

d. $2.0 \text{ M} \times \left(\frac{60.0 \text{ mL}}{80.0 \text{ mL}}\right) = 1.5 \text{ M}$

8.51 The presence of solute molecules decreases the ability of solvent molecules to escape.

8.53 It is a more concentrated solution and thus has a lower vapor pressure.

8.55 a. same b. greater than c. less than d. greater than

8.57 2 to 1

8.59 a. swell b. remain the same c. swell d. shrink

8.61 a. hemolyze b. remain unaffected c. hemolyze d. crenate

8.63 a. hypotonic b. isotonic c. hypotonic d. hypertonic

8.65 a. K^+ and Cl^- leave the bag.
b. K^+, Cl^-, and glucose leave the bag.

8.67 a. like (both soluble) b. unlike
c. unlike d. like (both insoluble)

8.69 3.50 qt solution x $\left(\frac{2.000 \text{ qt } H_2O}{100 \text{ qt solution}}\right)$ = 0.0700 qt H_2O

8.71 a. 0.400 M x $\left(\frac{2212 \text{ mL}}{1875 \text{ mL}}\right)$ = 0.472 M

b. 0.400 M x $\left(\frac{2212 \text{ mL}}{1250 \text{ mL}}\right)$ = 0.708 M

c. 0.400 M x $\left(\frac{2212 \text{ mL}}{853 \text{ mL}}\right)$ = 1.04 M

d. 0.400 M x $\left(\frac{2212 \text{ mL}}{553 \text{ mL}}\right)$ = 1.60 M

8.73 a. 4.00 M; both solutions are 4.00 M, so the mixture is 4.00 M.
b. (4.00 M x 352 mL) + (2.00 M x 225 mL) = (y M)(577 mL)
y = 3.22 M

Chemical Reactions Chapter 9

Solutions to Selected Problems

9.1 a. single replacement b. decompostition
c. double replacement d. combination

9.3. a. combination, single replacement, combustion
b. decomposition, single replacement
c. combination, decomposition, single replacement, double replacement, combustion
d. combination, decomposition, single replacement, double replacement, combustion

9.5 a. +2 b. +6 c. 0 d. +5

9.7 a. +3 b. +4 c. +6 d. +6 e. +6 f. +6 g. +6 h. +5

9.9 a. +3 P, –1 F b. +1 Na, –2 O, +1 H
c. +1 Na, +6 S, –2 O d. +4 C, –2 O

9.11 a. redox b. nonredux c. redox d. redox

9.13 a. H_2 oxidized, N_2 reduced b. KI oxidized, Cl_2 reduced
c. Fe oxidized, Sb_2O_3 reduced d. H_2SO_3 oxidized, HNO_3 reduced

9.15 a. N_2 oxidizing agent, H_2 reducing agent
b. Cl_2 oxidizing agent, KI reducing agent
c. Sb_2O_3 oxidizing agent, Fe reducing agent
d. HNO_3 oxidizing agent, H_2SO_3 reducing agent

9.17 because reactant molecules have greater freedom of movement

9.19 Molecular collisions are not effective if the activation energy requirement is not met.

9.21 a. exothermic b. endothermic c. endothermic d. exothermic

9.23

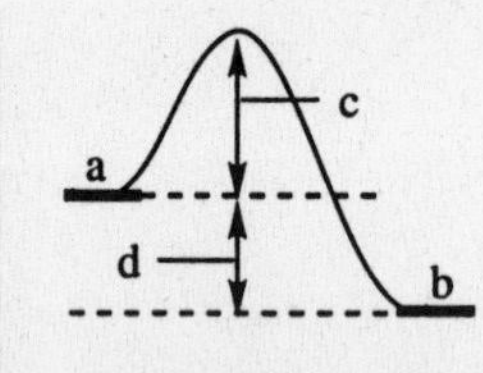

9.25 a. As temperature increases, so does the number of collisions per second.
b. A catalyst lowers the activation energy.

9.27 The concentration of O_2 has increased from 21% to 100%.

9.29

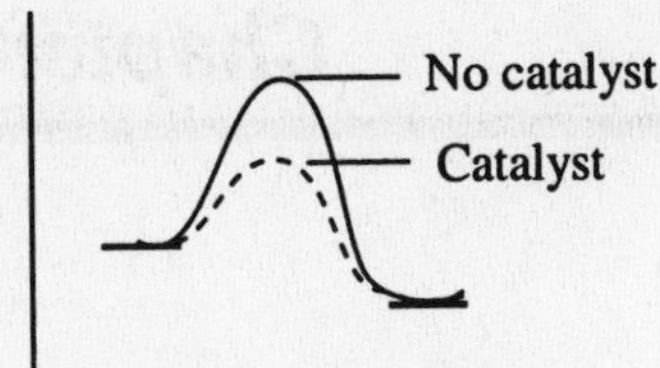

9.31 a. 1 b. 3 c. 4 d. 3

9.33 Rate of forward reaction is equal to rate of reverse reaction.

9.35

Products

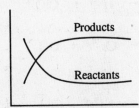

Reactants

9.37 a. $K_{eq} = \dfrac{[NO_2]^2}{[N_2O_4]}$ b. $K_{eq} = \dfrac{[Cl_2][CO]}{[COCl_2]}$

c. $K_{eq} = \dfrac{[H_2S]^2[CH_4]}{[H_2]^4[CS_2]}$ d. $K_{eq} = \dfrac{[SO_3]^2}{[O_2][SO_2]^2}$

9.39 a. $K_{eq} = [SO_3]$ b. $K_{eq} = \dfrac{1}{[Cl_2]}$

c. $K_{eq} = \dfrac{[NaCl]^2}{[Na_2SO_4][BaCl_2]}$ d. $K_{eq} = [O_2]$

9.41 $K_{eq} = \dfrac{[0.0032]^2}{[0.213]} = 4.8 \times 10^{-5}$

9.43 a. more products than reactants
b. essentially all reactants
c. significant amounts of both reactants and products
d. significant amounts of both reactants and products

9.45 a. right b. left c. left d. right

9.47 a. left b. left c. left d. left

9.49 a. shifts left b. no effect c. shifts right d. no effect

9.51 a. redox, single replacement b. redox, combustion
c. redox, decomposition d. nonredox, double replacement

9.53 a. gain b. reduction c. decrease d. increase

9.55 a. no b. no c. yes d. no

9.57 a. yes b. yes c. no d. yes

Acids, Bases, and Salts — Chapter 10

Solutions to Selected Problems

10.1 a. H^+ b. OH^-

10.3 a. Arrhenius acid b. Arrhenius base

10.5 a. $HI \xrightarrow{H_2O} H^+ + I^-$ b. $HClO \xrightarrow{H_2O} H^+ + ClO^-$

c. $LiOH \xrightarrow{H_2O} Li^+ + OH^-$ d. $CsOH \xrightarrow{H_2O} Cs^+ + OH^-$

10.7 a. acid b. base c. acid d. acid

10.9 a. $HClO + H_2O \rightarrow H_3O^+ + ClO^-$ b. $HClO_4 + NH_3 \rightarrow NH_4^+ + ClO_4^-$

c. $H_3O^+ + OH^- \rightarrow H_2O + H_2O$ d. $H_3O^+ + NH_2^- \rightarrow H_2O + NH_3$

10.11 a. HSO_3^- b. HCN c. $C_2O_4^{2-}$ d. $H_2PO_4^-$

10.13 a. $HS^- + H_2O \rightarrow H_3O^+ + S^{2-}$; $HS^- + H_2O \rightarrow H_2S + OH^-$

b. $HPO_4^{2-} + H_2O \rightarrow H_3O^+ + PO_4^{3-}$; $HPO_4^{2-} + H_2O \rightarrow H_2PO_4^- + OH^-$

c. $NH_3 + H_2O \rightarrow H_3O^+ + NH_2^-$; $NH_3 + H_2O \rightarrow NH_4^+ + OH^-$

d. $OH^- + H_2O \rightarrow H_3O^+ + O^{2-}$; $OH^- + H_2O \rightarrow H_2O + OH^-$

10.15 a. monoprotic b. diprotic c. monoprotic d. diprotic

10.17 $H_3C_6H_5O_7\ H_2O \rightarrow H_3O^+ + H_2C_6H_5O_7^-$

$H_2C_6H_5O_7^- + H_2O \rightarrow H_3O^+ + HC_6H_5O_7^{2-}$

$H_2C_6H_5O_7^{2-} + H_2O \rightarrow H_3O^+ + C_6H_5O_7^{3-}$

10.19 a. 1, 0 b. 2, 4 c. 1, 7 d. 0, 4

10.21 to show that it is a monoprotic acid

10.23 monoprotic; only one H atom is involved in a polar bond

10.25 a. strong b. weak c. weak d. strong

10.27 a. $K_a = \dfrac{[H^+][F^-]}{[HF]}$ b. $K_a = \dfrac{[H^+][C_2H_3O_2^-]}{[HC_2H_3O_2]}$

10.29 a. $K_b = \dfrac{[NH_4^+][OH^-]}{[NH_3]}$ b. $K_b = \dfrac{[C_6H_5NH_3^+][OH^-]}{[C_6H_5NH_2]}$

10.31 a. H_3PO_4 b. HF c. H_2CO_3 d. HNO_2

10.33 $[H_3O^+] = [A^-] = (0.12)(0.00300\text{ M}) = 0.00036\text{ M}$

$[HA] = (0.00300 - 0.00036)\text{ M} = 0.00264\text{ M}$

$$K_a = \frac{[0.00036][0.00036]}{[0.00264]} = 4.9 \times 10^{-5}$$

10.35 a. acid b. salt c. salt d. base e. salt f. base g. acid h. acid

10.37 a. $Ba(NO_3)_2 \xrightarrow{H_2O} Ba^{2+} + 2NO_3^-$

b. $Na_2SO_4 \xrightarrow{H_2O} 2Na^+ + SO_4^{2-}$

c. $CaBr_2 \xrightarrow{H_2O} Ca^{2+} + 2Br^-$

d. $K_2CO_3 \xrightarrow{H_2O} 2K^+ + CO_3^{2-}$

10.39 a. no b. yes c. yes d. no

10.41 a. 1 to 1 b. 1 to 2 c. 1 to 1 d. 2 to 1

10.43 a. $HCl + NaOH \rightarrow NaCl + H_2O$

b. $HNO_3 + KOH \rightarrow KNO_3 + H_2O$

c. $H_2SO_4 + 2LiOH \rightarrow Li_2SO_4 + 2H_2O$

d. $2H_3PO_4 + 3Ba(OH)_2 \rightarrow Ba_3(PO_4)_2 + 6H_2O$

10.45 a. $H_2SO_4 + 2LiOH \rightarrow Li_2SO_4 + 2H_2O$

b. $HCl + NaOH \rightarrow NaCl + H_2O$

c. $HNO_3 + KOH \rightarrow KNO_3 + H_2O$

d. $2H_3PO_4 + 3Ba(OH)_2 \rightarrow Ba_3(PO_4)_2 + 6H_2O$

10.47 a. $\frac{1.00 \times 10^{-14}\text{ M}}{3.00 \times 10^{-3}\text{ M}} = 3.3 \times 10^{-12}\text{ M}$ b. $\frac{1.00 \times 10^{-14}\text{ M}}{6.7 \times 10^{-3}\text{ M}} = 1.5 \times 10^{-9}\text{ M}$

c. $\frac{1.00 \times 10^{-14}\text{ M}}{9.1 \times 10^{-8}\text{ M}} = 1.1 \times 10^{-7}\text{ M}$ d. $\frac{1.00 \times 10^{-14}\text{ M}}{1.2 \times 10^{-11}\text{ M}} = 8.3 \times 10^{-4}\text{ M}$

10.49 a. acidic b. basic c. basic d. acidic

10.51 a. 4.00 b. 11.00
c. $[H_3O^+] = 1.0 \times 10^{-11}$; 11.00 d. $[H_3O^+] = 1.0 \times 10^{-7}$; 7.00

10.53 a. 7.68 b. 7.40
c. $[H_3O^+] = 1.4 \times 10^{-4}$; 3.85 d. $[H_3O^+] = 1.4 \times 10^{-12}$; 11.85

10.55 a. 1×10^{-2} M b. 1×10^{-6} M c. 1×10^{-8} M d. 1×10^{-10} M

10.57 a. 2.1×10^{-4} M b. 8.1×10^{-6} M c. 4.5×10^{-8} M d. 3.5×10^{-13} M

10.59 a. 3.35 b. 6.37 c. 7.21 d. 1.82

10.61 acid B

10.63 a. strong acid – strong base salt b. weak acid – strong base salt
c. strong acid – weak base salt d. strong acid – strong base salt

10.65 a. none b. $C_2H_3O_2^-$ c. NH_4^+ d. none

10.67 a. neutral b. basic c. acidic d. neutral

10.69 a. no b. yes c. no d. yes

10.71 a. HCN and CN^- b. H_3PO_4 and $H_2PO_4^-$
c. H_2CO_3 and HCO_3^- d. HCO_3^- and CO_3^{2-}

10.73 a. $F^- + H_3O^+ \rightarrow HF + H_2O$

b. $H_2CO_3 + OH^- \rightarrow HCO_3^- + H_2O$

c. $CO_3^{2-} + H_3O^+ \rightarrow HCO_3^- + H_2O$

d. $H_3PO_4 + OH^- \rightarrow H_2PO_4^- + H_2O$

10.75 $$pH = 6.72 + \log\left[\frac{0.500\text{ M}}{0.230\text{ M}}\right] = 7.06$$

10.77 $$pH = -\log\left[6.8 \times 10^{-6}\right] + \log\left[\frac{0.150\text{ M}}{0.150\text{ M}}\right] = 5.17$$

10.79 a. weak b. strong c. strong d. strong

10.81 a. $$5.00\text{ mL HNO}_3 \times \left(\frac{0.250\text{ mole HNO}_3}{1000\text{ mL HNO}_3}\right) \times \left(\frac{1\text{ mole NaOH}}{1\text{ mole HNO}_3}\right) = 0.00125\text{ mole NaOH}$$

$$\frac{0.00125\text{ mole NaOH}}{0.0250\text{ L solution}} = 0.0500\text{ M}$$

b. $$20.00\text{ mL H}_2\text{SO}_4 \times \left(\frac{0.500\text{ mole H}_2\text{SO}_4}{1000\text{ mL H}_2\text{SO}_4}\right) \times \left(\frac{2\text{ mole NaOH}}{1\text{ mole H}_2\text{SO}_4}\right) = 0.0200\text{ mole NaOH}$$

$$\frac{0.0200\text{ mole NaOH}}{0.0250\text{ L solution}} = 0.800\text{ M}$$

c. $23.76 \text{ mL HCl} \times \left(\frac{1.00 \text{ mole HCl}}{1000 \text{ mL HCl}}\right) \times \left(\frac{1 \text{ mole NaOH}}{1 \text{ mole HCl}}\right) = 0.0238 \text{ mole NaOH}$

$$\frac{0.0238 \text{ mole NaOH}}{0.0250 \text{ L solution}} = 0.952 \text{ M}$$

d. $10.00 \text{ mL } H_3PO_4 \times \left(\frac{0.100 \text{ mole } H_3PO_4}{1000 \text{ mL } H_3PO_4}\right) \times \left(\frac{3 \text{ mole NaOH}}{1 \text{ mole } H_3PO_4}\right) = 0.00300 \text{ mole NaOH}$

$$\frac{0.00300 \text{ mole NaOH}}{0.0250 \text{ L solution}} = 0.120 \text{ M}$$

10.83 a. yes b. no c. no d. yes

10.85 a. B b. A

10.87 HCl (strong acid), HCN (weak acid), KCl (salt that does not hydrolyze), NaOH (strong base)

10.89 CN^- ion undergoes hydrolysis to a greater extent than NH_4^+ ion, resulting in a basic solution. NH_4^+ ion and $C_2H_3O_2^-$ ion hydrolyze to an equal extent, resulting in a neutral solution.

Nuclear Chemistry

Chapter 11

Solutions to Selected Problems

11.1 a. $^{10}_{4}Be$, Be-10 b. $^{25}_{11}Na$, Na-25 c. $^{96}_{41}Nb$, Nb-96 d. $^{257}_{103}Lr$, Lr-257

11.3 a. $^{14}_{7}N$ b. $^{197}_{79}Au$ c. tin-121 d. boron-10

11.5 a. $^{4}_{2}\alpha$ b. $^{0}_{-1}\beta$ c. $^{0}_{0}\gamma$

11.7 2 protons and 2 neutrons

11.9 a. $^{200}_{84}Po \rightarrow ^{4}_{2}\alpha + ^{196}_{82}Pb$ b. $^{240}_{96}Cm \rightarrow ^{4}_{2}\alpha + ^{236}_{94}Pu$

c. $^{244}_{96}Cm \rightarrow ^{4}_{2}\alpha + ^{240}_{94}Pu$ d. $^{238}_{92}U \rightarrow ^{4}_{2}\alpha + ^{234}_{90}Th$

11.11 a. $^{10}_{4}Be \rightarrow ^{0}_{-1}\beta + ^{10}_{5}B$ b. $^{14}_{6}C \rightarrow ^{0}_{-1}\beta + ^{14}_{7}N$

c. $^{21}_{9}F \rightarrow ^{0}_{-1}\beta + ^{21}_{10}Ne$ d. $^{25}_{11}Na \rightarrow ^{0}_{-1}\beta + ^{25}_{12}Mg$

11.13 $A \rightarrow A-4,\ Z \rightarrow Z-2$

11.15 a. $^{0}_{-1}\beta$ b. $^{28}_{12}Mg$ c. $^{4}_{2}\alpha$ d. $^{200}_{80}Hg$

11.17 a. $^{4}_{2}\alpha$ b. $^{0}_{-1}\beta$

11.19 a. $\frac{1}{2^2} = \frac{1}{4}$ b. $\frac{1}{2^6} = \frac{1}{64}$ c. $\frac{1}{2^3} = \frac{1}{8}$ d. $\frac{1}{2^6} = \frac{1}{64}$

11.21 a. $\frac{1}{16} = \frac{1}{2^4}$; $\frac{5.4 \text{ days}}{4} = 1.4$ days b. $\frac{1}{64} = \frac{1}{2^6}$; $\frac{5.4 \text{ days}}{6} = 0.90$ day

c. $\frac{1}{256} = \frac{1}{2^8}$; $\frac{5.4 \text{ days}}{8} = 0.68$ day d. $\frac{1}{1024} = \frac{1}{2^{10}}$; $\frac{5.4 \text{ days}}{10} = 0.54$ day

11.23 $\frac{60.0 \text{ hr}}{15.0 \text{ hr}} = 4$; $\frac{1}{2^4} = \frac{1}{16}$; $\frac{1}{16} \times 4.00 \text{ g} = 0.250 \text{ g}$

11.25 2000

11.27 92, uranium

11.29 a. $^{4}_{2}\alpha$ b. $^{25}_{12}Mg$ c. $^{4}_{2}\alpha$ d. $^{1}_{1}p$

11.31 stable; termination of a decay series requires a stable isotope

11.33 $^{232}_{90}\text{Th} \rightarrow {}^{4}_{2}\alpha + {}^{228}_{88}\text{Ra}$

$^{228}_{88}\text{Ra} \rightarrow {}^{0}_{-1}\beta + {}^{228}_{89}\text{Ac}$

$^{228}_{89}\text{Ac} \rightarrow {}^{0}_{-1}\beta + {}^{288}_{90}\text{Th}$

$^{228}_{90}\text{Th} \rightarrow {}^{4}_{2}\alpha + {}^{224}_{88}\text{Ra}$

11.35 the electron and positive ion that are produced during an ionizing interaction between a molecule (or an atom) and radiation

11.37 a. yes b. no c. yes d. no

11.39 It continues on, interacting with other atoms and forming more ion pairs.

11.41 Alpha is stopped; beta and gamma go through.

11.43 alpha, 0.1 the speed of light; beta, up to 0.9 the speed of light; gamma, the speed of light

11.45 a. no detectable effects b. nausea, fatigue, lowered blood cell count

11.47 19% human-made, 81% natural sources

11.49 to monitor the extent of radiation exposure

11.51 so the radiation can be detected externally (outside the body)

11.53 a. bone tumors b. circulatory problems
c. iron metabolism d. intercellular space problems

11.55 They are usually α or β emitters instead of γ emitters.

11.57 a. fusion b. fusion c. both d. fission

11.59 a. fusion b. fission c. neither d. neither

11.61 a. $^{206}_{80}\text{Hg} \rightarrow {}^{0}_{-1}\beta + {}^{206}_{81}\text{Tl}$ b. $^{109}_{46}\text{Pd} \rightarrow {}^{0}_{-1}\beta + {}^{109}_{47}\text{Ag}$

c. $^{245}_{96}\text{Cm} \rightarrow {}^{4}_{2}\alpha + {}^{241}_{94}\text{Pu}$ d. $^{249}_{100}\text{Fm} \rightarrow {}^{4}_{2}\alpha + {}^{245}_{98}\text{Cf}$

11.63 a. $^{243}_{94}\text{Pu} + {}^{4}_{2}\alpha \rightarrow {}^{246}_{96}\text{Cm} + {}^{1}_{0}\text{n}$ b. $^{246}_{96}\text{Cm} + {}^{12}_{6}\text{C} \rightarrow {}^{254}_{102}\text{No} + 4{}^{1}_{0}\text{n}$

c. $^{27}_{13}\text{Al} + {}^{4}_{2}\alpha \rightarrow {}^{30}_{15}\text{P} + {}^{1}_{0}\text{n}$ d. $^{23}_{11}\text{Na} + {}^{2}_{1}\text{H} \rightarrow {}^{21}_{10}\text{Ne} + {}^{4}_{2}\alpha$

11.65 12 elements

11.67 A = 0 (negligible amount), B = 0 (negligible amount), C = 63 atoms, D = 937 atoms

Saturated Hydrocarbons — Chapter 12

Solutions to Selected Problems

12.1 a. false b. false c. true d. true

12.3 a. meets b. does not meet c. does not meet d. does not meet

12.5 A hydrocarbon contains only the elements carbon and hydrogen, and a hydrocarbon derivative contains at least one additional element besides carbon and hydrogen.

12.7 All bonds are single bonds in a saturated hydrocarbon, and at least one carbon–carbon multiple bond is present in an unsaturated hydrocarbon.

12.9 a. saturated b. unsaturated c. unsaturated d. unsaturated

12.11 a. 18 b. 4 c. 13 d. 22

12.13 a. $CH_3-CH_2-CH_2-CH_3$

b. 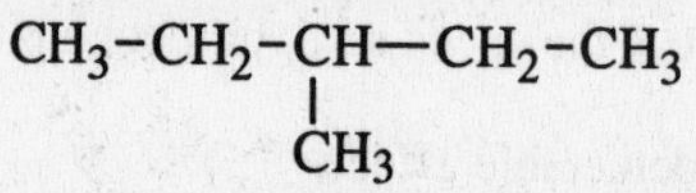
```
CH3-CH2-CH—CH2-CH3
        |
        CH3
```

c. 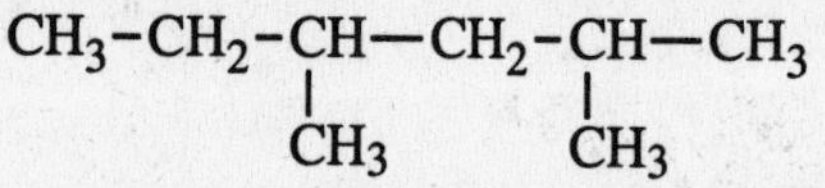
```
CH3-CH2-CH—CH2-CH—CH3
        |      |
        CH3    CH3
```

d. 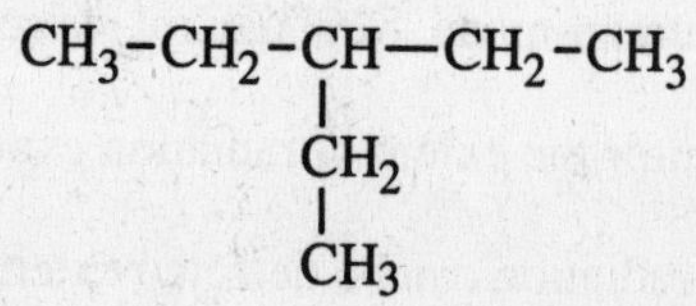
```
CH3-CH2-CH—CH2-CH3
        |
        CH2
        |
        CH3
```

12.15 a. 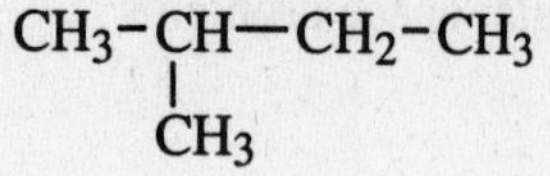
```
CH3-CH—CH2-CH3
    |
    CH3
```

b. 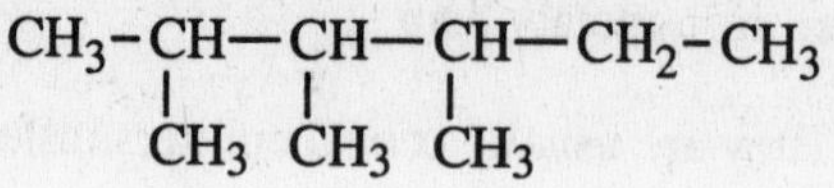
```
CH3-CH—CH—CH—CH2-CH3
    |  |  |
    CH3 CH3 CH3
```

c. $CH_3-CH_2-CH_2-CH_2-CH_2-CH_3$

d. 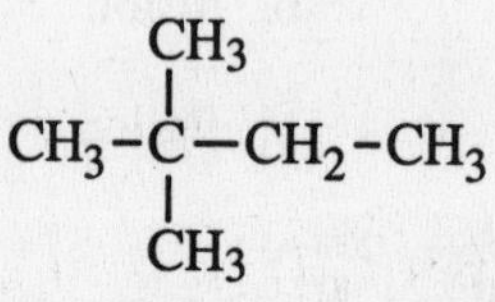
```
    CH3
    |
CH3-C-CH2-CH3
    |
    CH3
```

12.17 a. 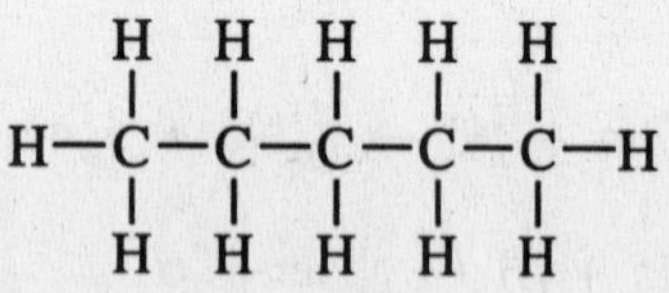
```
   H  H  H  H  H
   |  |  |  |  |
H—C—C—C—C—C—H
   |  |  |  |  |
   H  H  H  H  H
```

b. 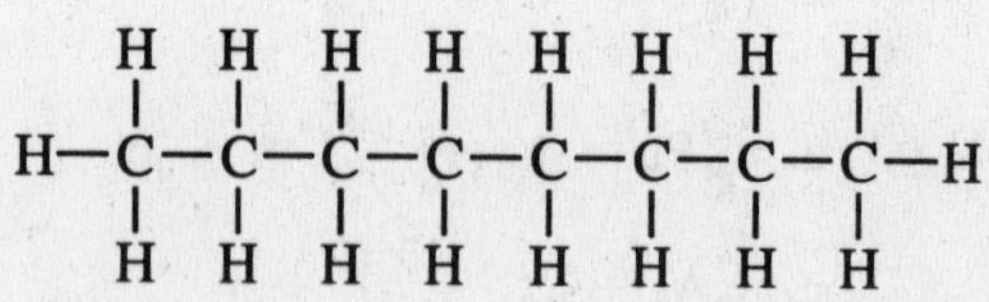
```
   H  H  H  H  H  H  H  H
   |  |  |  |  |  |  |  |
H—C—C—C—C—C—C—C—C—H
   |  |  |  |  |  |  |  |
   H  H  H  H  H  H  H  H
```

c. $CH_3\text{—}(CH_2)_8\text{—}CH_3$

d. C_6H_{14}

12.19 a. different compounds that are not structural isomers
b. different compounds that are structural isomers

c. different conformations of the same molecule
d. different compounds that are structural isomers

12.21 a. 7 b. 8 c. 8 d. 7

12.23 a. 2-methylpentane
b. 2,4,5-trimethylheptane
c. 3-ethyl-2,3-dimethylpentane
d. 3-ethyl-2,4-dimethylhexane
e. decane
f. 4-propylheptane

12.25 horizontal chain, because it has more substituents

12.27 a.
```
CH3-CH—CH2-CH3
    |
    CH3
```
b.
```
CH3-CH2-CH—CH—CH2-CH3
        |   |
        CH3 CH3
```
c.
```
           CH3
           |
CH3-CH2-C—CH2-CH3
           |
           CH2
           |
           CH3
```
d.
```
CH3-CH—CH—CH—CH—CH2-CH3
    |   |   |   |
    CH3 CH3 CH3 CH3
```
e.
```
CH3-CH2-CH—CH2-CH—CH2-CH2-CH3
        |       |
        CH2     CH2
        |       |
        CH3     CH3
```
f.
```
CH3-CH2-CH2-CH—CH2-CH2-CH2-CH2-CH3
            |
            CH2
            |
            CH2
            |
            CH3
```

12.29 a. 1, 1 b. 2, 2 c. 2, 2 d. 4, 4 e. 2, 2 f. 1, 1

12.31 a. carbon chain numbered from wrong end; 2-methylpentane
b. not based on longest carbon chain; 2,2-dimethylbutane
c. carbon chain numbered from wrong end; 2,2,3-trimethylbutane
d. not based on longest carbon chain; 3,3-dimethylhexane
e. carbon chain numbered from wrong end, substituents not alphabetical; 3-ethyl-4-methylhexane
f. like alkyl groups listed separately; 2,4-dimethylhexane

12.33 a. 3, 2, 1, 0 b. 5, 2, 3, 0 c. 5, 2, 1, 1 d. 5, 2, 3, 0 e. 2, 8, 0, 0 f. 3, 6, 1, 0

12.35 a. isopropyl b. isobutyl c. isopropyl d. *sec*-butyl

12.37 a.

```
CH3-CH2-CH2-CH2-CH-CH2-CH2-CH2-CH2-CH3
                |
                CH-CH3
                |
                CH2
                |
                CH3
```

b.

```
               CH3
               |
               CH-CH3
               |
CH3-CH2-CH2-C-CH2-CH2-CH2-CH3
               |
               CH-CH3
               |
               CH3
```

c.

```
CH3-CH-CH-CH2-CH-CH2-CH2-CH2-CH3
    |   |       |
   CH3 CH3     CH2
                |
                CH-CH3
                |
                CH3
```

d.

```
        CH3
        |
CH3-CH2-CH-CH-CH2-CH2-CH3
            |
       CH3-C-CH3
            |
            CH3
```

12.39 a. 16 b. 6 c. 5 d. 15

12.41 a. C_6H_{12} b. C_6H_{12} c. C_4H_8 d. C_7H_{14}

12.43 a. cyclohexane b. 1,2-dimethylcylobutane
c. methylcyclopropane d. 1,2-dimethylcyclopentane

12.45 a. must locate methyl groups with numbers
b. wrong numbering system for ring
c. no number needed
d. wrong numbering system for ring

12.47 a. (cyclobutane ring with substituent) $CH_2-CH_2-CH_3$

b. (cyclobutane ring with substituent)

```
CH3
|
CH-CH3
```

c. (cyclohexane ring) CH_2-CH_3, CH_2-CH_3, H, H

d. (cyclopentane ring) H, CH_2-CH_3, $H_3C-H_2C-CH_2$, H

12.49 a. not possible

b. 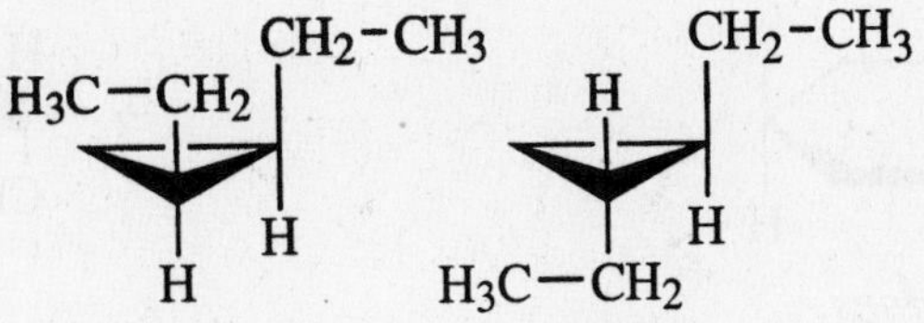 (cis: cyclopropane with CH_2-CH_3 and H_3C-CH_2 on the same side, H and H on the other; trans: CH_2-CH_3 up, H_3C-CH_2 down)

cis *trans*

c. not possible

d. 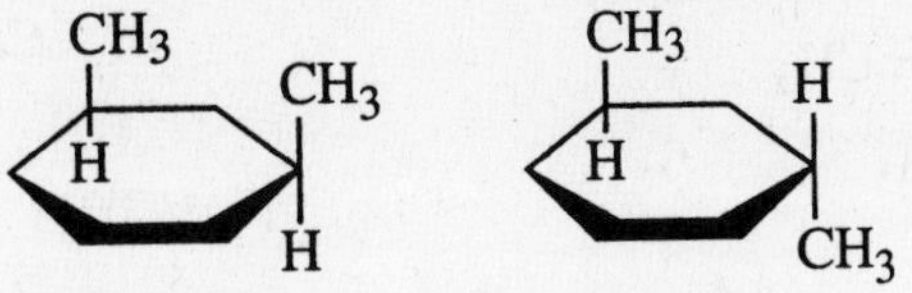 (cyclohexane with CH_3 and CH_3; cis: both CH3 up; trans: one CH3 up, one CH3 down)

cis *trans*

12.51 boiling point

12.53 a. octane b. cyclopentane c. pentane d. cyclopentane

12.55 a. different states b. same state c. same state d. same state

12.57 a. CO_2 and H_2O b. CO_2 and H_2O c. CO_2 and H_2O d. CO_2 and H_2O

12.59 CH_3Br, CH_2Br_2, $CHBr_3$, CBr_4

12.61 a. 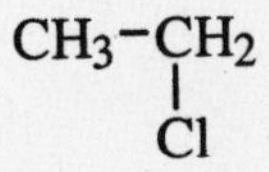 CH_3-CH_2-Cl

b. 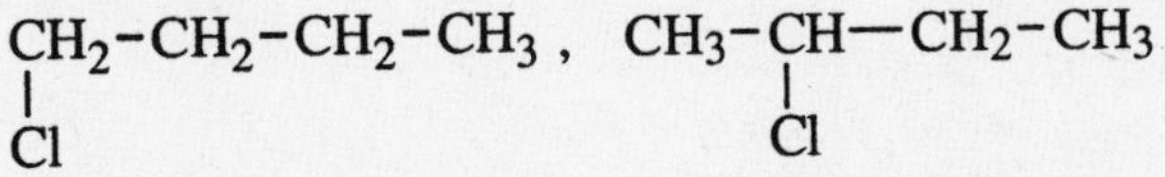 $CH_2(Cl)-CH_2-CH_2-CH_3$, $CH_3-CH(Cl)-CH_2-CH_3$

c. 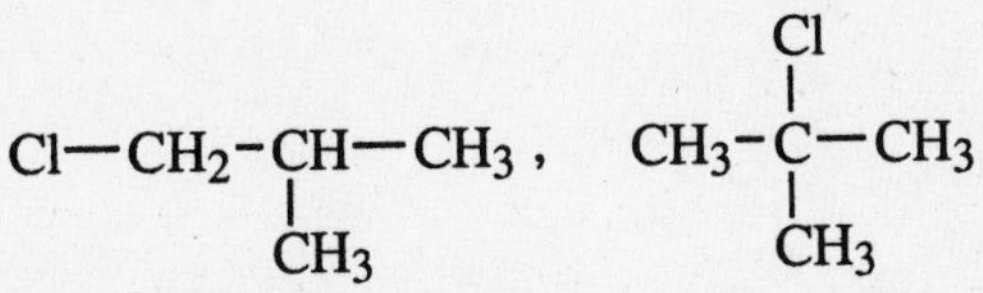 $Cl-CH_2-CH(CH_3)-CH_3$, $CH_3-C(Cl)(CH_3)-CH_3$

d. 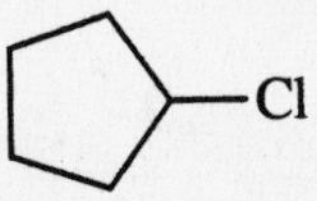 (cyclopentane)–Cl

12.63 a. iodomethane, methyl iodide
b. 1-chloropropane, propyl chloride
c. 2-fluorobutane, *sec*-butyl fluoride
d. chlorocyclobutane, cyclobutyl chloride

12.65 a. 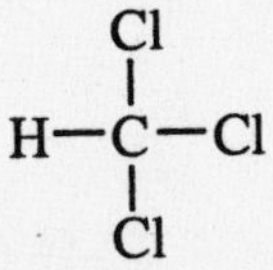 $H-C(Cl)_2-Cl$ (trichloromethane) b. $Cl-CF_2-CF_2-Cl$ c. $CH_3-CH(CH_3)-Br$ d.  (cyclobutane with Br and H on one carbon, H and Cl on the adjacent carbon; Br up, Cl down)

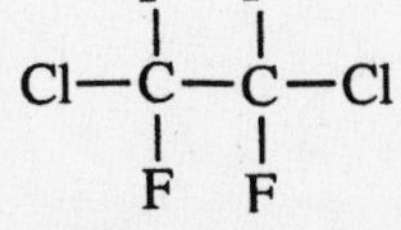

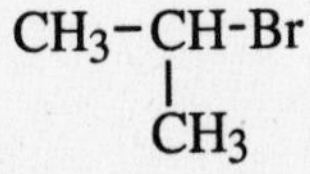

12.67 a. 16 b. 6 c. 5 d. 22
e. liquid f. less dense g. insoluble h. flammable

12.69 a. 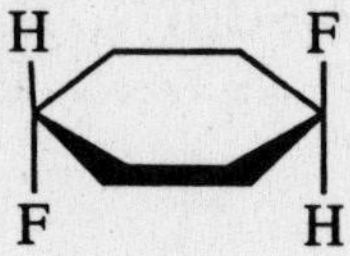 H F F H *trans* b. 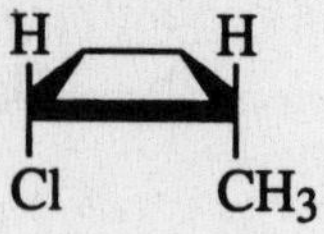 H H Cl CH_3 *cis*

c. 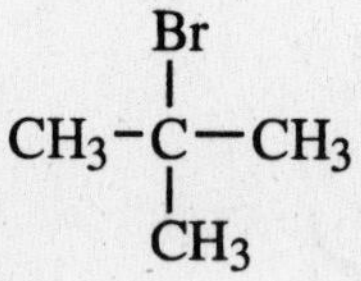
$$\begin{array}{c} Br \\ | \\ CH_3-C-CH_3 \\ | \\ CH_3 \end{array}$$

d. 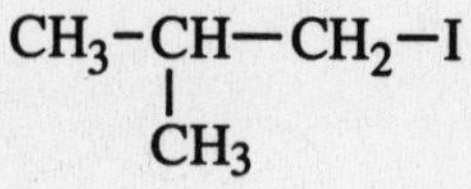
$$\begin{array}{l} CH_3-CH-CH_2-I \\ \qquad\ | \\ \qquad CH_3 \end{array}$$

12.71 a. $C_{18}H_{38}$ b. C_7H_{14} c. $C_7H_{14}F_2$ d. $C_6H_{10}Br_2$

12.73 a. alkane b. halogenated cycloalkane
c. halogenated alkane d. cycloalkane

12.75 a. 1,2-diethylcyclohexane
b. 3-methylhexane
c. 2,3-dimethyl-4-propylnonane
d. 1-isopropyl-3,5-dipropylcyclohexane

Unsaturated Hydrocarbons — Chapter 13

Solutions to Selected Problems

13.1 a. unsaturated, alkene with one double bond
b. saturated
c. unsaturated, alkene with one double bond
d. unsaturated, diene
e. unsaturated, triene
f. unsaturated, diene

13.3 a. C_4H_{10} b. C_5H_{10} c. C_5H_8 d. C_7H_{10}

13.5 a. C_nH_{2n-2} b. C_nH_{2n-2} c. C_nH_{2n-2} d. C_nH_{2n-6}

13.7 a. 2-butene
b. 2,4-dimethyl-2-pentene
c. cyclohexene
d. 1,3-cyclopentadiene
e. 2-ethyl-1-pentene
f. 2,4,6-octatriene

13.9 a. 2-pentene
b. pentane
c. 2,3,3-trimethyl-1-butene
d. 2-methyl-1,4-pentadiene
e. 1,3,5-hexatriene
f. 2,3-pentadiene

13.11

```
a.  H2C=CH−CH−CH2−CH3
           |
           CH3
```

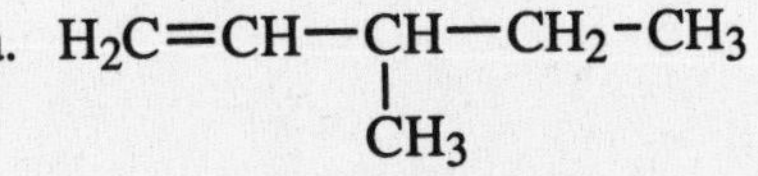

b. 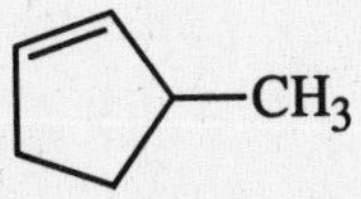 —CH3

```
c.  H2C=CH−CH=CH2
```

```
d.  H2C=CH−CH−CH2−CH3
           |
           CH2
           |
           CH3
```

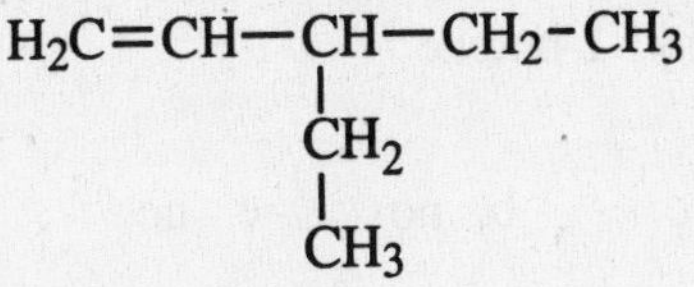

```
e.  CH3−CH=CH−CH−CH2−CH2−CH3
              |
              CH2
              |
              CH2
              |
              CH3
```

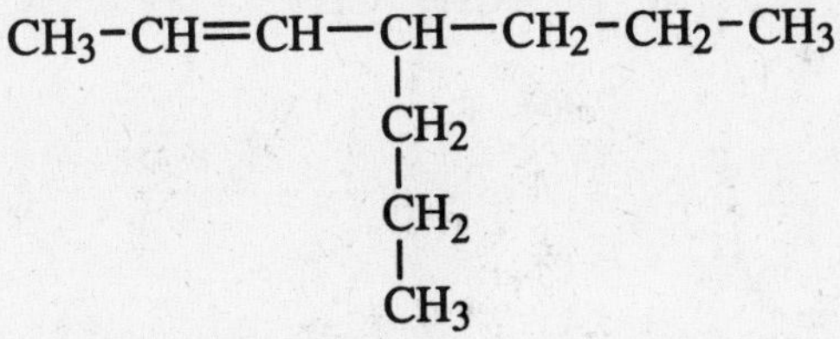

f. 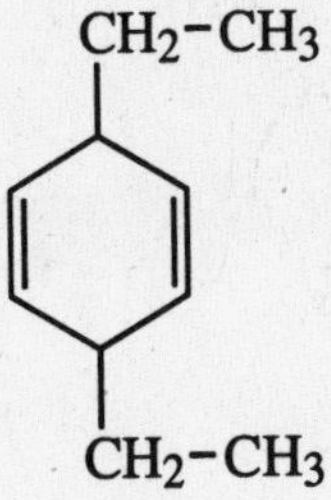 CH2−CH3

CH2−CH3

13.13 a. 3-methyl-3-hexene
b. 2,3-dimethyl-2-hexene
c. 1,3-cyclopentadiene
d. 4,5-dimethylcyclohexene

13.15

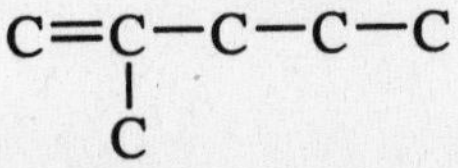

```
C=C−C−C−C−C          C−C=C−C−C−C
  1-hexene             2-hexene

C−C−C=C−C−C          C=C−C−C−C
  3-hexene             |
                       C
                     2-methyl-1-pentene
```

```
C=C-C-C-C
    |
    C
```
3-methyl-1-pentene

```
C=C-C-C-C
      |
      C
```
4-methyl-1-pentene

```
C-C=C-C-C
  |
  C
```
2-methyl-2-pentene

```
C-C=C-C-C
    |
    C
```
3-methyl-2-pentene

```
C-C=C-C-C
      |
      C
```
4-methyl-2-pentene

```
C=C-C-C
  | |
  C C
```
2,3-dimethyl-1-butene

```
    C
    |
C=C-C-C
    |
    C
```
3,3-dimethyl-1-butene

```
C-C=C-C
  | |
  C C
```
2,3-dimethyl-2-butene

```
C=C-C-C
  |
  C
  |
  C
```
2-ethyl-1-butene

13.17 a. no b. no c. no

d.
```
H3C-H2C      CH2-CH3        H3C-H2C      H
       \    /                      \    /
        C=C                         C=C
       /    \                      /    \
      H      H                    H      CH2-CH3
```
cis *trans*

e.
```
           CH3
           |
H3C        CH-CH3         H3C      H
   \      /                  \    /
    C=C                       C=C
   /    \                    /    \
  H      H                  H      CH-CH3
                                   |
                                   CH3
```
cis *trans*

f. 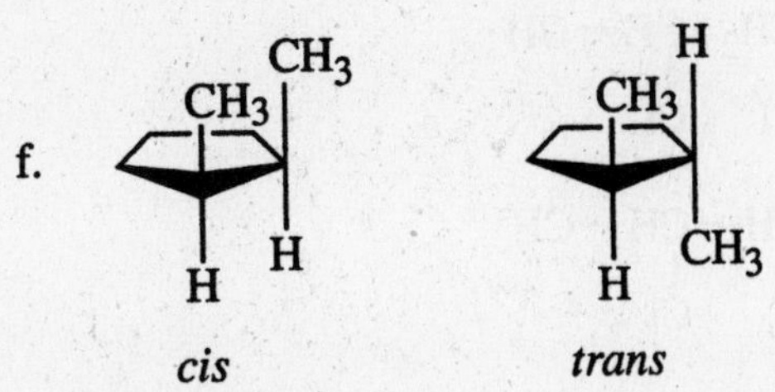

13.19 a. *cis*-2-pentene
b. *trans*-1-bromo-2-iodoethene
c. tetrafluoroethene
d. 2-methyl-2-butene

13.21 a. $\begin{matrix} H_3C{-}H_2C & & H \\ & C{=}C & \\ H_3C & & CH_2{-}CH_3 \end{matrix}$

b. $\begin{matrix} H_3C & & CH_2{-}CH_3 \\ & C{=}C & \\ H & & H \end{matrix}$

c. $\begin{matrix} H_3C & & H \\ & C{=}C & \\ H & & CH_2{-}\underset{CH_3}{CH}{-}CH_2{-}CH_3 \end{matrix}$

d. $\begin{matrix} H_2C{=}CH & & H \\ & C{=}C & \\ H & & CH_3 \end{matrix}$

13.23 Pheromones are compounds used by insects (and some animals) to transmit messages to other members of the same species.

13.25 because isoprene, the building block for terpenes, contains 5 carbon atoms

13.27 a. gas b. liquid c. liquid d. liquid

13.29 a. yes b. no c. yes d. no

13.31 a. $CH_2{=}CH_2 + Cl_2 \longrightarrow \underset{Cl}{CH_2}{-}\underset{Cl}{CH_2}$

b. $CH_2{=}CH_2 + HCl \longrightarrow CH_3{-}\underset{Cl}{CH_2}$

c. $CH_2{=}CH_2 + H_2 \xrightarrow{Ni} CH_3{-}CH_3$

d. $CH_2{=}CH_2 + HBr \longrightarrow CH_3{-}\underset{Br}{CH_2}$

13.33 a. $CH_2{=}CH{-}CH_3 + Cl_2 \longrightarrow CH_2Cl{-}CHCl{-}CH_3$

b. $CH_2{=}CH{-}CH_3 + HCl \longrightarrow CH_3{-}CHCl{-}CH_3$

c. $CH_2{=}CH{-}CH_3 + H_2 \xrightarrow{Ni} CH_3{-}CH_2{-}CH_3$

d. $CH_2{=}CH{-}CH_3 + HBr \longrightarrow CH_3{-}CHBr{-}CH_3$

13.35 a. $CH_3{-}CHCl{-}CHCl{-}CH_3$

b. $CH_3{-}CBr(CH_3){-}CH_3$

c. $CH_3{-}CH_2{-}CHCl{-}CH_3$

d. (cyclopentane)

e. (cyclopentene)

f. (cyclobutanol, HO on cyclobutane ring)

13.37 a. Br_2 b. H_2 + Ni catalyst c. HCl d. $H_2O + H_2SO_4$ catalyst

13.39 a. 2 b. 2 c. 2 d. 3

13.41 a. $F_2C{=}CF_2$

b. $H_2C{=}CCl{-}CH{=}CH_2$

c. $H_2C{=}CHCl$

d. $H_2C{=}CH{-}C_6H_5$ (styrene)

13.43 a. $-CH_2{-}CH_2{-}CH_2{-}CH_2{-}CH_2{-}CH_2-$

b. $-CH_2{-}CHCl{-}CH_2{-}CHCl{-}CH_2{-}CHCl-$

c. $-CHCl{-}CHCl{-}CHCl{-}CHCl{-}CHCl{-}CHCl-$

d. $-CH_2{-}CHCl{-}CH_2{-}CHCl{-}CH_2{-}CHCl-$

13.45 a. 1-hexyne
b. 4-methyl-2-pentyne
c. 2,2-dimethyl-3-heptyne
d. 2-butyne
e. 3-methyl-1,4-hexadiyne
f. 4-methyl-2-hexyne

13.47 a. CH_3-CH_3
b. 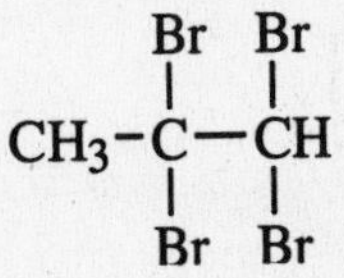 $CH_3-CBr_2-CHBr_2$
c. 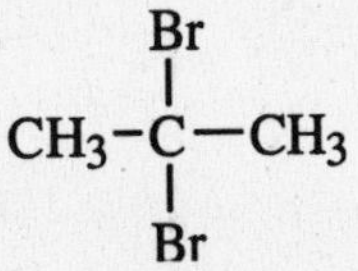 $CH_3-CBr_2-CH_3$
d. 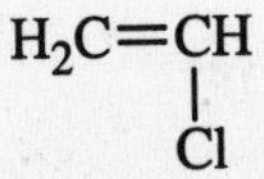 $H_2C=CH-Cl$
e. 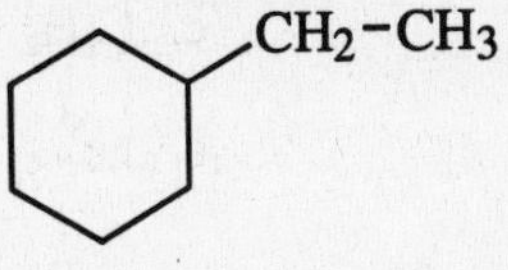 CH_2-CH_3 (on cyclohexane)
f. 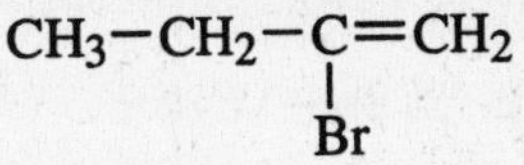 $CH_3-CH_2-C(Br)=CH_2$

13.49 a. 1,3-dibromobenzene
b. 1-chloro-2-fluorobenzene
c. 1-chloro-4-fluorobenzene
d. 3-chlorotoluene
e. 1-bromo-2-ethylbenzene
f. 4-bromotoluene

13.51 a. *m*-dibromobenzene
b. *o*-chlorofluorobenzene
c. *p*-chlorofluorobenzene
d. *m*-chlorotoluene
e. *o*-bromoethylbenzene
f. *p*-bromotoluene

13.53 a. 2,4-dibromo-1-chlorobenzene
b. 3-bromo-5-chlorotoluene
c. 1-bromo-3-chloro-2-fluorobenzene
d. 1,4-dibromo-2,5-dichlorobenzene

13.55 a. 2-phenylbutane
b. 3-phenyl-1-butene
c. 3-methyl-1-phenylbutane
d. 2,4-diphenylpentane

13.57 a. 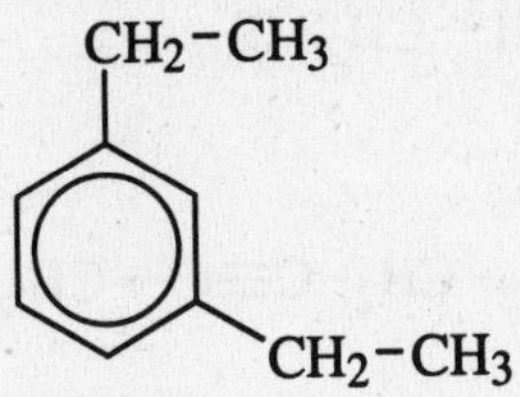 CH_2-CH_3, CH_2-CH_3
b. 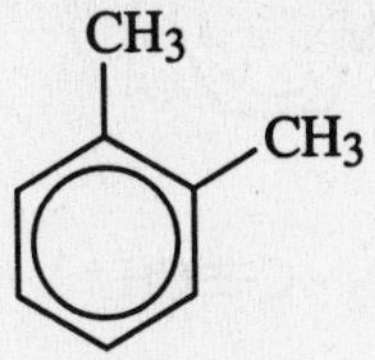 CH_3, CH_3
c. 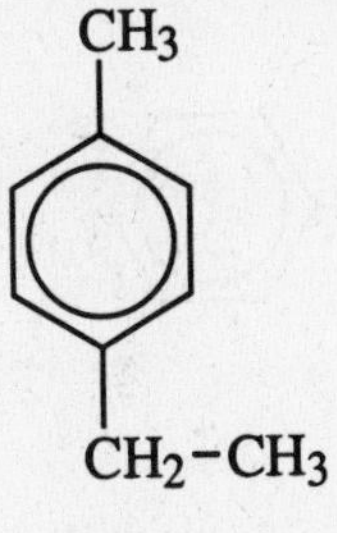 CH_3, CH_2-CH_3
d. 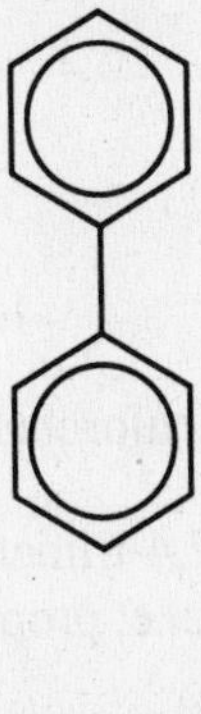
e. $-CH_2-CH_2-$
f. 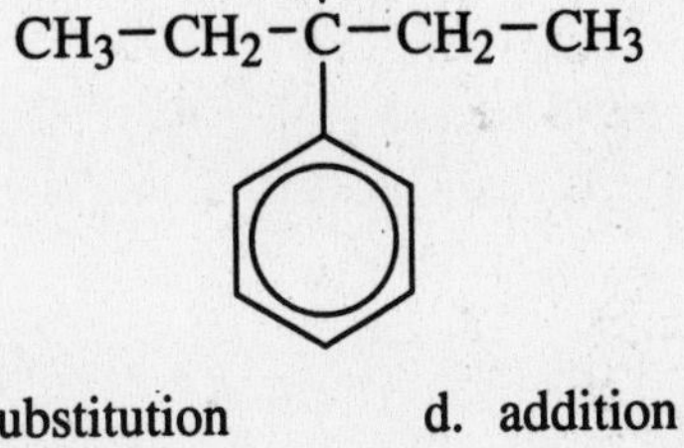 $CH_3-CH_2-C(CH_3)-CH_2-CH_3$

13.59 a. substitution
b. addition
c. substitution
d. addition

13.61 a. Br_2

b. (benzene ring with $CH(CH_3)_2$ substituent)

$$\begin{array}{l} CH_3 \\ | \\ CH-CH_3 \\ | \\ C_6H_5 \end{array}$$

c. CH_3-CH_2-Br

13.63 a. C_2H_4 b. C_3H_4 c. C_2H_2 d. CH_4

13.65 a. no b. yes c. no d. yes

13.67 a. $CH_3-C{\equiv}C-CH_2-\underset{\displaystyle CH_3}{\underset{|}{CH}}-CH_3$

b. $\underset{\displaystyle Cl}{\underset{|}{CH_2}}-CH{=}CH-CH_3$

c. $CH_3-C{\equiv}C-CH_2-\underset{\displaystyle CH_3}{\underset{|}{CH}}-\underset{\displaystyle CH_3}{\underset{|}{CH}}-CH_3$

d. $CH_2{=}CH-\underset{\displaystyle \underset{\displaystyle CH_3}{\underset{|}{CH-CH_3}}}{\underset{|}{CH}}-CH_2-CH_2-CH_3$

e. $CH_2{=}CH-CH_2-CH_2-CH_2-CH{=}CH_2$

f. $HC{\equiv}C-\underset{\displaystyle CH_3}{\underset{|}{CH}}-C{\equiv}CH$

13.69 a. $CH_2{=}CH-C_6H_5$ (benzene ring)

b. $CH_2{=}CH-CH_2-Cl$

c. $CH_3-CH_2-CH_2-C{\equiv}CH$

d. $CH_3-CH_2-CH_2-C{\equiv}C-CH_2-CH_2-CH_3$

e. (benzene ring with two adjacent CH_3 groups)

f. (biphenyl with H_3C on one ring)

13.71 Each carbon atom in 1,2-dichlorobenzene has only one substituent.

13.73 1,2,3-trimethylbenzene; 1,2,4-trimethylbenzene; 1,3,5-trimethylbenzene; 2-ethyltoluene; 3-ethyltoluene; 4-ethyltoluene; propylbenzene; isopropylbenzene

Alcohols, Phenols, and Ethers

Chapter 14

Solutions to Selected Problems

14.1 a. 2 b. 1 c. 4 d. 1

14.3 R–OH

14.5 R–O–H versus H–O–H

14.7 a. 2-pentanol b. ethanol c. 3-methyl-2-butanol
d. 2-ethyl-1-pentanol e. 2-butanol f. 3,3-dimethyl-1-butanol

14.9 a. $CH_3-CH_2-\underset{OH}{CH}-CH_2-CH_3$

b. $CH_3-CH_2-\overset{CH_2-CH_3}{\underset{OH}{C}}-CH_2-CH_2-CH_3$

c. $\underset{OH}{CH_2}-\underset{CH_3}{CH}-CH_3$

d. $CH_3-\underset{OH}{CH}-CH_2-\underset{CH_3}{CH}-CH_3$

e. 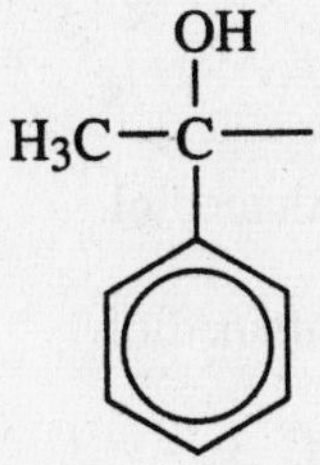 $H_3C-\overset{OH}{C}-CH_3$

f. 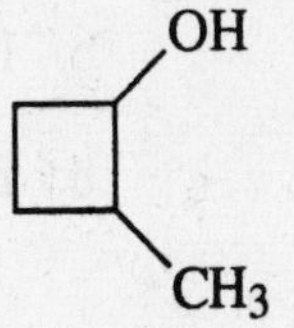 OH, CH_3

14.11 a. $CH_3-CH_2-CH_2-CH_2-CH_2-OH$
1-pentanol

b. $CH_3-CH_2-CH_2-OH$
1-propanol

c. $CH_3-\underset{CH_3}{CH}-CH_2-OH$
2-methyl-1-propanol

d. $CH_3-CH_2-\underset{CH_3}{CH}-OH$
2-butanol

14.13 a. 1,2-propanediol b. 1,4-pentanediol
c. 1,3-pentanediol d. 3-methyl-1,2,4-butanetriol

14.15 a. cyclohexanol b. *trans*-3-chlorocyclohexanol
c. *cis*-2-methylcyclohexanol d. 1-methylcyclobutanol

14.17 a. $CH_3-\underset{OH}{CH}-CH_2-CH=CH_2$

b. $HC\equiv C-\underset{OH}{CH}-CH_2-CH_3$

c. $CH_3-CH-C=CH_2$
OH CH_3

d. $HO-CH_2$ CH_3
$C=C$
H H

14.19 a. $CH_2-CH-CH_3$
OH CH_2
CH_3
2-methyl-1-butanol

b. $CH_3-CH-CH_2-CH_2$
OH OH
1,3-butanediol

c. $CH_3-CH-CH-CH_3$
CH_3 OH
3-methyl-2-butanol

d. HO OH
1,3-cyclopentanediol

14.21 a. ethanol with all traces of water removed b. ethanol
c. 70% solution of isopropyl alcohol d. ethanol

14.23 a. glycerol b. ethanol c. methanol d. methanol

14.25 Alcohols can hydrogen-bond to each other; alkanes cannot.

14.27 a. 1-heptanol b. 1-propanol c. 1,2-ethanediol

14.29 a. 1-butanol b. 1-pentanol c. 1,2-butanediol

14.31 a. 3 b. 3 c. 3 d. 3

14.33 a. CH_3-CH_2
OH

b. $CH_3-CH_2-CH_2$
OH

c. OH
$CH_3-CH_2-C-CH_3$
CH_3

d. OH
$CH_3-CH_2-CH-CH_2-CH_3$

14.35 a. $CH_2=CH-CH_3$

b. $CH_3-C=CH_2$
CH_3

c. $CH_2=CH$
CH_3

d. $CH_3-CH_2-CH_2-O-CH_2-CH_2-CH_3$

14.37 a. $CH_3-CH(OH)-CH(CH_3)-CH_3$

b. $CH_3-CH_2-CH_2(OH)$ or $CH_3-CH(CH_3)-OH$

c. CH_3-CH_2-OH

d. $CH_3-CH(CH_3)-OH$

14.39 a. $CH_3-CH_2-CH(OH)-CH_3$

b. $CH_3-CH_2-CH_2-OH$

c. $CH_3-CH_2-CH_2-OH$

d. cyclopentyl$-CH_2-OH$

14.41 a. $CH_3-CH_2-CH_2-Cl$

b. 1-methylcyclopentene ($-CH_3$ on ring double bond carbon)

c. $CH_3-C(=O)-CH_2-CH_3$

d. $CH_3-CH_2-CH(Cl)-CH_2-CH_3$

e. $CH_3-CH_2-O-CH_2-CH_3$

f. $CH_2(Br)-CH_2(Br)$

14.43 $\left(CH(OH)-CH(OH) \right)_n$

14.45 In a phenol the –OH group must be directly attached to the benzene ring.

14.47 a. 3-ethylphenol b. 2-chlorophenol c. *o*-cresol
d. hydroquinone e. 2-bromophenol f. 2-bromo-3-ethylphenol

14.49 a. benzene ring with OH and Cl (para)

b. benzene ring with OH and CH_2-CH_3 (ortho)

c. benzene ring with OH, Br (ortho) and Br (para)

d. benzene ring with OH and CH_3 (meta)

e. (benzene ring with OH at positions 1 and 3)

f. (benzene ring with OH; H_3C-H_2C and CH_2-CH_3 on either side of OH; CH_3 para to OH)

14.51 An antiseptic kills microorganisms on living tissue; a disinfectant kills microorganisms on inanimate objects.

14.53 (phenol, C_6H_5OH) + $H_2O \rightleftharpoons$ (phenoxide, $C_6H_5O^-$) + H_3O^+

14.55 a. yes b. no c. yes d. yes

14.57 a. 1-methoxypropane b. 1-ethoxypropane
c. 2-methoxypropane d. methoxybenzene
e. cyclohexoxycyclohexane f. ethoxybenzene

14.59 a. methyl propyl ether b. ethyl propyl ether
c. isopropyl methyl ether d. methyl phenyl ether
e. dicyclohexyl ether f. cyclobutyl ethyl ether

14.61 a. $CH_3-CH(CH_3)-O-CH_2-CH_2-CH_3$ b. $CH_3-CH_2-O-C_6H_5$

c. (benzene ring with $O-CH_3$ and CH_3 in meta positions) d. $CH_3-CH(O-CH_2-CH_3)-CH_2-CH_2-CH_3$

e. (cyclobutane ring with $O-CH_2-CH_3$) f. $CH_3-O-CH_2-C(CH_3)_2-CH_3$

14.63 There is no hydrogen bonding in the ether; there is hydrogen bonding in the alcohol.

14.65 flammability and peroxide formation

14.67 There are no hydrogen atoms bonded to oxygen atoms.

14.69 a. noncyclic ether b. noncyclic ether c. cyclic ether
d. cyclic ether e. noncyclic ether f. nonether

14.71 R–S–H versus R–O–H

14.73 a. CH_3-SH

b. $CH_3-CH-CH_3$ (SH on the middle carbon)

c. $CH_2-CH_2-CH_2-CH_3$ (SH on the first carbon)

d. $CH_2-CH_2-CH-CH_2-CH_3$ (SH on the first carbon; CH_3 on the third carbon)

e. 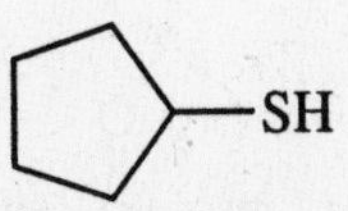 –SH

f. CH_2-CH_2 (SH on each carbon)

14.75 a. methyl mercaptan b. propyl mercaptan
c. *sec*-butyl mercaptan d. isobutyl mercaptan

14.77 Alcohol oxidation produces aldehydes and ketones; thiol oxidation produces disulfides.

14.79 a. methylthioethane, ethyl methyl sulfide
b. 2-methylthiopropane, isopropyl methyl sulfide
c. methylthiocyclohexane, cyclohexyl methyl sulfide
d. cyclohexylthiocyclohexane, dicyclohexyl sulfide
e. 1-(methylthio)-2-propene, allyl methyl sulfide
f. 2-methylthiobutane, *sec*-butyl methyl sulfide

14.81 a. secondary b. secondary c. primary d. primary e. tertiary f. secondary

14.83 a. 2-hexanol b. 3-pentanol c. 3-phenoxy-1-propene
d. 2-methyl-1-propanol e. 2-methyl-2-propanol f. ethoxyethane

14.85 1-pentanol

14.87 a. disulfide b. thiol, thioalcohol
c. alcohol d. peroxide
e. alcohol, thiol, thioalcohol f. ether, sulfide, thioether

Aldehydes and Ketones — Chapter 15

Solutions to Selected Problems

15.1 a. yes b. no c. yes d. yes e. no f. no

15.3 similarity: both have σ and π components
difference: C=O is polar, C=C is nonpolar

15.5 a. neither b. aldehyde c. ketone d. neither e. aldehyde f. aldehyde

15.7 a. $H-\overset{O}{\overset{\|}{C}}-H$ b. $CH_3-\overset{O}{\overset{\|}{C}}-H$ c. $CH_3-\overset{O}{\overset{\|}{C}}-CH_3$ d. $CH_3-\overset{O}{\overset{\|}{C}}-CH_2-CH_3$

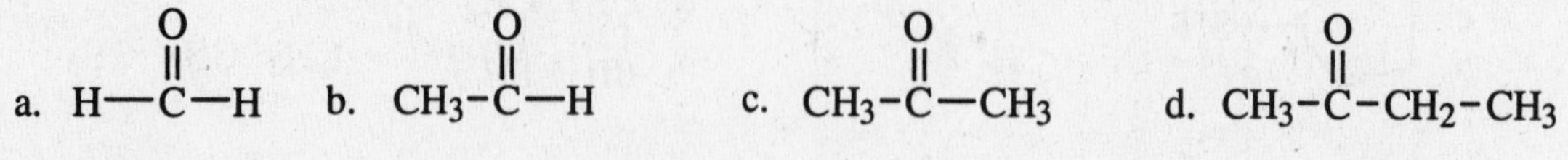

15.9 a. neither b. aldehyde c. neither d. ketone e. ketone f. aldehyde

15.11 a. butanal b. 2-methylbutanal c. 4-methylheptanal
d. 3-phenylpropanal e. propanal f. 3,3-dimethylbutanal

15.13 a. $CH_3-CH_2-\underset{CH_3}{\underset{|}{CH}}-CH_2-\overset{O}{\overset{\|}{C}}-H$ b. $CH_3-CH_2-CH_2-CH_2-\underset{CH_3}{\underset{|}{\underset{CH_2}{\underset{|}{CH}}}}-\overset{O}{\overset{\|}{C}}-H$

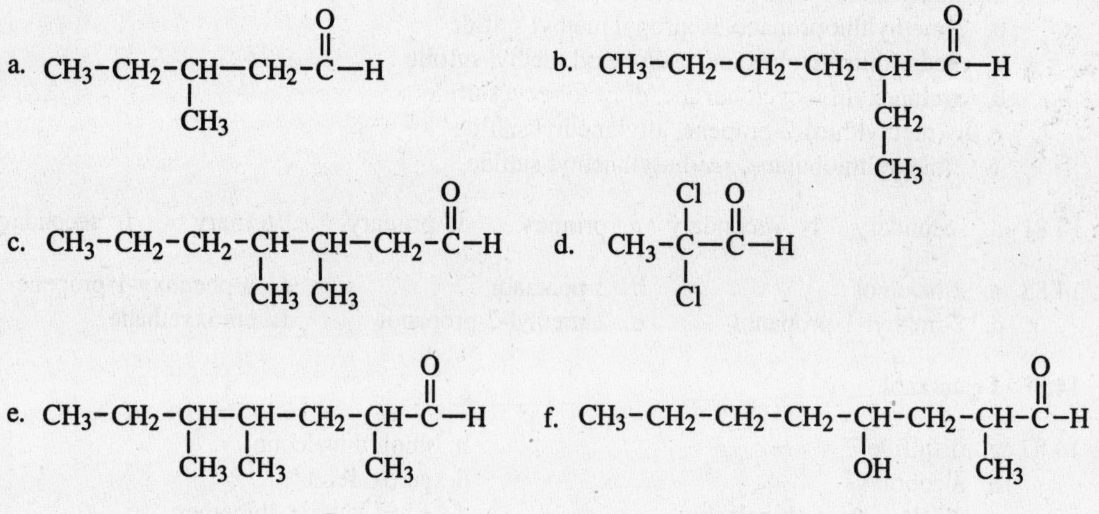

c. $CH_3-CH_2-CH_2-\underset{CH_3}{\underset{|}{CH}}-\underset{CH_3}{\underset{|}{CH}}-CH_2-\overset{O}{\overset{\|}{C}}-H$ d. $CH_3-\underset{Cl}{\underset{|}{\overset{Cl}{\overset{|}{C}}}}-\overset{O}{\overset{\|}{C}}-H$

e. $CH_3-CH_2-\underset{CH_3}{\underset{|}{CH}}-\underset{CH_3}{\underset{|}{CH}}-CH_2-\underset{CH_3}{\underset{|}{CH}}-\overset{O}{\overset{\|}{C}}-H$ f. $CH_3-CH_2-CH_2-CH_2-\underset{OH}{\underset{|}{CH}}-CH_2-\underset{CH_3}{\underset{|}{CH}}-\overset{O}{\overset{\|}{C}}-H$

15.15 a. $H-\overset{O}{\overset{\|}{C}}-H$ b. $CH_3-CH_2-\overset{O}{\overset{\|}{C}}-H$

c. $\underset{Cl}{\underset{|}{CH_2}}-\overset{O}{\overset{\|}{C}}-H$ d. $CH_3-CH_2-\underset{Cl}{\underset{|}{CH}}-\overset{O}{\overset{\|}{C}}-H$

e. (benzene ring with CHO and ortho CH_3) C(=O)H, CH_3

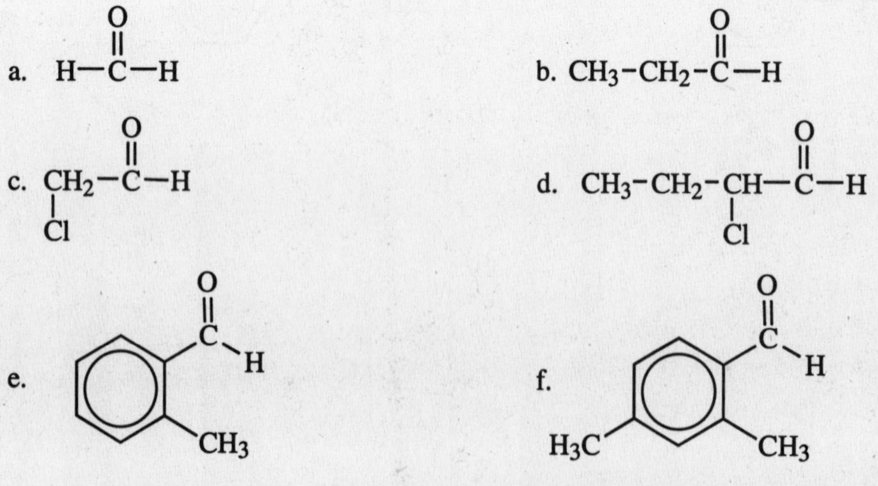

f. (benzene ring with CHO, CH_3 ortho and H_3C para) C(=O)H, H_3C, CH_3

15.17 a. propionaldehyde b. 2-methylpropionaldehyde
c. butyraldehyde d. dichloroacetaldehyde
e. 2-chlorobenzaldehyde f. 3-chloro-4-hydroxybenzaldehyde

15.19 a. 2-butanone b. 2,4,5-trimethyl-3-hexanone
c. 6-methyl-3-heptanone d. 2-octanone
e. 1,5-dichloro-3-pentanone f. 1,1-dichloro-2-butanone

15.21 a. cyclohexanone b. 3-methylcyclohexanone
c. 2-methylcyclohexanone d. 3-chlorocyclopentanone

15.23 a. $CH_3-\overset{O}{\overset{\|}{C}}-\underset{CH_3}{\underset{|}{CH}}-CH_2-CH_3$ b. $CH_3-CH_2-\overset{O}{\overset{\|}{C}}-CH_2-CH_2-CH_3$

c. $\square\!=\!O$ d. $CH_3-\underset{CH_3}{\underset{|}{CH}}-\overset{O}{\overset{\|}{C}}-\underset{CH_3}{\underset{|}{CH}}-CH_3$

e. $\underset{Cl}{\underset{|}{CH_2}}-\overset{O}{\overset{\|}{C}}-CH_3$ f. $\underset{Cl}{\underset{|}{CH_2}}-\overset{O}{\overset{\|}{C}}-\underset{Cl}{\underset{|}{CH_2}}$

15.25 a. $CH_3-CH_2-\overset{O}{\overset{\|}{C}}-CH_2-CH_3$ b. $CH_3-\overset{O}{\overset{\|}{C}}-CH_3$

c. $CH_3-\underset{CH_3}{\underset{|}{CH}}-\overset{O}{\overset{\|}{C}}-CH_2-CH_2-CH_3$ d. $\underset{Cl}{\underset{|}{CH_2}}-\overset{O}{\overset{\|}{C}}-CH_3$

e. $CH_3-\overset{O}{\overset{\|}{C}}-C_6H_5$ f. $CH_3-\overset{O}{\overset{\|}{C}}-C_6H_5$

15.27 Dipole-dipole attractions between molecules raise the boiling point.

15.29 2

15.31 ethanal, because it has a short carbon chain

15.33 a. $CH_3-CH_2-CH_2-CH_2-\overset{O}{\overset{\|}{C}}-H$ b. $CH_3-CH_2-\overset{O}{\overset{\|}{C}}-CH_3$

c. 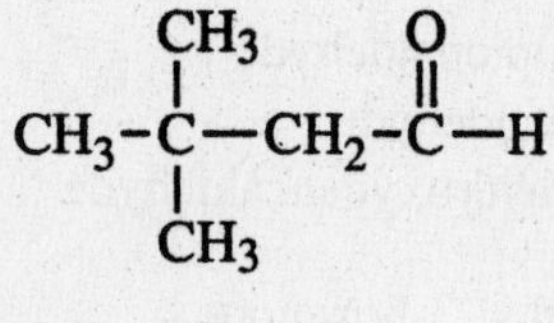 $CH_3-C(CH_3)_2-CH_2-C(=O)-H$

d. 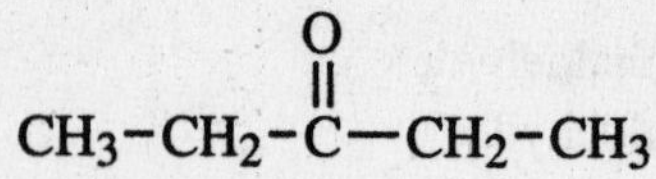 $CH_3-CH_2-C(=O)-CH_2-CH_3$

e. 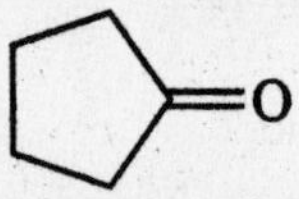 cyclopentanone (five-membered ring with =O)

f. 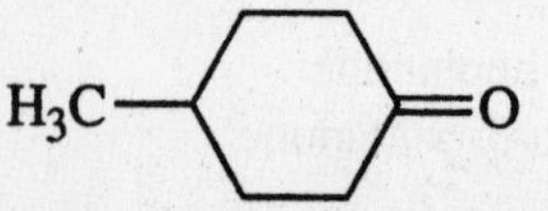 H_3C–cyclohexanone (4-methylcyclohexanone, ring with =O)

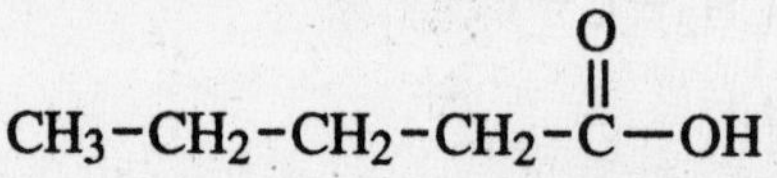

15.35 a. CH_3-CH_2-OH

b. 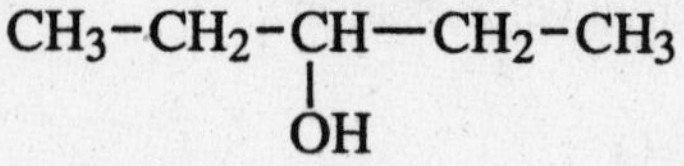 $CH_3-CH_2-CH(OH)-CH_2-CH_3$

c. 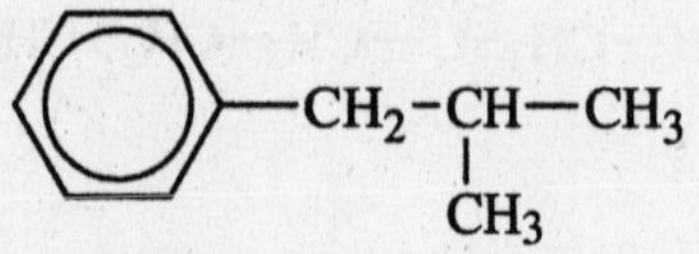 $C_6H_5-CH_2-CH(CH_3)-CH_3$

d. CH_3-CH_2-OH

e. 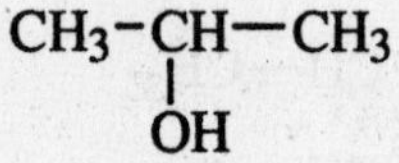 $CH_3-CH(OH)-CH_3$

f. 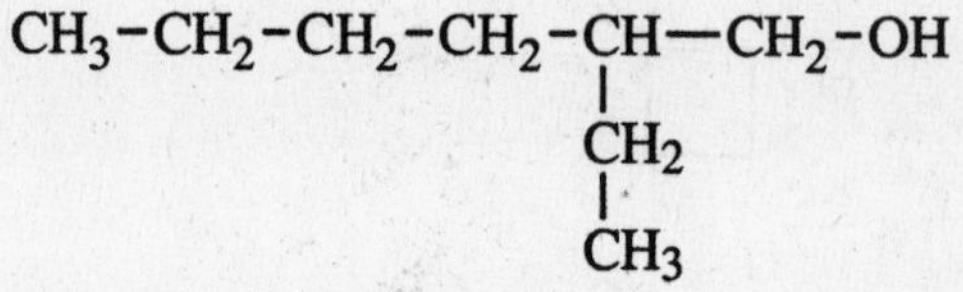 $CH_3-CH_2-CH_2-CH_2-CH(CH_2CH_3)-CH_2-OH$

15.37 a. 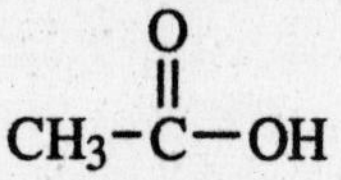 $CH_3-C(=O)-OH$

b. $CH_3-CH_2-CH_2-CH_2-C(=O)-OH$

c. $H-C(=O)-OH$

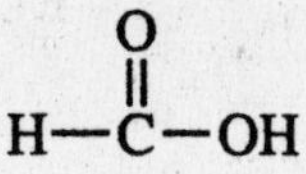

d. 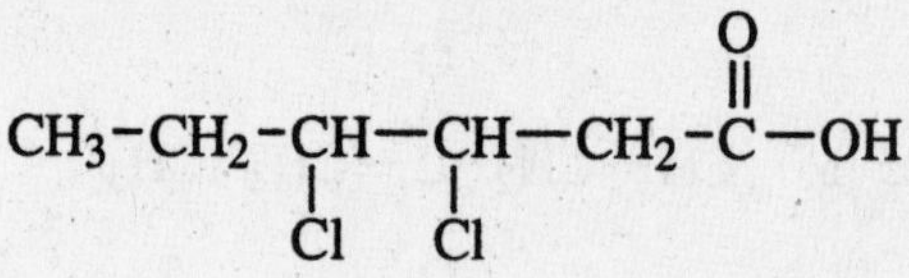 $CH_3-CH_2-CH(Cl)-CH(Cl)-CH_2-C(=O)-OH$

15.39 appearance of a silver mirror

15.41 Cu^{2+} ion

15.43 a. no b. yes c. yes d. no

15.45 a. 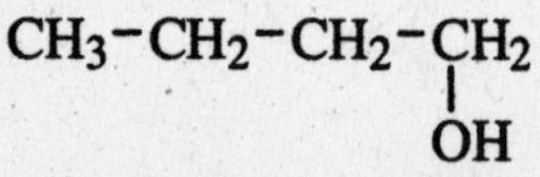 $CH_3-CH_2-CH_2-CH_2-OH$

b. 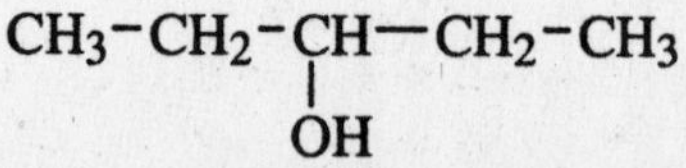 $CH_3-CH_2-CH(OH)-CH_2-CH_3$

c. 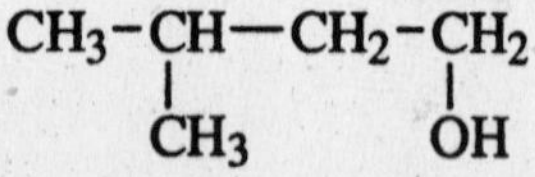 $CH_3-CH(CH_3)-CH_2-CH_2-OH$

d. 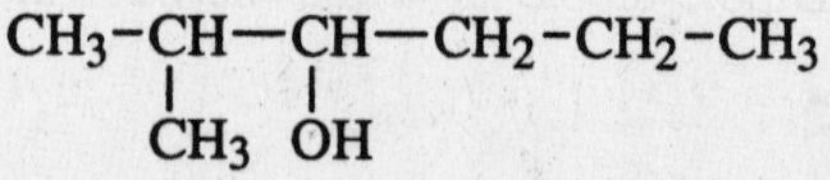 $CH_3-CH(CH_3)-CH(OH)-CH_2-CH_2-CH_3$

15.47 R–O– and H–

15.49 a. no b. yes c. no d. yes e. yes f. no

15.51 a.

$$\begin{array}{c} \quad\;\; OH \\ \quad\;\; | \\ CH_3-CH-O-CH_2-CH_3 \end{array}$$

b.

$$\begin{array}{l} \qquad\quad OH \\ \qquad\quad\; | \\ CH_3-C-CH_2-CH_2-CH_3 \\ \qquad\quad\; | \\ \qquad\quad O-CH_3 \end{array}$$

c.

$$\begin{array}{l} \qquad\qquad\qquad\qquad OH \\ \qquad\qquad\qquad\qquad\; | \\ CH_3-CH_2-CH_2-CH \\ \qquad\qquad\qquad\qquad\; | \\ \qquad\qquad\qquad\qquad O-CH_2-CH_3 \end{array}$$

d.

$$\begin{array}{l} \qquad\quad OH \\ \qquad\quad\; | \\ CH_3-C-CH_3 \\ \qquad\quad\; | \\ \qquad\quad O-CH-CH_3 \\ \qquad\qquad\quad\; | \\ \qquad\qquad\quad CH_3 \end{array}$$

15.53 a.

$$\begin{array}{l} CH_3-(CH_2)_2-CH-O-CH_2-CH_3 \\ \qquad\qquad\qquad\;\; | \\ \qquad\qquad\qquad OH \end{array}$$

b.

$$\begin{array}{l} \qquad\qquad\qquad\;\; O \\ \qquad\qquad\qquad\;\; \| \\ CH_3-CH_2-C-H \end{array}$$

c.

$$\begin{array}{l} \qquad\qquad\qquad OH \\ \qquad\qquad\qquad\; | \\ CH_3-CH_2-C-CH_3 \\ \qquad\qquad\qquad\; | \\ \qquad\qquad\qquad O-CH_3 \end{array}$$

d. (ring structure) CH_2-OH; O; $-OH$; OH

15.55 a. yes b. yes c. no d. yes

15.57 a. CH_3-OH

b.

$$\begin{array}{l} CH_3-CH-O-CH_3 \\ \qquad\quad\; | \\ \qquad\quad OH \end{array}$$

c.

$$\begin{array}{l} CH_3-CH-O-CH_3 \\ \qquad\quad\; | \\ \qquad\quad O-CH-CH_3 \\ \qquad\qquad\quad\; | \\ \qquad\qquad\quad CH_3 \end{array}$$

d.

$$\begin{array}{l} CH_3-CH-O-CH_3, \; CH_3-OH \\ \qquad\quad\; | \\ \qquad\quad OH \end{array}$$

15.59 a.

$$\begin{array}{l} \qquad\quad\;\; O \\ \qquad\quad\;\; \| \\ CH_3-C-H, \quad 2\; CH_3-OH \end{array}$$

b.

$$\begin{array}{l} \qquad\quad\;\; O \\ \qquad\quad\;\; \| \\ CH_3-C-CH_3, \quad 2\; CH_3-OH \end{array}$$

c.

$$\begin{array}{l} \qquad\qquad\qquad\; O \\ \qquad\qquad\qquad\; \| \\ CH_3-CH_2-C-CH_2-CH_3, \quad CH_3-OH, \quad CH_3-CH_2-OH \end{array}$$

d.

$$\begin{array}{l} \qquad\qquad\qquad\qquad\qquad\qquad\quad\; O \\ \qquad\qquad\qquad\qquad\qquad\qquad\quad\; \| \\ CH_3-CH_2-CH_2-CH_2-C-H, \quad 2\; CH_3-OH \end{array}$$

15.61 a. dimethyl acetal of ethanal
b. dimethyl acetal of propanone
c. ethyl methyl acetal of 3-pentanone
d. dimethyl acetal of pentanal

15.63 a. By definition the carbonyl carbon atom is numbered 1 in an aldehyde; therefore, the number does not have to specified in the name.
b. There is only one possible location for the carbonyl group in propanone; therefore, its location does not have to be specified.

15.65 a. heptanal b. 2-heptanone, 3-heptanone, 4-heptanone

15.67

Structure	Name
$CH_3-CH_2-CH_2-CH_2-CH_2-\overset{O}{\overset{\|}{C}}-H$	hexanal
$CH_3-CH_2-CH_2-\overset{CH_3}{\overset{\|}{CH}}-\overset{O}{\overset{\|}{C}}-H$	2-methylpentanal
$CH_3-CH_2-\overset{CH_3}{\overset{\|}{CH}}-CH_2-\overset{O}{\overset{\|}{C}}-H$	3-methylpentanal
$CH_3-\overset{CH_3}{\overset{\|}{CH}}-CH_2-CH_2-\overset{O}{\overset{\|}{C}}-H$	4-methylpentanal
$CH_3-CH_2-\underset{\underset{CH_3}{\|}}{\overset{CH_3}{\overset{\|}{C}}}-\overset{O}{\overset{\|}{C}}-H$	2,2-dimethylbutanal
$CH_3-\underset{\underset{CH_3}{\|}}{\overset{CH_3}{\overset{\|}{C}}}-CH_2-\overset{O}{\overset{\|}{C}}-H$	3,3-dimethybutanal
$CH_3-\overset{CH_3}{\overset{\|}{CH}}-\overset{CH_3}{\overset{\|}{CH}}-\overset{O}{\overset{\|}{C}}-H$	2,3-dimethylbutanal
$CH_3-\overset{O}{\overset{\|}{C}}-CH_2-CH_2-CH_2-CH_3$	2-hexanone
$CH_3-CH_2-\overset{O}{\overset{\|}{C}}-CH_2-CH_2-CH_3$	3-hexanone
$CH_3-\overset{O}{\overset{\|}{C}}-\overset{CH_3}{\overset{\|}{CH}}-CH_2-CH_3$	3-methyl-2-pentanone
$CH_3-\overset{O}{\overset{\|}{C}}-CH_2-\overset{CH_3}{\overset{\|}{CH}}-CH_3$	4-methyl-2-pentanone
$CH_3-\overset{CH_3}{\overset{\|}{CH}}-\overset{O}{\overset{\|}{C}}-CH_2-CH_3$	2-methyl-3-pentanone
$CH_3-\overset{O}{\overset{\|}{C}}-\underset{\underset{CH_3}{\|}}{\overset{CH_3}{\overset{\|}{C}}}-CH_3$	3,3-dimethyl-2-butanone

15.69 a. ketone, alkene b. aldehyde, alcohol, ether
c. ketone, alkyne d. aldehyde, ketone

Carboxylic Acids and Esters — Chapter 16

Solutions to Selected Problems

16.1 a. yes b. no c. yes d. yes e. no f. yes

16.3 a. butanoic acid b. heptanoic acid
c. 2,3-dimethypentanoic acid d. 4-bromopentanoic acid
e. 3-methylpentanoic acid f. chloroethanoic acid

16.5 a. $CH_3-CH_2-CH(CH_2CH_3)-C(=O)-OH$

b. $CH_3-CH(CH_3)-CH_2-CH_2-CH(CH_3)-C(=O)-OH$

c. $CH_3-CH(CH_3)-C(=O)-OH$

d. $CH_3-CH(Cl)-C(=O)-OH$

e. $CH_3-CH_2-CH_2-CH(Cl)-CH_2-CH(Br)-CH_2-C(=O)-OH$

f. $CH_3-CH(CH_3)-CH(CH_3)-C(=O)-OH$

16.7 a. butanedioic acid b. propanedioic acid
c. 3-methylpentanedioic acid d. 2-chlorobenzoic acid
e. *m*-toluic acid f. 2-bromo-4-chlorobenzoic acid

16.9 a. $CH_3-CH_2-C(CH_3)_2-C(=O)-OH$

b. $HO-C(=O)-C(CH_3)_2-CH_2-C(=O)-OH$

c. $HO-C(=O)-C(CH_3)_2-CH_2-CH_2-C(=O)-OH$

d. Benzene ring with $-C(=O)-OH$ and Br (ortho)

e. Benzene ring with $-C(=O)-OH$, Cl (ortho) and Cl (para)

f. Benzene ring with $-C(=O)-OH$ and H_3C (para)

16.11 a. $CH_3-CH_2-CH_2-CH_2-\overset{O}{\overset{\|}{C}}-OH$ b. $CH_3-CH_2-\overset{O}{\overset{\|}{C}}-OH$

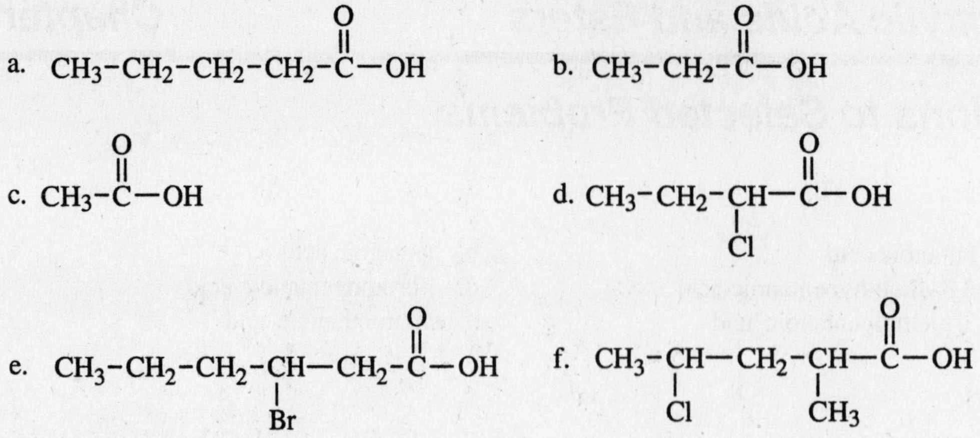

c. $CH_3-\overset{O}{\overset{\|}{C}}-OH$ d. $CH_3-CH_2-\underset{Cl}{\underset{|}{CH}}-\overset{O}{\overset{\|}{C}}-OH$

e. $CH_3-CH_2-CH_2-\underset{Br}{\underset{|}{CH}}-CH_2-\overset{O}{\overset{\|}{C}}-OH$ f. $CH_3-\underset{Cl}{\underset{|}{CH}}-CH_2-\underset{CH_3}{\underset{|}{CH}}-\overset{O}{\overset{\|}{C}}-OH$

16.13 a. $OH-\overset{O}{\overset{\|}{C}}-CH_2-\overset{O}{\overset{\|}{C}}-OH$ b. $OH-\overset{O}{\overset{\|}{C}}-CH_2-CH_2-\overset{O}{\overset{\|}{C}}-OH$

c. $OH-\overset{O}{\overset{\|}{C}}-(CH_2)_4-\overset{O}{\overset{\|}{C}}-OH$ d. $OH-\overset{O}{\overset{\|}{C}}-(CH_2)_2-\underset{Br}{\underset{|}{CH}}-(CH_2)_2-\overset{O}{\overset{\|}{C}}-OH$

e. $OH-\overset{O}{\overset{\|}{C}}-\underset{CH_3}{\underset{|}{CH}}-(CH_2)_2-\overset{O}{\overset{\|}{C}}-OH$ f. $OH-\overset{O}{\overset{\|}{C}}-\underset{Br}{\underset{|}{\overset{Cl}{\overset{|}{C}}}}-CH_2-\overset{O}{\overset{\|}{C}}-OH$

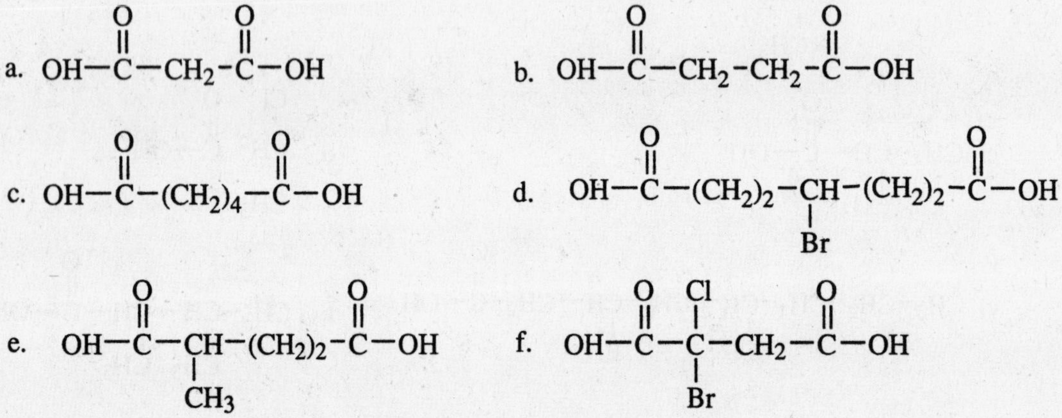

16.15 a. 3 b. 1 c. 2 d. 1

16.17 a. carbon-carbon double bond b. hydroxyl group
c. carbon-carbon double bond d. hydroxyl group

16.19 a. propenoic acid b. 2-hydroxypropanoic acid
c. *cis*-butenedioic acid d. 2-hydroxyethanoic acid

16.21 a. $CH_3-CH_2-\overset{O}{\overset{\|}{C}}-CH_2-\overset{O}{\overset{\|}{C}}-OH$ b. $CH_3-CH_2-\overset{OH}{\overset{|}{CH}}-\overset{O}{\overset{\|}{C}}-OH$

c. $H_3C(H)C=C(H)-CH_2-CH_2-\overset{O}{\overset{\|}{C}}-OH$ (H₃C and H on one carbon; H and CH_2-CH_2-COOH on the other, H₃C and CH₂ trans)

d. $OH-\overset{O}{\overset{\|}{C}}-\underset{OH}{\underset{|}{CH}}-\underset{OH}{\underset{|}{CH}}-CH_2-\overset{O}{\overset{\|}{C}}-OH$

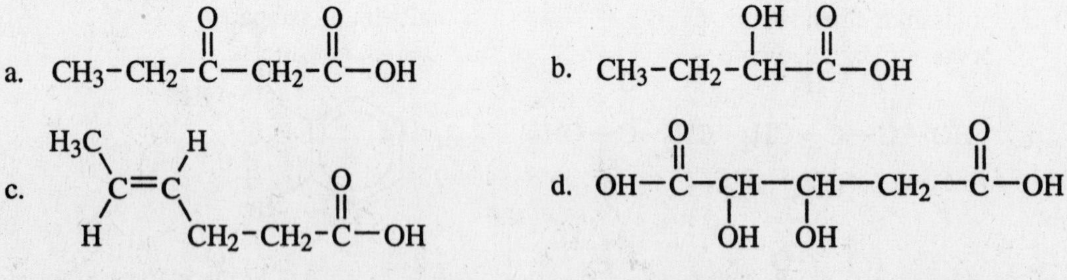

16.23 a. propionic acid b. propionic acid c. succinic acid d. glutaric acid

16.25 a. hydroxy, carboxy b. hydroxy, carboxy
c. keto, carboxy d. hydroxy, carboxy

16.27 a. 2 b. 5

16.29 a. solid b. solid c. liquid d. solid

16.31 a. $CH_3-\overset{O}{\overset{\|}{C}}-OH$ b. $CH_3-\overset{O}{\overset{\|}{C}}-OH$

c. $CH_3-CH_2-\underset{\underset{CH_3}{|}}{CH}-CH_2-\overset{O}{\overset{\|}{C}}-OH$ d. $C_6H_5-\overset{O}{\overset{\|}{C}}-OH$

16.33 a. 1 b. 3 c. 2 d. 2

16.35 a. –1 b. –3 c. –2 d. –2

16.37 a. pentanoate ion b. citrate ion c. succinate ion d. oxalate ion

16.39 a. $CH_3-\overset{O}{\overset{\|}{C}}-OH + H_2O \longrightarrow H_3O^+ + CH_3-\overset{O}{\overset{\|}{C}}-O^-$

b. $OH-\overset{O}{\overset{\|}{C}}-CH_2-\overset{OH}{\overset{|}{\underset{\underset{\underset{OH}{|}}{\underset{C=O}{|}}}{C}}}-CH_2-\overset{O}{\overset{\|}{C}}-OH + 3H_2O \longrightarrow 3H_3O^+ + {}^-O-\overset{O}{\overset{\|}{C}}-CH_2-\overset{OH}{\overset{|}{\underset{\underset{\underset{O^-}{|}}{\underset{C=O}{|}}}{C}}}-CH_2-\overset{O}{\overset{\|}{C}}-O^-$

c. $CH_3-\overset{O}{\overset{\|}{C}}-OH + H_2O \longrightarrow H_3O^+ + CH_3-\overset{O}{\overset{\|}{C}}-O^-$

d. $CH_3-CH_2-\underset{\underset{CH_3}{|}}{CH}-\overset{O}{\overset{\|}{C}}-OH + H_2O \longrightarrow H_3O^+ + CH_3-CH_2-\underset{\underset{CH_3}{|}}{CH}-\overset{O}{\overset{\|}{C}}-O^-$

16.41 a. potassium ethanoate b. calcium propanoate
c. potassium butanedioate d. sodium pentanoate

16.43 a. $CH_3-\overset{O}{\overset{\|}{C}}-OH + KOH \longrightarrow CH_3-\overset{O}{\overset{\|}{C}}-O^-\ K^+ + H_2O$

b. $2\ CH_3-CH_2-\overset{O}{\overset{\|}{C}}-OH + Ca(OH)_2 \longrightarrow \left(CH_3-CH_2-\overset{O}{\overset{\|}{C}}-O^-\right)_2 Ca^{2+} + 2H_2O$

c. $HO-\overset{O}{\overset{\|}{C}}-CH_2-CH_2-\overset{O}{\overset{\|}{C}}-OH + 2\ KOH \longrightarrow K^{+\,-}O-\overset{O}{\overset{\|}{C}}-CH_2-CH_2-\overset{O}{\overset{\|}{C}}-O^-\ K^+ + 2H_2O$

d. $CH_3-CH_2-CH_2-CH_2-\overset{O}{\overset{\|}{C}}-OH + NaOH \longrightarrow CH_3-CH_2-CH_2-CH_2-\overset{O}{\overset{\|}{C}}-O^- Na^+ + H_2O$

16.45 a. $CH_3-CH_2-CH_2-\overset{O}{\overset{\|}{C}}-O^- Na^+ + HCl \longrightarrow CH_3-CH_2-CH_2-\overset{O}{\overset{\|}{C}}-OH + NaCl$

b. $K^+ \, ^-O-\overset{O}{\overset{\|}{C}}-\overset{O}{\overset{\|}{C}}-O^- K^+ + 2HCl \longrightarrow HO-\overset{O}{\overset{\|}{C}}-\overset{O}{\overset{\|}{C}}-OH + 2KCl$

c. $\left(^-O-\overset{O}{\overset{\|}{C}}-CH_2-\overset{O}{\overset{\|}{C}}-O^-\right) Ca^{2+} + 2HCl \longrightarrow HO-\overset{O}{\overset{\|}{C}}-CH_2-\overset{O}{\overset{\|}{C}}-OH + CaCl_2$

d. $C_6H_5-\overset{O}{\overset{\|}{C}}-O^- Na^+ + HCl \longrightarrow C_6H_5-\overset{O}{\overset{\|}{C}}-OH + NaCl$

16.47 a. yes b. yes c. no d. yes e. yes f. no

16.49 a. $CH_3-CH_2-\overset{O}{\overset{\|}{C}}-O-CH_3$

b. $CH_3-\overset{O}{\overset{\|}{C}}-O-CH_2-CH_2-CH_3$

c. $CH_3-CH_2-\underset{CH_3}{\underset{|}{CH}}-\overset{O}{\overset{\|}{C}}-O-\underset{CH_3}{\underset{|}{CH}}-CH_3$

d. $CH_3-CH_2-CH_2-CH_2-\overset{O}{\overset{\|}{C}}-O-\underset{CH_3}{\underset{|}{CH}}-CH_2-CH_3$

16.51 a. $CH_3-CH_2-\overset{O}{\overset{\|}{C}}-OH$, CH_3-CH_2-OH b. $CH_3-CH_2-CH_2-\overset{O}{\overset{\|}{C}}-OH$, CH_3-OH

c. $CH_3-CH_2-CH_2-\overset{O}{\overset{\|}{C}}-OH$, CH_3-OH d. $CH_3-\overset{O}{\overset{\|}{C}}-OH$, C_6H_5-OH

e. $C_6H_5-\overset{\displaystyle O}{\overset{\|}{C}}-OH$, CH_3-OH

f. $CH_3-\underset{\displaystyle Cl}{\underset{|}{CH}}-\overset{\displaystyle O}{\overset{\|}{C}}-OH$, CH_3-CH_2-OH

16.53 a. methyl propanoate
b. methyl methanoate
c. methyl ethanoate
d. propyl ethanoate
e. isopropyl propanoate
f. ethyl benzoate

16.55 a. methyl propionate
b. methyl formate
c. methyl acetate
d. propyl acetate
e. isopropyl propionate
f. ethyl benzoate

16.57 a. $H-\overset{\displaystyle O}{\overset{\|}{C}}-O-CH_3$

b. $CH_3-\overset{\displaystyle O}{\overset{\|}{C}}-O-CH_2-CH_2-CH_3$

c. $CH_3-(CH_2)_8-\overset{\displaystyle O}{\overset{\|}{C}}-O-(CH_2)_7-CH_3$

d. $C_6H_5-CH_2-\overset{\displaystyle O}{\overset{\|}{C}}-O-CH_2-CH_3$

e. $CH_3-\overset{\displaystyle O}{\overset{\|}{C}}-O-\underset{\displaystyle CH_3}{\underset{|}{CH}}-CH_3$

f. $CH_3-\overset{\displaystyle O}{\overset{\|}{C}}-O-CH_2-\underset{\displaystyle Br}{\underset{|}{CH}}-CH_3$

16.59 a. ethyl ethanoate
b. methyl ethanoate
c. ethyl butanoate
d. propyl 2-hydroxypropanoate
e. pentyl pentanoate
f. 1-methylpropyl hexanoate

16.61 No oxygen-hydrogen bonds are present within an ester.

16.63 Ester molecules cannot hydrogen-bond to each other; acid molecules can hydrogen-bond to each other.

16.65 a. $CH_3-CH_2-\overset{\displaystyle O}{\overset{\|}{C}}-OH$, CH_3-CH_2-OH

b. $CH_3-\overset{\displaystyle O}{\overset{\|}{C}}-OH$, CH_3-CH_2-OH

c. $CH_3-\underset{\displaystyle CH_3}{\underset{|}{CH}}-\overset{\displaystyle O}{\overset{\|}{C}}-OH$, C_6H_5-OH

d. $CH_3-CH_2-CH_2-\overset{\displaystyle O}{\overset{\|}{C}}-OH$, CH_3-OH

e. $H-\overset{\displaystyle O}{\overset{\|}{C}}-OH$, CH_3-CH_2-OH

f. $C_6H_5-\overset{\displaystyle O}{\overset{\|}{C}}-OH$, $CH_3-\underset{\displaystyle CH_3}{\underset{|}{CH}}-OH$

16.67 a. $CH_3-CH_2-\overset{O}{\overset{\|}{C}}-O^- Na^+$, CH_3-CH_2-OH b. $CH_3-\overset{O}{\overset{\|}{C}}-O^- Na^+$; CH_3-CH_2-OH

c. $CH_3-\underset{CH_3}{\underset{|}{CH}}-\overset{O}{\overset{\|}{C}}-O^- Na^+$; C_6H_5-OH d. $CH_3-CH_2-CH_2-\overset{O}{\overset{\|}{C}}-O^- Na^+$; CH_3-OH

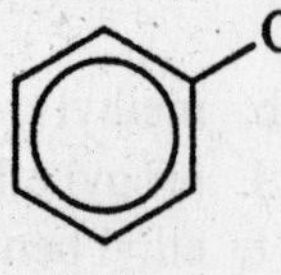

e. $H-\overset{O}{\overset{\|}{C}}-O^- Na^+$; CH_3-CH_2-OH f. 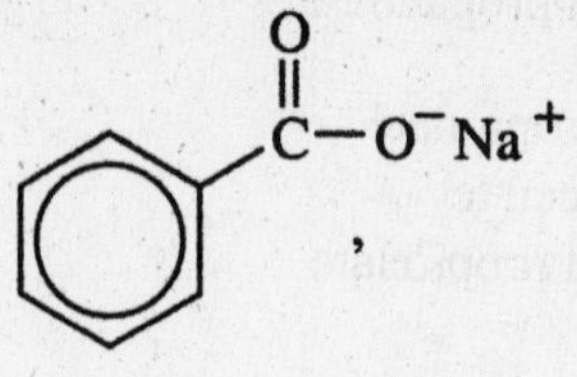 $C_6H_5-\overset{O}{\overset{\|}{C}}-O^- Na^+$, $CH_3-\underset{CH_3}{\underset{|}{CH}}-OH$

16.69 a. $CH_3-\underset{CH_3}{\underset{|}{CH}}-\overset{O}{\overset{\|}{C}}-OH$, CH_3-CH_2-OH

b. $CH_3-\underset{CH_3}{\underset{|}{CH}}-\overset{O}{\overset{\|}{C}}-O^- Na^+$; CH_3-CH_2-OH

c. $H-\overset{O}{\overset{\|}{C}}-OH$, $CH_3-CH_2-CH_2-CH_2-OH$

d. $CH_3-\overset{O}{\overset{\|}{C}}-O^- Na^+$; $CH_3-\underset{CH_3}{\underset{|}{CH}}-\underset{CH_3}{\underset{|}{CH}}-CH_2-OH$

16.71 a. $CH_3-\overset{O}{\overset{\|}{C}}-S-CH_2-CH_3$ b. $CH_3-(CH_2)_8-\overset{O}{\overset{\|}{C}}-S-CH_3$

c. 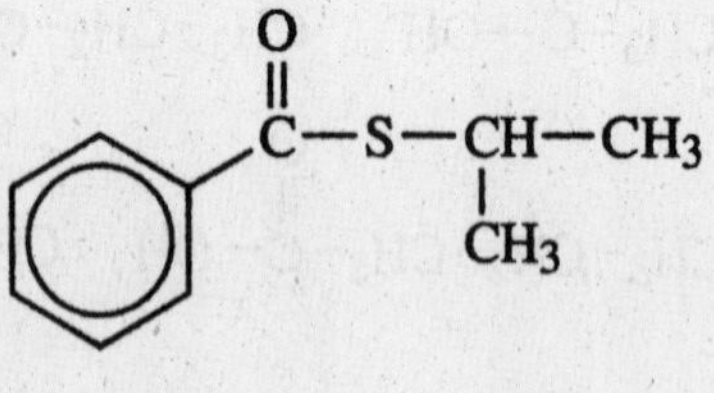 $C_6H_5-\overset{O}{\overset{\|}{C}}-S-\underset{CH_3}{\underset{|}{CH}}-CH_3$ d. $H-\overset{O}{\overset{\|}{C}}-S-CH_2-CH_2-CH_3$

16.73 $-\overset{O}{\overset{\|}{C}}-\overset{O}{\overset{\|}{C}}-O-(CH_2)_3-O-\overset{O}{\overset{\|}{C}}-\overset{O}{\overset{\|}{C}}-O-(CH_2)_3-O-$

16.75 $HO-\overset{O}{\overset{\|}{C}}-(CH_2)_2-\overset{O}{\overset{\|}{C}}-OH$, $HO-(CH_2)_3-OH$

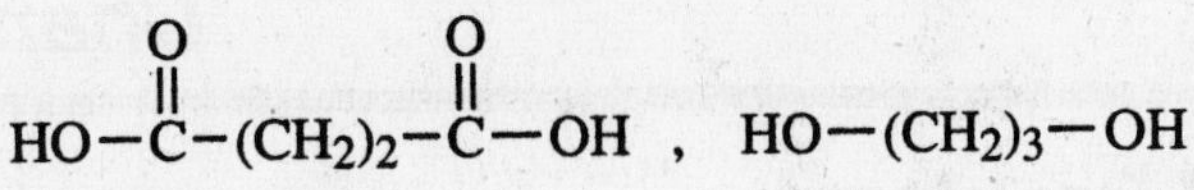

16.77 a. 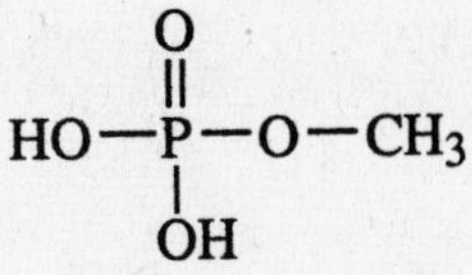 $HO-\overset{O}{\overset{\|}{P}}(OH)-O-CH_3$

b. 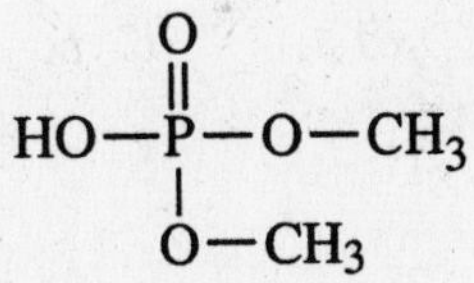 $HO-\overset{O}{\overset{\|}{P}}(O-CH_3)-O-CH_3$

c. 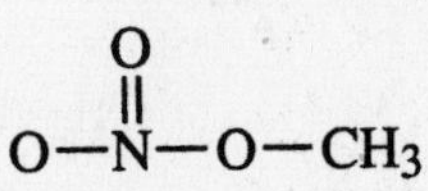 $O-\overset{O}{\overset{\|}{N}}-O-CH_3$

d. 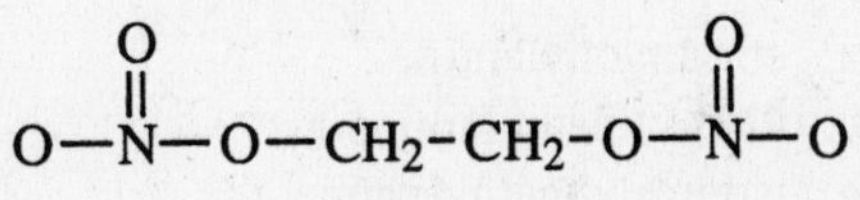 $O-\overset{O}{\overset{\|}{N}}-O-CH_2-CH_2-O-\overset{O}{\overset{\|}{N}}-O$

16.79 H_3PO_4 is a triprotic acid, and H_2SO_4 is a diprotic acid.

16.81 a. 2, 2 b. 7, 1 c. 7, 1 d. 6, 3 e. 3, 1 f. 2, 1

16.83 $C_nH_{2n-2}O_2$

16.85 methyl propanoate; ethyl ethanoate; propyl methanoate; isopropyl methanoate

16.87 $CH_3-\overset{O}{\overset{\|}{C}}-O-CH_2-CH_3$

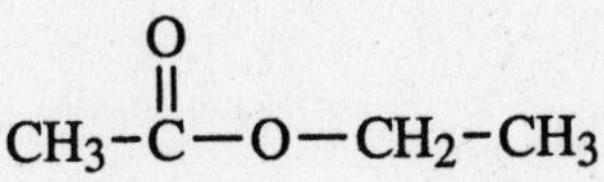

Amines and Amides — Chapter 17

Solutions to Selected Problems

17.1 a. yes b. yes c. no d. yes e. no f. yes

17.3 a. 1° b. 1° c. 2° d. 2° e. 1° f. 3°

17.5 a. 2° b. 3° c. 3° d. 1° e. 2° f. 2°

17.7 a. ethylmethylamine b. propylamine
b. diethylmethylamine d. diphenylamine
e. isopropylmethylamine f. diisopropylamine

17.9 a. 3-pentanamine b. 2-methyl-3-pentanamine
c. *N*-methyl-3-pentanamine d. 1,5-pentanediamine
e. 2,3-butanediamine f. *N*,*N*-dimethyl-1-butanamine

17.11 a. 2-bromoaniline b. *N*-isopropylaniline
c. *N*-ethyl-*N*-methylaniline d. *N*-methyl-*N*-phenylaniline
e. *N*-ethyl-*N*-methylaniline f. *N*-(1-chloroethyl)aniline

17.13 a. $CH_3-CH_2-NH_2$

b. 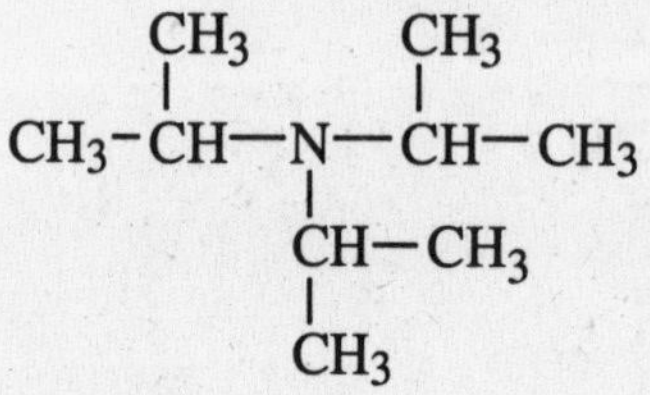 $CH_3-CH(CH_3)-N(CH(CH_3)-CH_3)-CH(CH_3)-CH_3$

c. 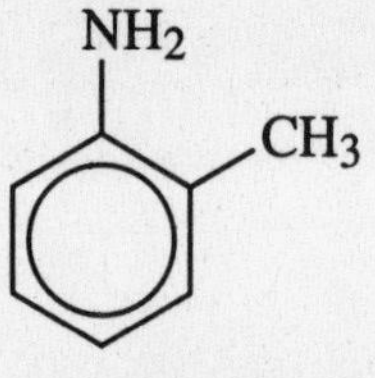 NH_2, CH_3 (benzene ring)

d. 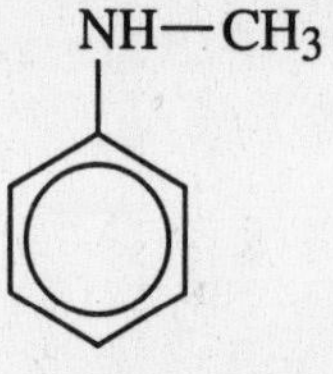 $NH-CH_3$ (benzene ring)

e. 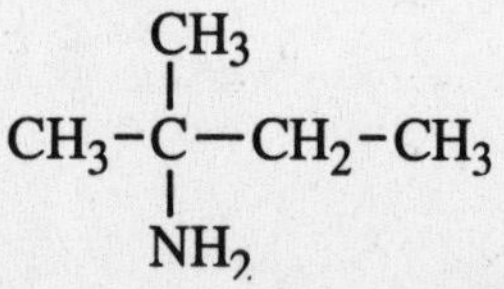 $CH_3-C(CH_3)(NH_2)-CH_2-CH_3$

f. $H_2N-CH_2-CH_2-CH_2-CH_2-CH_2-CH_2-NH_2$

g. 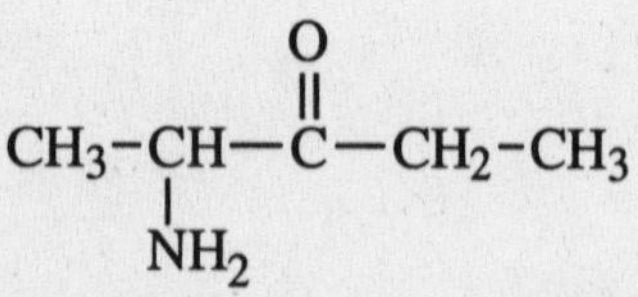 $CH_3-CH(NH_2)-C(=O)-CH_2-CH_3$

h. 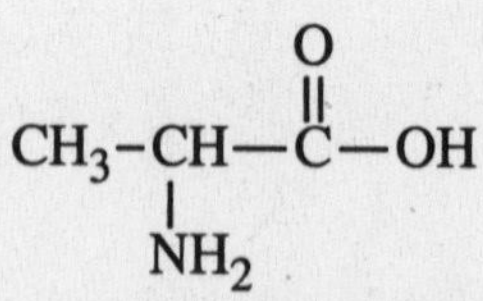 $CH_3-CH(NH_2)-C(=O)-OH$

17.15 a. liquid b. gas c. gas d. liquid

17.17 a. 3 b. 3

17.19 Hydrogen bonding is possible for the amine.

17.21 a. $CH_3–CH_2–NH_2$; it has a shorter carbon chain.
b. $H_2N–CH_2–CH_2–CH_2–NH_2$; it has two amino groups rather than one.

17.23 a. $CH_3-CH_2-\overset{+}{N}H_3$
b. OH^-
c. $CH_3-CH(CH_3)-NH-CH_3$
d. $CH_3-CH_2-\overset{+}{(NH)_2}-CH_2-CH_3 + OH^-$

17.25 a. dimethylammonium ion
b. triethylammonium ion
c. *N,N*-diethylanilinium ion
d. dimethylpropylammonium ion
e. propylammonium ion
f. *N*-isopropylanilinium ion

17.27 a. $CH_3-NH-CH_3$
b. $CH_3-CH_2-N(CH_2-CH_3)-CH_2-CH_3$
c. 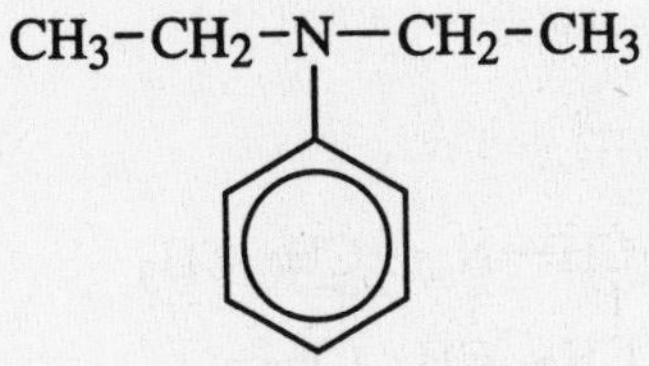 $CH_3-CH_2-N(C_6H_5)-CH_2-CH_3$ (N bonded to a benzene ring)
d. $CH_3-N(CH_3)-CH_2-CH_2-CH_3$
e. $CH_3-CH_2-CH_2-NH_2$
f. 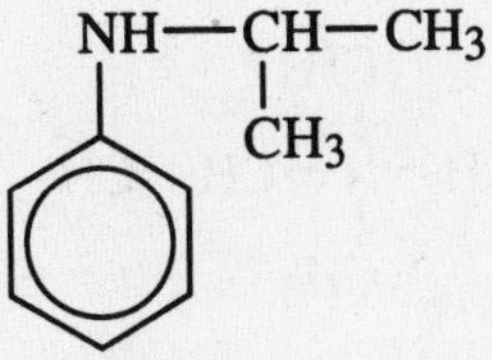 $C_6H_5-NH-CH(CH_3)-CH_3$ (NH bonded to a benzene ring)

17.29 a. $CH_3-CH_2-\overset{+}{N}H_3\ Cl^-$
b. 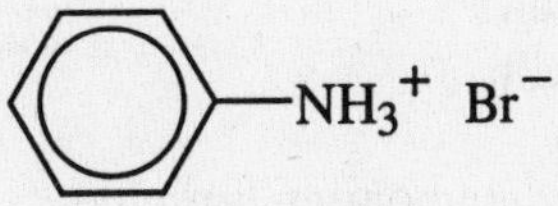 $C_6H_5-NH_3^+\ Br^-$ (benzene ring bonded to NH_3^+)
c. $CH_3-C(CH_3)_2-NH_2$
d. HCl

17.31 a. $CH_3-CH(CH_3)-NH_2$
b. $CH_3-\overset{+}{N}H_2-CH_3\ Cl^-$
c. 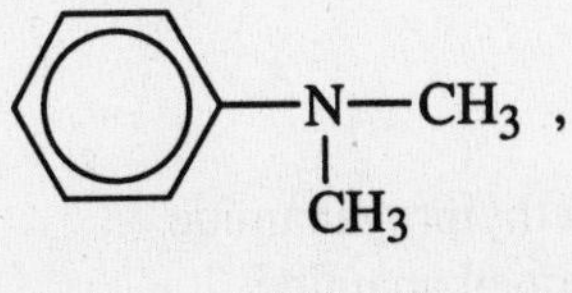 $C_6H_5-N(CH_3)-CH_3$, NaBr (benzene ring bonded to N)
d. $CH_3-NH-CH_3$

17.33 a. propylammonium chloride
b. methylpropylammonium chloride
c. ethyldimethylammonium bromide
d. *N,N*-dimethylanilinium bromide

17.35 to increase the water solubility of the drug

17.37 ethylmethylamine hydrochloride

17.39 a. $CH_3-CH_2-CH_2-NH_2$, NaCl, H_2O

b. 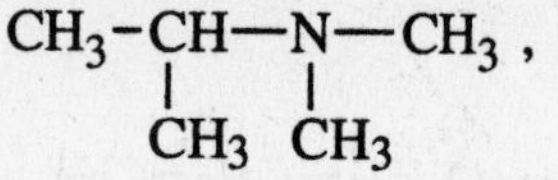 $CH_3-CH(CH_3)-N(CH_3)-CH_3$, NaBr, H_2O

c. $CH_3-CH_2-NH-CH_2-CH_3$, NaCl, H_2O

d. 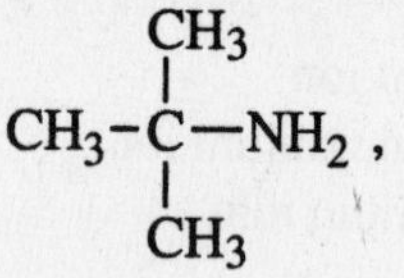 $CH_3-C(CH_3)_2-NH_2$, NaBr, H_2O

17.41 ethylmethylamine, propyl chloride
ethylpropylamine, methyl chloride
methylpropylamine, ethyl chloride

17.43 a. 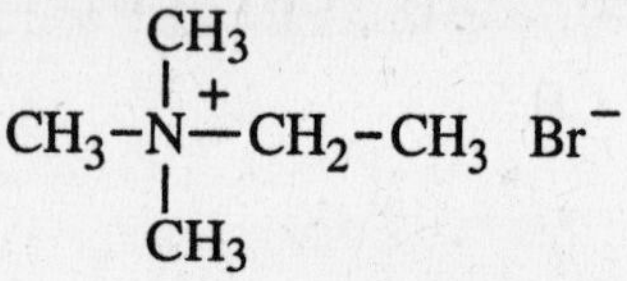 $CH_3-\overset{+}{N}(CH_3)_2-CH_2-CH_3$ Br^-

b. 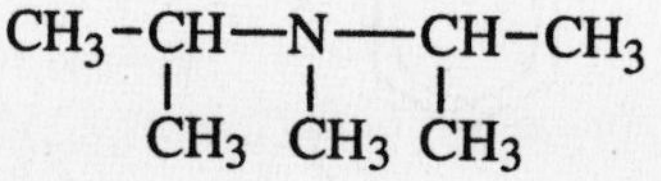 $CH_3-CH(CH_3)-N(CH_3)-CH(CH_3)-CH_3$

c. 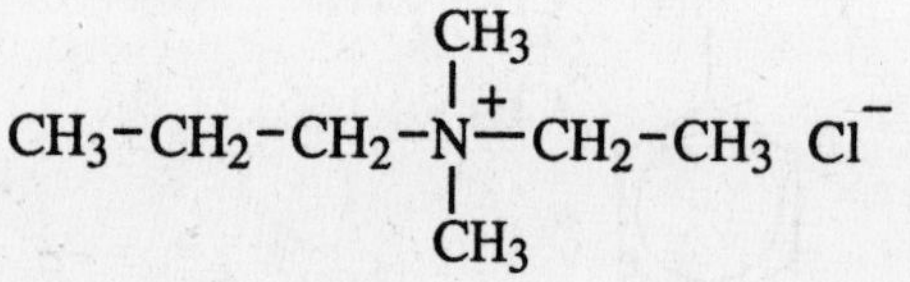 $CH_3-CH_2-CH_2-\overset{+}{N}(CH_3)_2-CH_2-CH_3$ Cl^-

d. $CH_3-CH_2-NH-CH_2-CH_3$

17.45 a. amine salt
b. quaternary ammonium salt
c. amine salt
d. quaternary ammonium salt

17.47 a. trimethylammonium bromide
b. tetramethylammonium chloride
c. ethylmethylammonium bromide
d. diethyldimethylammonium chloride

17.49 a. purine b. pyrrole c. imidazole d. indole

17.51 a. true b. false c. true d. false e. false f. false

17.53 a. yes b. yes c. no d. yes e. no f. yes

17.55 a. monosubstituted
b. disubstituted
c. unsubstituted
d. monosubstituted

17.57 a. 2° b. 3° c. 1° d. 2°

17.59 a. *N*-ethylethanamide
b. *N,N*-dimethylpropanamide
c. butanamide
d. *N*-methylmethanamide
e. 2-chloropropanamide
f. 2,*N*-dimethylpropanamide

17.61 a. *N*-ethylacetamide
b. *N,N*-dimethylpropionamid
c. butyramide
d. *N*-methylformamide
e. 2-chloropropionamide
f. 2,*N*-dimethylpropionamide

17.63 a. $CH_3-\overset{O}{\overset{\|}{C}}-\underset{CH_3}{\underset{|}{N}}-CH_3$

b. $CH_3-CH_2-\underset{CH_3}{\underset{|}{CH}}-\overset{O}{\overset{\|}{C}}-NH_2$

c. $CH_3-\underset{CH_3}{\underset{|}{CH}}-CH_2-\overset{O}{\overset{\|}{C}}-NH-CH_3$

d. $H-\overset{O}{\overset{\|}{C}}-NH_2$

e. $C_6H_5-\overset{O}{\overset{\|}{C}}-NH-C_6H_5$

f. $H-\overset{O}{\overset{\|}{C}}-NH_2$

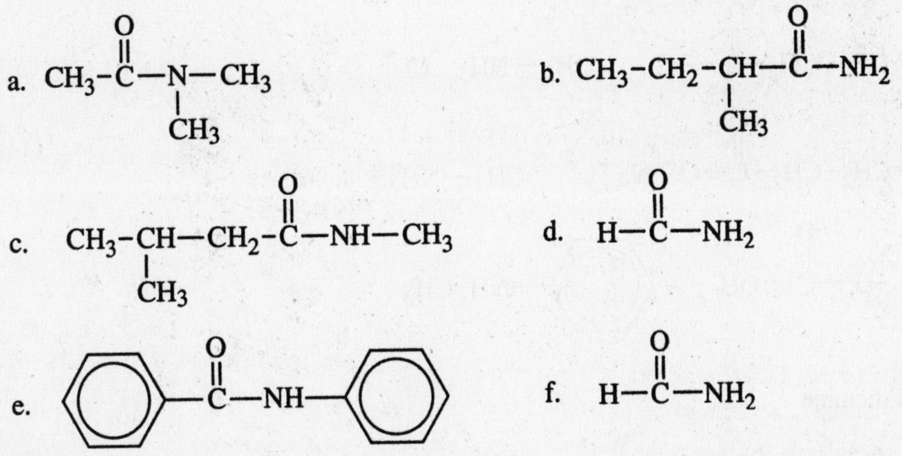

17.65 An electronegativity effect induced by the carbonyl oxygen atom makes the lone pair of electrons on the nitrogen atom unavailable.

17.67 a. 5 b. 5

17.69 a. CH_3-NH_2

b. $CH_3-\overset{CH_3}{\underset{CH_3}{C}}-\overset{O}{\overset{\|}{C}}-\underset{CH_3}{\underset{|}{N}}-CH_3$

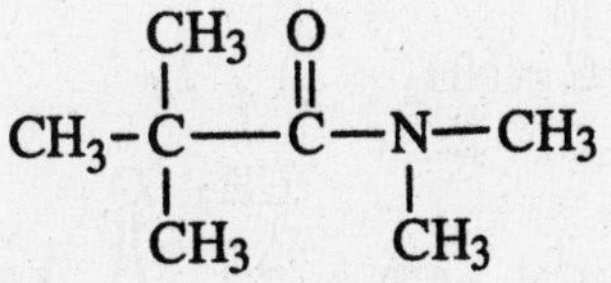

c. NH_3

d. $C_6H_5-\overset{O}{\overset{\|}{C}}-OH$

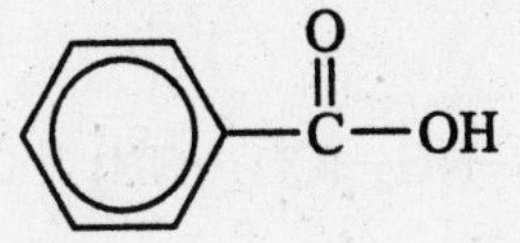

17.71 a. $CH_3-\overset{O}{\overset{\|}{C}}-OH$, $CH_3-NH-\underset{CH_3}{\underset{|}{CH}}-CH_3$

b. $CH_3-CH_2-CH_2-CH_2-\overset{O}{\overset{\|}{C}}-OH$, CH_3-NH_2

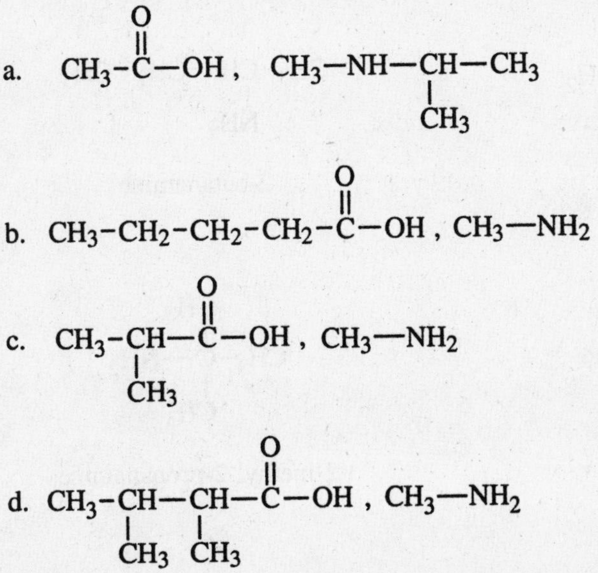

c. $CH_3-\underset{CH_3}{\underset{|}{CH}}-\overset{O}{\overset{\|}{C}}-OH$, CH_3-NH_2

d. $CH_3-\underset{CH_3}{\underset{|}{CH}}-\underset{CH_3}{\underset{|}{CH}}-\overset{O}{\overset{\|}{C}}-OH$, CH_3-NH_2

17.73 a. $CH_3-CH_2-CH_2-\overset{\overset{\displaystyle O}{\|}}{C}-OH$, CH_3-NH_2

b. $CH_3-CH_2-CH_2-\overset{\overset{\displaystyle O}{\|}}{C}-OH$, $CH_3-\overset{+}{N}H_3\ Cl^-$

c. $CH_3-CH_2-CH_2-\overset{\overset{\displaystyle O}{\|}}{C}-O^-\ Na^+$, CH_3-NH_2

d. 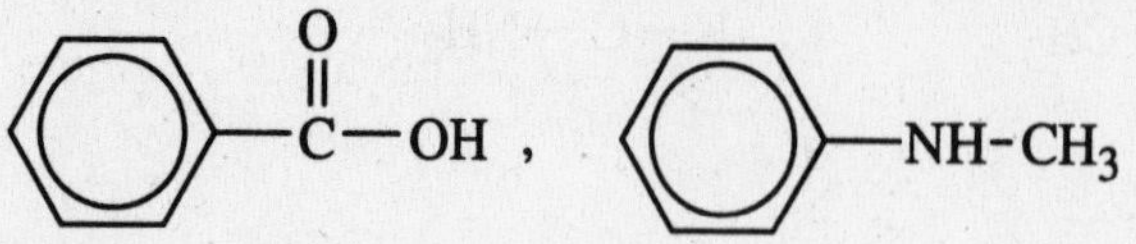 $C_6H_5-\overset{\overset{\displaystyle O}{\|}}{C}-OH$, $C_6H_5-NH-CH_3$

17.75 diacid and diamine

17.77 $\left(\overset{\overset{\displaystyle O}{\|}}{C}-CH_2-CH_2-\overset{\overset{\displaystyle O}{\|}}{C}-\overset{\overset{\displaystyle H}{|}}{N}-(CH_2)_4-\overset{\overset{\displaystyle H}{|}}{N}\right)_n$

17.79 a. $H-\overset{\overset{\displaystyle O}{\|}}{C}-NH_2$

b. $CH_3-\underset{\underset{\displaystyle NH_2}{|}}{CH}-CH_2-CH_2-CH_3$

c. $CH_3-CH_2-CH_2-\overset{\overset{\displaystyle CH_3}{|}}{CH}-\overset{\overset{\displaystyle O}{\|}}{C}-NH_2$

d. $CH_3-\overset{\overset{\displaystyle O}{\|}}{C}-NH-\overset{\overset{\displaystyle CH_3}{|}}{CH}-CH_3$

e. $CH_3-CH_2-\overset{+}{N}H_2-CH_2-CH_3\ Cl^-$

f. 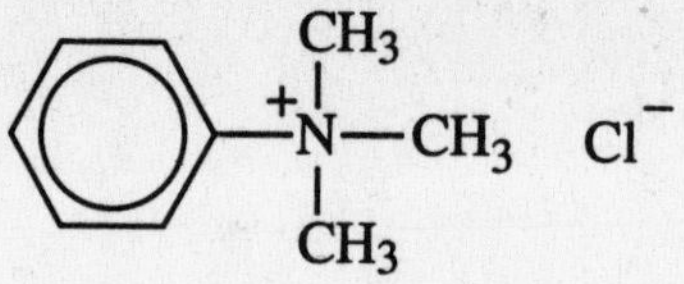 $C_6H_5-\overset{+}{N}(CH_3)_3\ Cl^-$ (phenyl–N⁺ bearing three CH_3 groups)

17.81 $CH_3-CH_2-CH_2-CH_2-NH_2$

1-butanamine

$CH_3-\underset{\underset{\displaystyle NH_2}{|}}{CH}-CH_2-CH_3$

2-butanamine

$CH_3-\underset{\underset{\displaystyle CH_3}{|}}{CH}-CH_2-NH_2$

2-methyl-1-propanamine

$CH_3-\overset{\overset{\displaystyle CH_3}{|}}{\underset{\underset{\displaystyle CH_3}{|}}{C}}-NH_2$

2-methyl-2-propanamine

$CH_3-CH_2-CH_2-NH-CH_3$

N-methyl-1-propanamine

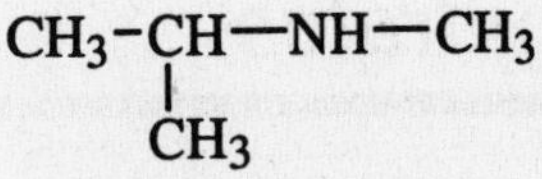

$$CH_3-\underset{\displaystyle CH_3}{\underset{|}{CH}}-NH-CH_3$$

N-methyl-2-propanamine

$CH_3-CH_2-NH-CH_2-CH_3$

N-ethylethanamine

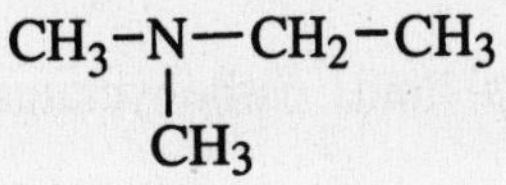

$$CH_3-\underset{\displaystyle CH_3}{\underset{|}{N}}-CH_2-CH_3$$

N,*N*-dimethylethanamine

17.83 $$CH_3-\overset{\displaystyle CH_3}{\overset{+|}{\underset{\displaystyle CH_3}{\underset{|}{N}}}}-CH_2-CH_3 \quad Cl^-$$

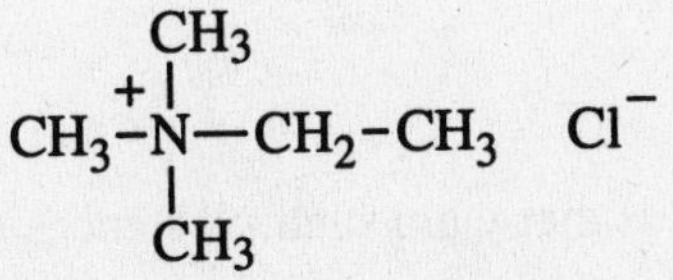

17.85 a. amide b. amine c. amide d. amine e. amine f. amide

Carbohydrates — Chapter 18

Solutions to Selected Problems

18.1 Biochemistry is the study of the chemical substances found in living systems and the chemical interactions of these substances with each other.

18.3 proteins, lipids, carbohydrates, nucleic acids

18.5 $CO_2 + H_2O + \text{solar energy} \xrightarrow[\text{plant enzymes}]{\text{chlorophyll}} \text{carbohydrates} + O_2$

18.7 serve as structural elements, provide energy reserves

18.9 polyhydroxyaldehydes or polyhydroxyketones or compounds that yield such substances upon hydrolysis

18.11 a. one monosaccharide unit versus a few monosaccharide units
b. two monosaccharide units versus four monosaccharide units

18.13 Superimposable objects have parts that coincide exactly at all points when the objects are laid upon each other.

18.15 a. drill bit b. hand, foot, ear c. pop, peep

18.17 a. no b. no c. yes d. yes

18.19 a. $CH_2(Cl)-{}^{*}CH(Cl)-Br$

b. $CH_2(Br)-{}^{*}C(Cl)(Br)-{}^{*}CH(Cl)(Br)$

c. $CH_2(OH)-{}^{*}CH(OH)-{}^{*}CH(OH)-{}^{*}CH(OH)-C(=O)-H$

d. $CH_2(OH)-{}^{*}CH(OH)-{}^{*}CH(OH)-{}^{*}CH(OH)-{}^{*}CH(OH)-CH_2(OH)$

18.21 a. zero b. two c. zero d. zero

18.23 Structural isomers have different connectivity of atoms; stereoisomers have the same connectivity of atoms with different arrangements of atoms in space.

18.25 (Fischer projections: top / left / right / bottom)

a. top H; left Br; right Cl; bottom CH_3

b. top CH_3; left Br; right Cl; bottom H

c. top CH_3; left Br; right H; bottom Cl

d. top CH_3; left H; right Br; bottom Cl

18.27

a.
```
      CHO
  H ──┼── OH
 HO ──┼── H
 HO ──┼── H
      CH2OH
```

b.
```
      CH2OH
       ══O
 HO ──┼── H
  H ──┼── OH
 HO ──┼── H
      CH2OH
```

c.
```
      CHO
 HO ──┼── H
 HO ──┼── H
 HO ──┼── H
  H ──┼── OH
      CH2OH
```

d.
```
      CHO
 HO ──┼── H
  H ──┼── OH
 HO ──┼── H
  H ──┼── OH
      CH2OH
```

18.29 a. D-enantiomer b. D-enantiomer c. L-enantiomer d. L-enantiomer

18.31 a. diastereomers b. neither enantiomers nor diastereomers c. enantiomers d. diastereomers

18.33 effect on plane-polarized light

18.35 a. same b. different c. same d. different

18.37 a. aldose b. ketose c. ketose d. ketose

18.39 a. aldohexose b. ketohexose c. ketotriose d. ketotetrose

18.41 a. D-galactose b. D-psicose c. dihydroxyacetone d. L-erythrulose

18.43 a. carbon 4 b. carbons 1 and 2 c. carbons 1 and 2 d. carbon 2

18.45 a. 1, 3, and 4 b. 3 c. 3 d. 1

18.47

a.
```
      CHO
  H ──┼── OH
 HO ──┼── H
  H ──┼── OH
  H ──┼── OH
      CH2OH
```

b.
```
      CHO
  H ──┼── OH
      CH2OH
```

c.
```
      CH2OH
       ══O
 HO ──┼── H
  H ──┼── OH
  H ──┼── OH
      CH2OH
```

d.
```
      CHO
 HO ──┼── H
  H ──┼── OH
  H ──┼── OH
 HO ──┼── H
      CH2OH
```

18.49 a. D-fructose b. D-glucose c. D-galactose

18.51 a. 1 and 5 b. 1 and 5 c. 2 and 5 d. 1 and 4

18.53 the hydroxyl group orientation on carbon 1

18.55 The fructose cyclization process involves carbons 2 and 5, and the ribose cyclization process involves carbons 1 and 4; both give five-membered rings.

18.57 The cyclic and noncyclic forms interconvert; an equilibrium exists between forms.

18.59 a. α-D-monosaccharide b. α-D-monosaccharide c. β-D-monosaccharide d. α-D-monosaccharide

18.61 a. yes b. yes c. yes d. yes

18.63

a.	b.	c.	d.
CHO	CHO	CHO	CH_2OH
H—OH	H—OH	HO—H	=O
HO—H	HO—H	HO—H	H—OH
H—OH	HO—H	H—OH	HO—H
H—OH	H—OH	H—OH	H—OH
CH_2OH	CH_2OH	CH_2OH	CH_2OH

18.65 a. α-D-glucose b. α-D-galactose c. β-D-mannose d. α-D-sorbose

18.67 a. b. c. d.

18.69 a. reducing sugar b. reducing sugar c. reducing sugar d. reducing sugar

18.71 The aldehyde group in glucose is oxidized to a carboxylic acid group; the Ag^{+} ion in the Tollens solution is reduced to Ag.

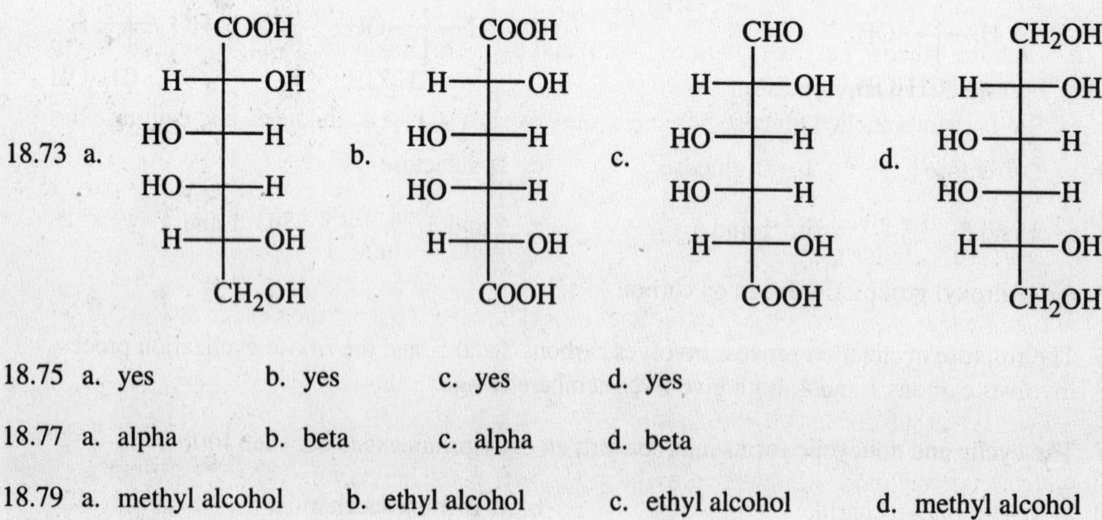

18.73

a.	b.	c.	d.
COOH	COOH	CHO	CH_2OH
H—OH	H—OH	H—OH	H—OH
HO—H	HO—H	HO—H	HO—H
HO—H	HO—H	HO—H	HO—H
H—OH	H—OH	H—OH	H—OH
CH_2OH	COOH	COOH	CH_2OH

18.75 a. yes b. yes c. yes d. yes

18.77 a. alpha b. beta c. alpha d. beta

18.79 a. methyl alcohol b. ethyl alcohol c. ethyl alcohol d. methyl alcohol

18.81 A glycoside is an acetal formed from a cyclic form of a monosaccharide; a glucoside is a glycoside in which the monosaccharide is glucose.

18.83 a. 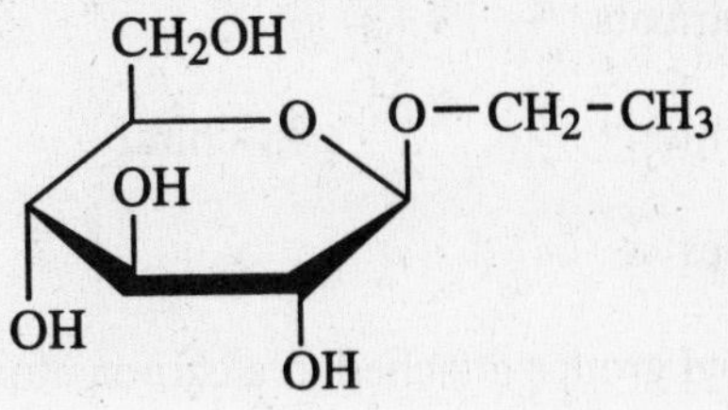 CH_2OH; O; $O-CH_2-CH_3$; OH; OH; OH   b. 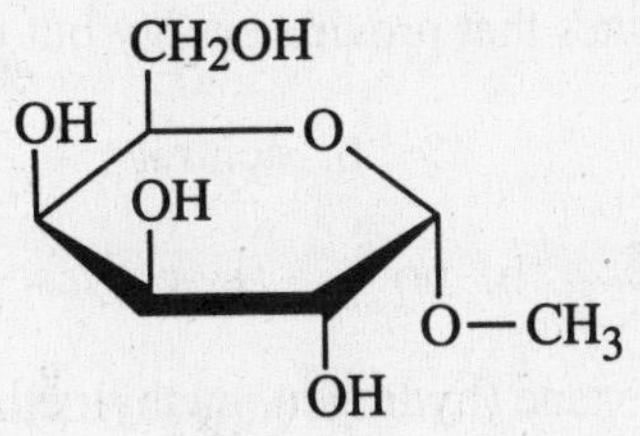 CH_2OH; OH; O; OH; $O-CH_3$; OH

18.85 a. 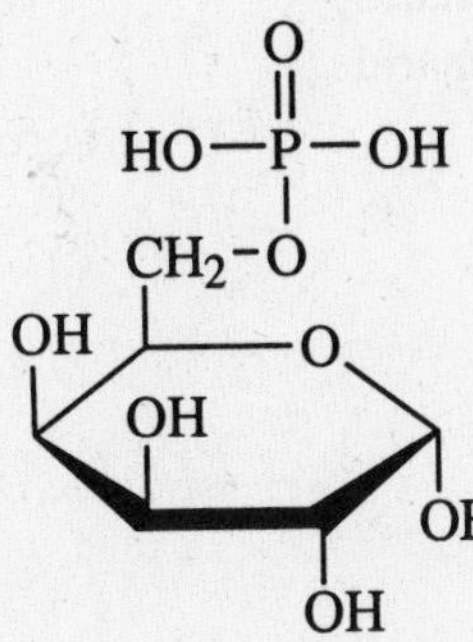 O; HO—P—OH; CH_2-O; OH; O; OH; OH; OH   b. 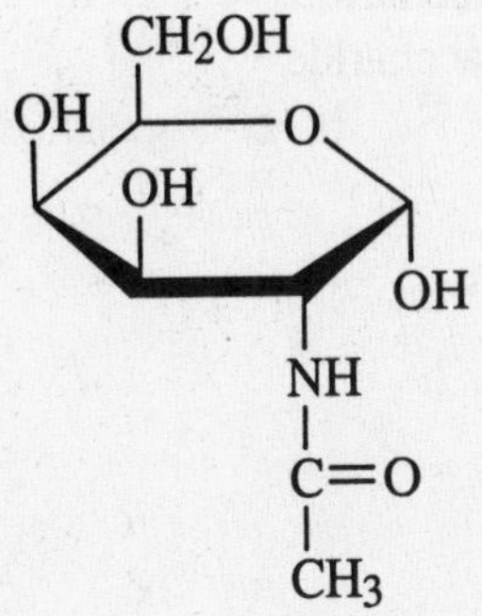 CH_2OH; OH; O; OH; OH; NH; C=O; CH_3

18.87 a. glucose and fructose b. glucose c. glucose and galactose d. glucose

18.89 The glucose part of the lactose structure has a hemiacetal carbon atom.

18.91 a. negative b. positive c. positive d. positive

18.93 a. $\alpha(1 \rightarrow 6)$ b. $\beta(1 \rightarrow 4)$ c. $\alpha(1 \rightarrow 4)$ d. $\alpha(1 \rightarrow 4)$

18.95 a. alpha b. beta c. alpha d. beta

18.97 a. reducing sugar b. reducing sugar c. reducing sugar d. reducing sugar

18.99 a. glucose b. galactose and glucose c. glucose and altrose d. glucose

18.101 a. Both are glucose polymers with $\alpha(1 \rightarrow 4)$ and $\alpha(1 \rightarrow 6)$ linkages; glycogen is more highly branched than amylopectin.
b. Both are unbranched glucose polymers; amylose has $\alpha(1 \rightarrow 4)$ linkages, and cellulose has $\beta(1 \rightarrow 4)$ linkages.

18.103 a. amylopectin, glycogen b. amylopectin, amylose, glycogen, cellulose
c. amylose, cellulose, chitin d. cellulose, chitin

18.105 The human body possesses enzymes for $\alpha(1 \rightarrow 4)$ linkages (starch) but not for $\beta(1 \rightarrow 4)$ linkages (cellulose).

18.107 connective tissue associated with joints

18.109 both contain the monosaccharide derivative *N*-acetyl-β-D-glucosamine

18.111 Simple carbohydrates are the mono- and disaccharides, and complex carbohydrates are the polysaccharides.

18.113 carbohydrates that provide energy but few other nutrients

18.115 a. chiral b. achiral c. chiral d. chiral

18.117 a. no b. no c. yes d. yes

18.119 3-methylhexane (hydrogen, methyl, ethyl, and propyl groups attached to a carbon atom)

18.121 a. homopolysaccharide b. homopolysaccharide
c. heteropolysaccharide d. homopolysaccharide

Solutions to Selected Problems

19.1 insoluble or only sparingly soluble in water

19.3 a. insoluble b. soluble c. insoluble d. soluble

19.5 energy-storage lipids, membrane lipids, emulsification lipids, messenger lipids, and protective-coating lipids

19.7 a. long-chain b. medium-chain c. long-chain d. medium-chain

19.9 a. saturated b. polyunsaturated c. polyunsaturated d. monounsaturated

19.11 In a SFA, there are no double bonds in the carbon chain; in a MUFA, there is one carbon-carbon double bond in the carbon chain.

19.13 a. neither b. omega-3 c. omega-3 d. neither

19.15

```
          H       H H       H
           \     /   \     /
            C===C     C===C
           /     \   /     \
CH3—(CH2)4—H2C    CH2       CH2—(CH2)7—COOH
```

19.17 There are fewer attractions between fatty-acid carbon chains because of bends in the chains caused by the presence of double bonds.

19.19 a. 18:1 acid b. 18:3 acid c. 14:0 acid d. 18:1 acid

19.21 a. tetradecanoic acid b. *cis*-9-hexadecenoic acid

19.23 a glycerol molecule and three fatty acid molecules

19.25 one - ester

19.27

```
              O
              ||
H2C—O—C—(CH2)14—CH3
 |            O
 |            ||
HC—O—C—(CH2)14—CH3
 |            O
 |            ||
H2C—O—C—(CH2)14—CH3
```

19.29

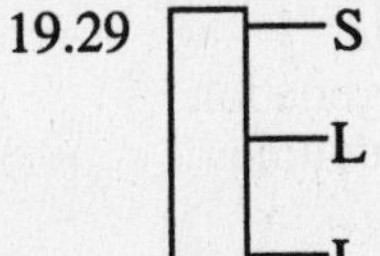

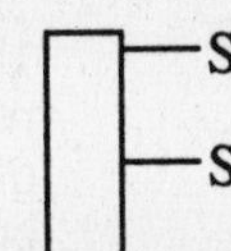

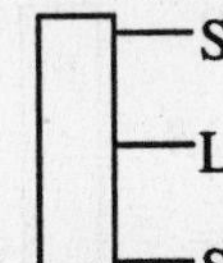

19.31 a. palmitic acid, myristic acid, oleic acid b. oleic acid, palmitic acid, palmitoleic acid

19.33 (top) 16 carbon atoms and 1 oxygen atom; (middle) 14 carbon atoms and 1 oxygen atom; (bottom) 18 carbon atoms and 1 oxygen atom

19.35 a. no difference
b. A triacylglycerol may be a solid or a liquid; a fat is a triacylglycerol that is a solid.
c. A triacylglycerol can have fatty acid residues that are all the same, or two or more different kinds may be present; in a mixed triacylglycerol two or more different fatty acid residues must be present.
d. A fat is a triacylglycerol that is a solid; an oil is a triacylglycerol that is a liquid.

19.37 a. not correct b. not correct

19.39 a. correct b. not correct

19.41 a. nonessential fatty acid b. essential fatty acid
c. nonessential fatty acid d. nonessential fatty acid

19.43 a. glycerol and three fatty acids b. glycerol and three fatty acid salts

19.45 $CH_2(OH){-}CH(OH){-}CH_2(OH)$ $CH_3{-}(CH_2)_{14}{-}COOH$

$CH_3{-}(CH_2)_{12}{-}COOH$ $CH_3{-}(CH_2)_7{-}CH{=}CH{-}(CH_2)_7{-}COOH$

19.47 glycerol, palmitic acid, myristic acid, oleic acid

19.49 $CH_2(OH){-}CH(OH){-}CH_2(OH)$ $CH_3{-}(CH_2)_{14}{-}COO^-\ Na^+$

$CH_3{-}(CH_2)_{12}{-}COO^-\ Na^+$ $CH_3{-}(CH_2)_7{-}CH{=}CH{-}(CH_2)_7{-}COO^-\ Na^+$

19.51 glycerol, sodium palmitate, sodium myristate, sodium oleate

19.53 Hydrogenation requires the presence of double bonds.

19.55 six

19.57 a.

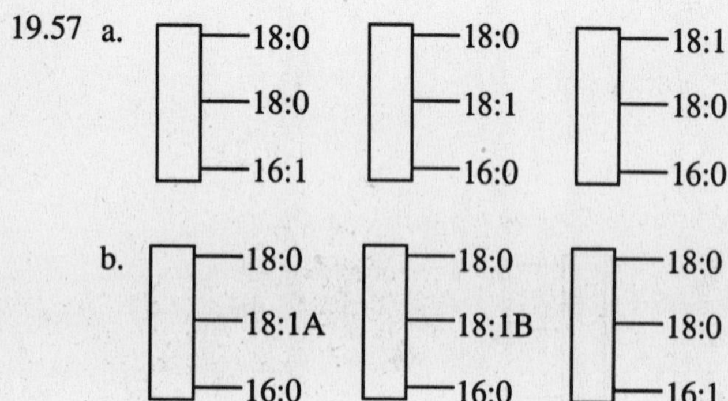

18:1A and 18:1B denote, respectively, the hydrogenation of the first and second double bonds in the 18:2 acid.

19.59 Rancidity results from hydrolysis of ester linkages and oxidation of carbon-carbon double bonds.

19.61 glycerol and sphingosine

19.63

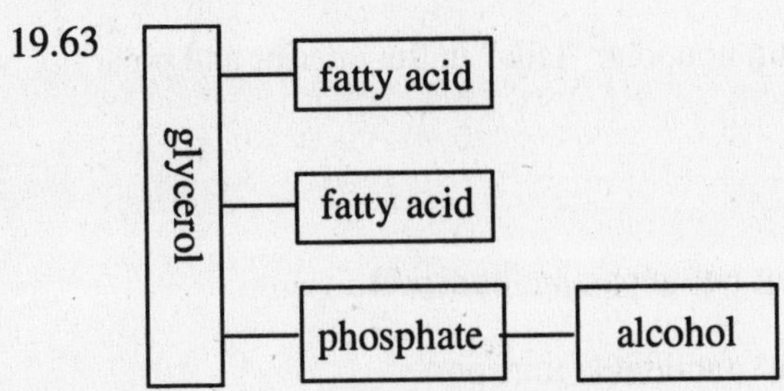

19.65 $HO-CH_2-CH_2-\overset{+}{N}(CH_3)_3$

choline

$HO-CH_2-CH_2-\overset{+}{N}H_3$

ethanolamine

$HO-CH_2-\underset{\displaystyle COO^-}{CH}-\overset{+}{N}H_3$

serine

19.67 The two tails are the carbon chain of sphingosine and the fatty acid carbon chain; the head is the phosphate-alcohol portion of the molecule.

19.69 the two tails

19.71 a. four b. three

19.73 the identity of the amino alcohol group; choline in a lecithin and serine in a phosphatidylserine.

19.75

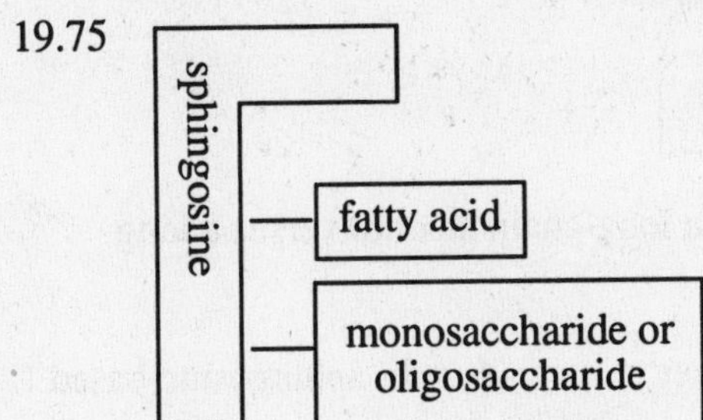

19.77 a carbohydrate group versus a phosphate-alcohol group

19.79

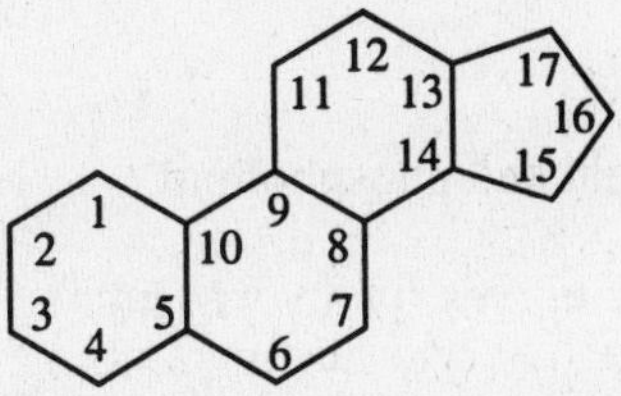

19.81 –OH on carbon 3, $-CH_3$ on carbons 10 and 13, hydrocarbon chain on carbon 17

19.83 "Good cholesterol" is that present in HDLs, and "bad cholesterol" is that present in LDLs.

19.85 phospholipids, glycolipids, and cholesterol

19.87 a two-layer-thick structure of lipid molecules with nonpolar "tails" in the interior and polar "heads" on the exterior

19.89 creates "open" areas in the lipid bilayer

19.91 Protein help is required in facilitated transport but not in passive transport.

19.93 a. active transport b. facilitated transport c. active transport d. passive transport and facilitated transport

19.95 tri- or dihydroxy versus monohydroxy; oxidized side chain amidified to an amino acid versus nonoxidized side chain

19.97 amino acid glycine versus amino acid taurine

19.99 bile fluid

19.101 gall bladder

19.103 sex hormones and adrenocortical hormones

19.105 Estradiol has an –OH group on carbon 3 and testosterone has a ketone group on carbon 3; testosterone has an extra $-CH_3$ group at carbon 10.

19.107 Prostaglandins have a bond between carbons 8 and 12, which creates a cyclopentane ring.

19.109 inflammatory process, pain and fever production, blood pressure regulation, induction of blood clotting, control of some reproductive functions, regulation of sleep/wake cycle

19.111 

19.113 mixture of esters involving a long-chain fatty acid and a long-chain alcohol versus a long-chain alkane mixture

19.115 a. neither b. glycerol-based c. neither d. sphingosine-based e. neither f. neither

19.117 a. sphingomyelins b. triacylglycerols c. steroids d. leukotrienes e. prostaglandins f. cerebrosides

19.119 a. glycerolipid b. sphingolipid c. glycerolipid and phospholipid d. sphingolipid and phospholipid

19.121 a. no b. yes c. yes d. yes e. yes f. no

Proteins — Chapter 20

Solutions to Selected Problems

20.1 a. yes b. no c. no d. yes

20.3 the identity of the R group (side chain)

20.5 a. phenylalanine, tyrosine, tryptophan b. methionine, cysteine
c. aspartic acid, glutamic acid d. serine, threonine, tyrosine

20.7 An amino group is part of the side chain.

20.9 The side chain is part of a cyclic structure.

20.11 a. alanine b. leucine c. methionine d. tryptophan

20.13 asparagine, glutamine, isoleucine, tryptophan

20.15 a. polar neutral b. polar acidic c. nonpolar d. polar neutral

20.17 L-family

20.19

```
a.       COOH           b.     COOH          c.     COOH          d.       COOH
          |                     |                    |                      |
    H2N---+---H             H---+---NH2          H---+---NH2          H2N---+---H
          |                     |                    |                      |
         CH2                   CH2                  CH3                    CH2
          |                     |                                           |
         OH                    OH                                          CH—CH3
                                                                            |
                                                                           CH3
```

20.21 They exist as charged species (zwitterions).

20.23

```
a.       COO⁻           b.       COO⁻           c.       COO⁻           d.       COO⁻
    +     |                 +     |                 +     |                 +     |
    H3N---+---H             H3N---+---H             H3N---+---H             H3N---+---H
          |                       |                       |                       |
         CH2                     CH—CH3                  CH2                      H
          |                       |                       |
         CH—CH3                  CH2                     SH
          |                       |
         CH3                     CH3
```

20.25

```
a.       COO⁻           b.       COOH           c.      COO⁻           d.       COOH
    +     |                 +     |                        |                +     |
    H3N---+---H             H3N---+---H             H2N---+---H             H3N---+---H
          |                       |                       |                       |
         CH2                     CH2                     CH2                     CH2
          |                       |                       |                       |
         OH                      OH                      OH                      OH
```

20.27 the pH at which zwitterion concentration in a solution is maximized

20.29 Two –COOH groups are present, which deprotonate at different times.

20.31 a. toward positive electrode b. isoelectric
c. toward negative electrode d. toward positive electrode

20.33 Aspartic acid migrates toward the positive electrode, histidine migrates toward the negative electrode, and valine does not migrate.

20.35 They react with each other to produce a covalent disulfide bond.

20.37 carboxyl group and amino group

20.39

```
            O  H        O  H
 +          ||  |       ||  |
H3N—CH—C—N—CH—C—N—CH—COO⁻
     |          |           |
     CH—CH3     CH2         CH2
     |          |           |
     CH3      (C6H5)        SH
```

20.41 Ser is the N-terminal end in Ser–Cys, and Cys is the N-terminal end in Cys–Ser.

20.43 Ser–Val–Gly Val–Ser–Gly Gly–Val–Ser
Ser–Gly–Val Val–Gly–Ser Gly–Ser–Val

20.45 a. Ser–Ala–Cys b. Asp–Thr–Asn

20.47 a. two b. two

20.49 peptide bonds and α-carbon –CH groups

20.51 a. Both are nonapeptides with six of the residues held in the form of a loop by a disulfide bond.
b. They differ in the identity of the amino acid present at two positions in the nonapeptide.

20.53 They bind at the same sites.

20.55 Glu is bonded to Cys through the side-chain carboxyl group rather than through the α-carbon carboxyl group.

20.57 the sequence of amino acids in the protein chain

20.59 α-helix, β-pleated sheet, triple helix

20.61 Intermolecular involves two separate chains, and intramolecular involves a single chain bending back on itself.

20.63 Yes, both α-helix and β-pleated sheet can occur at different locations in the same molecule.

20.65 Secondary-structure hydrogen bonding involves C=O····H–N interactions; tertiary-structure hydrogen bonding involves R-group interactions.

20.67 a. hydrophobic b. electrostatic
c. disulfide bond d. hydrogen bonding

20.69 a multi chain protein

20.71 A simple protein contains only amino acids; a conjugated protein has one or more other chemical components besides amino acids.

20.73 a. fibrous: generally water-insoluble; globular: generally water-soluble
b. fibrous: support an external protection; globular: involvement in metabolic reactions

20.75 a. fibrous b. fibrous c. globular d. globular

20.77 α-Keratin has a double-helix structure and collagen a triple-helix structure.

20.79 Yes, both Ala and Val are products in each case.

20.81 Drug hydrolysis would occur in the stomach.

20.83 Ala–Gly–Met–His–Val–Arg

20.85 five: Ala–Gly–Ser, Gly–Ser–Tyr, Ala–Gly, Gly–Ser, Ser–Tyr

20.87 secondary, tertiary, and quaternary

20.89 The primary structure is the same.

20.91 4-hydroxyproline and 5-hydroxylysine

20.93 They are involved with cross-linking.

20.95 An antigen is a substance foreign to the human body, and an antibody is a substance that defends against an invading antigen.

20.97 four polypeptide chains that have constant and variable amino acid regions, two chains are longer thant the other two; 1% to 12% carbohydrate present by mass; long and short chains are connected through disulfide linkages

20.99 suspend and transport lipids in the bloodstream

20.101 a. tertiary b. tertiary c. secondary d. primary

20.103 a. +1 b. +1 c. +1 d. +3

20.105

$$\begin{array}{ccc} & COOH & \\ H_2N- & C & -H \\ H_3C- & C & -H \\ & | & \\ & CH-CH_3 & \\ & | & \\ & CH_3 & \end{array} \qquad \begin{array}{ccc} & COOH & \\ H- & C & -NH_2 \\ H- & C & -CH_3 \\ & | & \\ & CH-CH_3 & \\ & | & \\ & CH_3 & \end{array} \qquad \begin{array}{ccc} & COOH & \\ H_2N- & C & -H \\ H- & C & -CH_3 \\ & | & \\ & CH-CH_3 & \\ & | & \\ & CH_3 & \end{array} \qquad \begin{array}{ccc} & COOH & \\ H- & C & -NH_2 \\ H_3C- & C & -H \\ & | & \\ & CH-CH_3 & \\ & | & \\ & CH_3 & \end{array}$$

20.107 a.

$$\begin{array}{c} H_3\overset{+}{N}-CH-COOH \\ | \\ CH_2 \\ | \\ OH \end{array} \; ; \quad \begin{array}{c} H_3\overset{+}{N}-CH-COOH \\ | \\ CH_3 \end{array} \; ; \quad \begin{array}{c} H_3\overset{+}{N}-CH-COOH \\ | \\ CH_2 \\ | \\ SH \end{array}$$

b.

$$\begin{array}{c} H_2N-CH-COO^- \\ | \\ CH_2 \\ | \\ OH \end{array} \; ; \quad \begin{array}{c} H_2N-CH-COO^- \\ | \\ CH_3 \end{array} \; ; \quad \begin{array}{c} H_2N-CH-COO^- \\ | \\ CH_2 \\ | \\ SH \end{array}$$

Enzymes and Vitamins — Chapter 21

Solutions to Selected Problems

21.1 catalyst

21.3 more efficient, more specific

21.5 a. yes b. no c. yes d. yes

21.7 a. add a carboxyl group to pyruvate b. remove H_2 from an alcohol
c. reduce an L-amino acid d. hydrolyze maltose

21.9 a. sucrase b. pyruvate decarboxylase
c. glucose isomerase d. lactate dehydrogenase

21.11 a. pyruvate b. galactose
c. an alcohol d. an L-amino acid

21.13 a. isomerase b. lyase c. ligase d. transferase

21.15 a. isomerase b. lyase c. transferase d. hydrolase

21.17 a. decarboxylase b. lipase c. phosphatase d. dehydrogenase

21.19 a. conjugated b. conjugated c. simple d. conjugated

21.21 A cofactor may be inorganic or organic; a coenzyme is an organic cofactor.

21.23 to provide additional functional groups

21.25 the portion of an enzyme actually involved in the catalysis process

21.27 The substrate must have the same shape as the active site.

21.29 interactions with amino acid R groups

21.31 a. accepts only one substrate
b. accepts substrate with a particular type of bond

21.33 absolute specificity and stereochemical specificity

21.35 a. absolute b. stereochemical

21.37 The rate increases until enzyme denaturation occurs.

21.39 Enzymes vary in the number of acidic and basic groups present.

21.41
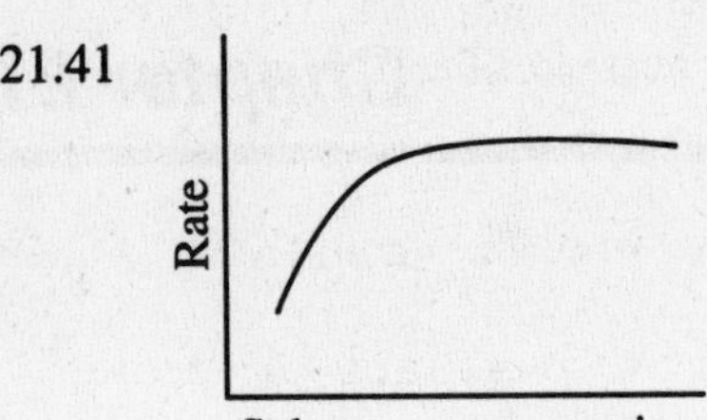

21.43 nothing; the rate remains the same

21.45 no; only one molecule may occupy the active site at a given time

21.47 a. reversible competitive
b. reversible noncompetitive, irreversible
c. reversible noncompetitive, irreversible
d. reversible noncompetitive

21.49 enzyme that has quaternary structure and more than one binding site

21.51 The product of a subsequent reaction in a series of reactions inhibits a prior reaction.

21.53 A zymogen is the inactive form of a proteolytic enzyme.

21.55 so they will not destroy the tissues that produce them

21.57 competitive inhibition of the enzyme necessary for conversion of PABA to folic acid

21.59 It has absolute specificity for bacterial transpeptidase.

21.61 the threat of biological weapon use by terrorists

21.63 dietary organic compounds needed by the body in trace amounts

21.65 a. fat-soluble b. water-soluble c. water-soluble d. water-soluble

21.67 a. likely b. unlikely c. unlikely d. unlikely

21.69 serves as a cosubstrate in the formation of collagen and as a general antioxidant

21.71 coenzymes

21.73 a. no b. yes c. yes d. no

21.75 alcohol, aldehyde, acid

21.77 Cell differentiation is the process whereby immature cells change in structure and function to become specialized cells. Vitamin A binds to protein receptors in the process.

21.79 They differ only in the identity of the side chain present.

21.81 to maintain normal blood levels of calcium and phosphorus ions so that bones can absorb these minerals

21.83 alpha-tocopherol

21.85 antioxidant effect

21.87 Vitamin K_1 and vitamin K_2 differ structurally in the length and degree of unsaturation of a side chain.

21.89 Menaquinones are forms of vitamin K_1, and phylloquinones are forms of vitamin K_2.

21.91 a. An apoenzyme is the protein portion of a conjugated enzyme; a proenzyme is an inactive precursor of an enzyme.
b. A simple enzyme contains only protein; an allosteric enzyme has two or more protein chains and two binding sites.
c. A coenzyme is an organic cofactor; an isoenzyme is one of several similar forms of an enzyme.
d. A conjugated enzyme has both a protein and a nonprotein portion; holoenzyme is another name for a conjugated enzyme.

21.93 a. no b. no c. no d. yes e. yes f. yes

21.95 a. oxidation-reduction reactions
b. addition of a group to or removal of a group from a double bond in a manner that does not involve hydrolysis or oxidation/reduction
c. conversion of a compound into another compound isomeric with it
d. bonding together of two molecules with the involvement of ATP
e. hydrolysis reactions
f. transfer of functional groups between two molecules

21.97 a. ethanol b. zinc ion
c. protein molecule d. alcohol dehydrogenase

Nucleic Acids

Chapter 22

Solutions to Selected Problems

22.1 At carbon 2, ribose has an –H atom and an –OH group and deoxyribose has two –H atoms.

22.3	a. pyrimidine	b. pyrimidine	c. purine	d. purine
22.5	a. one	b. four	c. one	
22.7	a. adenine	b. guanine	c. thymine	d. uracil
22.9	a. ribose	b. 2-deoxyribose	c. 2-deoxyribose	d. ribose
22.11	a. false	b. false	c. false	d. false

22.13

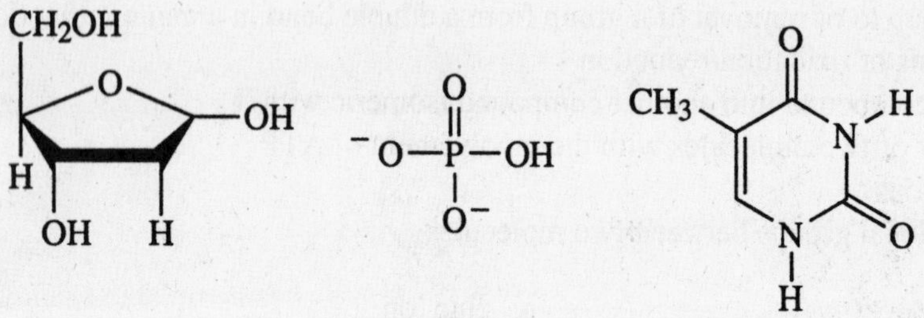

22.15 a pentose sugar and a phosphate

22.17 base sequence

22.19 The 5′ end has a phosphate group attached to the 5′ carbon; the 3′ end has a hydroxyl group attached to the 3′ carbon.

22.21 a phosphate group and two pentose sugars

22.23

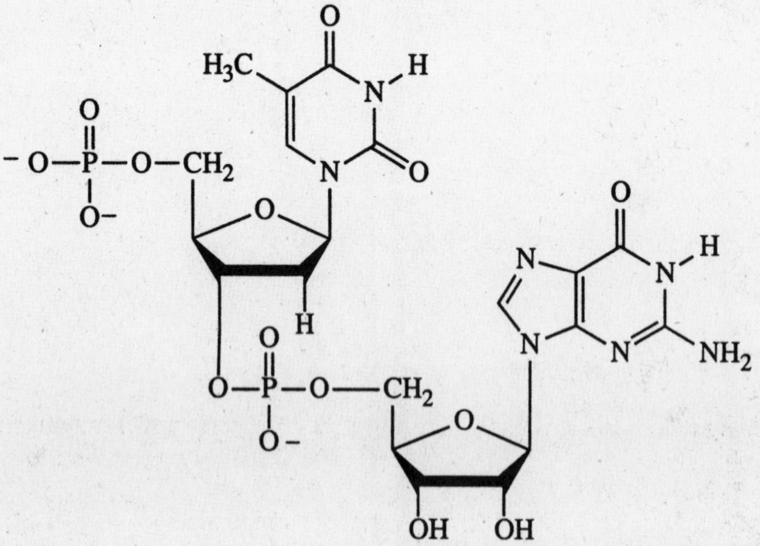

22.25 a. Two polynuleotide chains are coiled around each other in a helical fashion.
b. The nucleic acid backbones are the outside, and the nitrogen-containing bases are on the inside.

22.27 a. 36% b. 14% c. 14%

22.29 G–C pairing involves 3 hydrogen bonds, and A–T pairing involves 2 hydrogen bonds.

22.31 They are the same.

22.33 5′ TAGCC 3′

22.35 a. 3′ TGCATA 5′ b. 3′ AATGGC 5′ c. 5′ CGTATT 3′ d. 3′ TTGACC 5′

22.37 20 hydrogen bonds; 3 per G–C and 2 per A–T

22.39 catalyzes the unwinding of the double helix structure

22.41 a. 3′ TGAATC 5′ b. 3′ TGAATC 5′ c. 5′ ACTTAG 3′

22.43 The two strands are antiparallel (5′ → 3′ and 3′ → 5′), and only the 5′ → 3′ strand can grow continuously.

22.45 a DNA molecule bound to a group of proteins

22.47 (1) RNA contains ribose instead of deoxyribose, (2) RNA contains the base U instead of T, (3) RNA is single-stranded rather than double-stranded, and (4) RNA has a lower molecular mass.

22.49 a. hnRNA b. tRNA c. tRNA d. hnRNA

22.51 a. nuclear region b. extranuclear region
c. extranuclear region d. both nuclear and extranuclear regions

22.53 a strand of DNA

22.55 causes a DNA helix to unwind; links aligned ribonucleotides together

22.57 A–U, C–G, G–C, T–A

22.59 3′ UACGAAU 5′

22.61 3′ AAGCGTC 5′

22.63 Exons are the parts of genes that convey genetic information.

22.65 3′ AGUCAAGU 5′

22.67 a. hnRNA b. snRNA

22.69 a mechanism by which a number of proteins that are variations of a basic structural motif can be produced from a single gene

22.71 a three-nucleotide sequence in mRNA that codes for a specific amino acid

22.73 a. Leu b. Asn c. Ser d. Gly

22.75 a. CUC, CUA, CUG, UUA, UUG b. AAC
c. AGC, UCU, UCC, UCA, UCG d. GGU, GGC, GGA

22.77 The base T cannot be present in a codon.

22.79 Met–Lys–Glu–Asp–Leu

22.81 a cloverleaf shape with three hairpin loops and one open side

22.83 covalent bond

22.85 a. UCU b. GCA c. AAA d. GUU

22.87 a. Thr b. Leu c. Pro d. Ser

22.89 (1) activation of tRNA, (2) initiation, (3) elongation, (4) termination, and (5) post-translational processing

22.91 A site

22.93 gly: GGU, GGC, GGA or GGG; ala: GCU, GCC, GCA or GCG; cys: UGU or UGC; val: GUU, GUC, GUA or GUG; tyr: UAU or UAC

22.95 a. Gly–Tyr–Ser–Ser–Pro b. Gly–Tyr–Ser–Ser–Pro c. Gly–Tyr–Ser–Ser–Thr

22.97 a DNA or RNA molecule with a protein coating

22.99 (1) attaches itself to cell membrane, (2) opens a hole in the membrane, and (3) injects itself into the cell

22.101 contains a "foreign" gene

22.103 host for a "foreign" gene

22.105 Recombinant DNA is incorporated into a host cell.

22.107

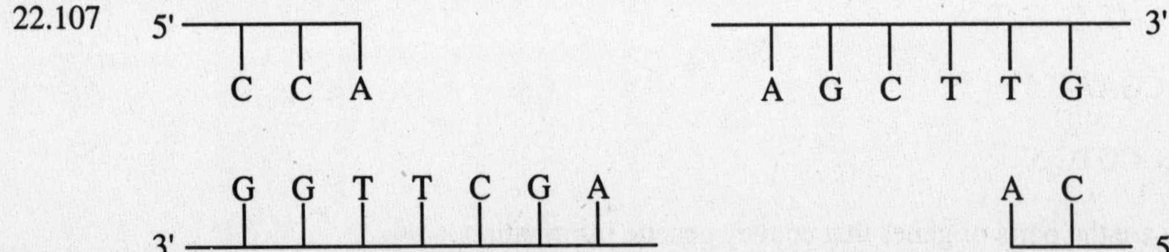

22.109 Many copies of a specific DNA sequence can be produced in a relatively short time.

22.111 a short nucleotide chain bound to the template DNA strand to which new nucleotides can be attached

22.113 dATP stands for an ATP in which deoxyribose is present; ddATP stands for an ATP in which dideoxyribose is present.

22.115 a. 5′ GCCGACTACT 3′ b. 5′ AGTAGTCGGC 3′

22.117 a. Thymine is a methyl uracil; the methyl group is on C–5′.
b. Adenine is the 6-amino derivative of purine, and guanine is the 2-amino-6-oxo derivative.

22.119 a. 1 b. 4 c. 2 d. 4

22.121 a. 3 b. 1 c. 4 d. 1, 2, 3, and 4

22.123 A

Solutions to Selected Problems

23.1 anabolism – synthetic; catabolism – degradative

23.3 a series of consecutive biochemical reactions

23.5 Large molecules are broken down to smaller molecules; energy is produced.

23.7 Prokaryotic cells have no nucleus, and the DNA is usually a single circular molecule; eukaryotic cells have their DNA in a membrane-enclosed nucleus.

23.9 a small structure within the cell cytosol that carries out a specific cellular function

23.11 inner membrane

23.13 region between the inner and outer membranes

23.15 adenosine triphosphate

23.17

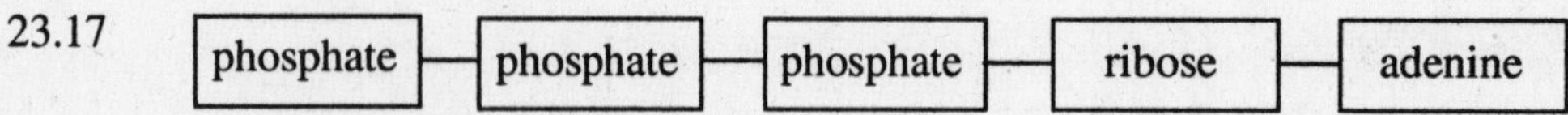

23.19 three phosphates versus one phosphate

23.21 adenine versus guanine

23.23 $$ATP \xrightarrow{H_2O} ADP + P_i$$

23.25 flavin adenine dinucleotide

23.27 flavin — ribitol — ADP

23.29

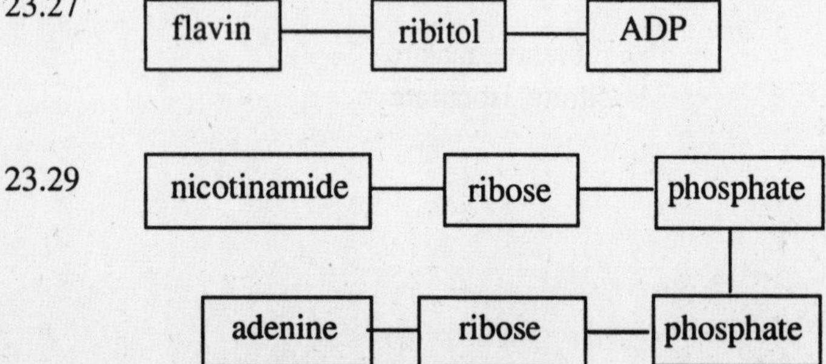

23.31 nicotinamide subunit

23.33 a. $FADH_2$ b. NAD^+

23.35 a. nicotinamide b. riboflavin

23.37 | 2-aminoethanethiol | — | pantothenic acid | — | phosphorylated ADP |

23.39 a compound with a greater free energy of hydrolysis than is typical for a compound

23.41 free monophosphate species

23.43 a. phosphoenolpyruvate b. creatine phosphate
c. 1,3-diphosphoglycerate d. AMP

23.45 (1) digestion, (2) acetyl group formation, (3) citric acid cycle, (4) electron transport chain and oxidative phosphorylation

23.47 tricarboxylic acid cycle, Krebs cycle

23.49 acetyl CoA

23.51 a. 2 (Steps 3, 4) b. 1 (Step 6) c. 2 (Steps 3, 8) d. 2 (Steps 2, 7)

23.53 a. Steps 3, 4, 6, 8 b. Step 2 c. Step 7

23.55 $^{-}OOC{-}CH_2{-}CH_2{-}COO^{-}$

succinate

$^{-}OOC{-}CH{=}CH{-}COO^{-}$

fumarate

$^{-}OOC{-}CH(OH){-}CH_2{-}COO^{-}$

malate

$^{-}OOC{-}C({=}O){-}CH_2{-}COO^{-}$

oxaloacetate

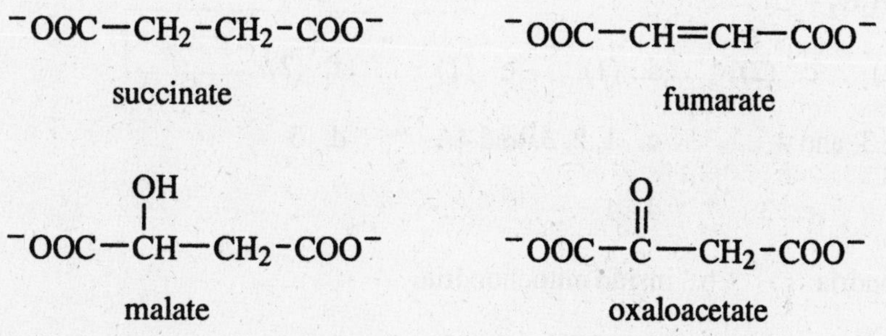

23.57 oxidation and decarboxylation

23.59 a. NAD^+ b. FAD

23.61 a. isocitrate, α-ketoglutarate b. fumarate, malate
c. malate, oxaloacetate d. citrate, isocitrate

23.63 respiratory chain

23.65 O_2

23.67 a. oxidized form of flavin mononucleotide
b. cytochrome
c. iron/sulfur protein
d. reduced form of nicotinamide adenine dinucleotide

23.69 a. mobile b. fixed c. fixed d. mobile

23.71 NADH, FeSP, cyt c_1, cyt a_3

23.73 a. oxidation b. reduction c. oxidation d. reduction

23.75 a. $FADH_2$, CoQ, $2Fe^{3+}$ b. $FMNH_2$, 2Fe(II)SP, $CoQH_2$

23.77 a. FMN + 2Fe(II)SP b. CoQ, $CoQH_2$

23.79 ATP synthesis from ADP using energy from the electron transport chain

23.81 protons (H^+ ions)

23.83 intermembrane space

23.85 ATP synthase

23.87 Protons flow through the ATP synthase complex.

23.89 2.5 molecules ATP

23.91 They enter the ETC at different stages.

23.93 reactive oxygen species

23.95 a. $2O_2 + NADPH \rightarrow 2O_2\cdot + NADP^+ + H^+$
b. $2O_2\cdot + 2H^+ \rightarrow H_2O_2 + O_2$

23.97 a. (1) b. (2) c. (2) d. (1) e. (1) f. (2)

23.99 a. 4 b. 2, 3, and 4 c. 1, 2, 3, and 4 d. 3

23.101 a. 4 b. 1 c. 3 d. 1

23.103 a. inside mitochondria b. inside mitochondria

23.105 Reduced coenzymes are oxidized and ADP is phosphorylated.

Carbohydrate Metabolism

Chapter 24

Solutions to Selected Problems

24.1 mouth; salivary α-amylase

24.3 small intestine; pancreas

24.5 outer membranes of intestinal mucosal cells; hydrolysis of sucrose

24.7 glucose, galactose, fructose

24.9 glucose

24.11 NAD^+

24.13 formation of glucose 6-phosphate, a species that cannot cross cell membranes

24.15 dihydroxyacetone phosphate, glyceraldehyde 3-phosphate

24.17 two

24.19 two

24.21 Steps 1, 3, and 6

24.23 cytosol

24.25 a. glucose 6-phosphate
b. 2-phosphoglycerate
c. phosphoglyceromutase
d. ADP

24.27 a. Step 10 b. Step 1 c. Step 8 d. Step 6

24.29 a. +2 b. +4

24.31 a.

$$\begin{array}{c} COOH \\ | \\ C{=}O \\ | \\ CH_3 \end{array} \qquad \begin{array}{c} COO^- \\ | \\ C{=}O \\ | \\ CH_3 \end{array}$$

b.

$$\begin{array}{c} CH_2{-}OH \\ | \\ C{=}O \\ | \\ CH_2{-}OH \end{array} \qquad \begin{array}{c} CH_2{-}OH \\ | \\ C{=}O \\ | \\ CH_2{-}O{-}\text{Ⓟ} \end{array}$$

c. 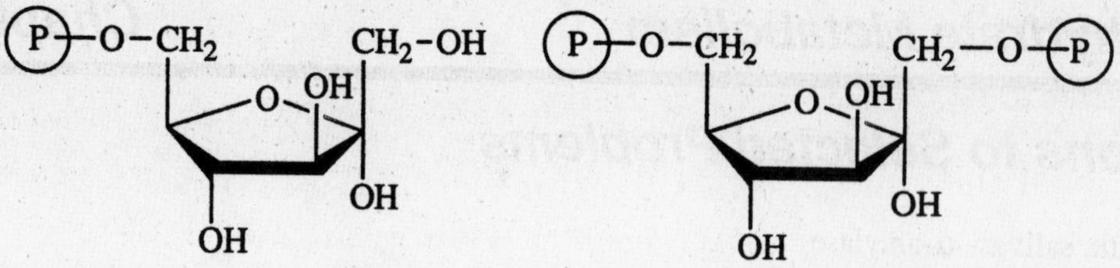 (P)–O–CH_2 … CH_2–OH (furanose ring with OH groups); (P)–O–CH_2 … CH_2–O–(P) (furanose ring with OH groups)

OH OH OH | OH OH OH

d. COOH | CH–OH | CH_2–OH   CHO | CH–OH | CH_2–OH

24.33 1 CH_2—O—(P) | 2 C=O | 3 CH_2–OH   4 CHO | 5 CH—OH | 6 CH_2—O—(P)

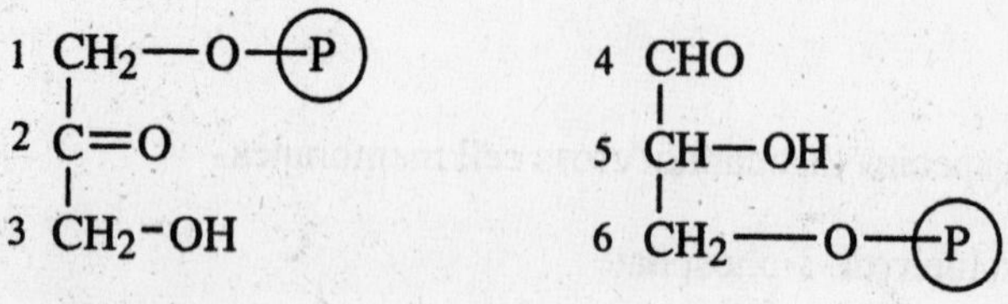

24.35 acetyl CoA, lactate, ethanol

24.37 pyruvate + CoA + NAD^+ → acetyl CoA + NADH + CO_2

24.39 NADH is oxidized to NAD^+, a substance needed for glycolysis.

24.41 CO_2

24.43 glucose + 2ADP + $2P_i$ → 2 lactate + 2ATP

24.45 decreases ATP production by 2

24.47 30 ATP versus 2 ATP

24.49 two

24.51 Glycogenesis converts glucose to glycogen, and glycogenolysis is the reverse process.

24.53 glucose 6-phosphate

24.55 UTP

24.57 UDP + ATP → UTP + ADP

24.59 Step 2

24.61 In liver cells the product is glucose, and in muscle cells it is glucose 6-phosphate.

24.63 as glucose 6-phosphate

24.65 the liver

24.67 2-step pathway for Step 10; different enzymes for Steps 1 and 3

24.69 oxaloacetate

24.71 converted to glucose in the liver

24.73 glucose 6-phosphate

24.75 NADPH is consumed in its reduced form, NADH in its oxidized form (NAD^+).

24.77 glucose 6-phosphate + $2NADP^+$ + H_2O $\rightarrow$ ribulose 5-phosphate + CO_2 + 2NADPH + $2H^+$

24.79 CO_2

24.81 increases the rate of glucogen synthesis

24.83 increases blood glucose levels

24.85 pancreas

24.87 Epinephrine attaches to cell membrane and stimulates the production of cAMP.

24.89 glucagon (liver cells) and epinephrine (muscle cells)

24.91 a. all four
b. glycogenesis and glycogenolysis
c. glycolysis and gluconeogenesis
d. gluconeogenesis

24.93 a. glycolysis b. glycogenesis c. glycogenolysis d. glyconeogenesis

24.95 a. 2 b. 2 c. 3 d. 4

24.97 a. 12 moles b. 4 moles c. 4 moles d. 60 moles

Lipid Metabolism

Chapter 25

Solutions to Selected Problems

25.1 98%

25.3 no effect

25.5 because lipids have a long residence time in the stomach

25.7 acts as an emulsifier

25.9 Hydrolysis results in the release of only two of the three fatty acid units, producing a monoglyceride + two free fatty acids.

25.11 reassembled into triacylglycerols; converted to chylomicrons

25.13 They have a large storage capacity for triacylglycerols.

25.15 hydrolysis of triacylglycerols in adipose tissue; entry of hydrolysis products into bloodstream

25.17 activates hormone-sensitive lipase

25.19 glycerol 3-phosphate, dihydroxyacetone phosphate

25.21 one

25.23 outer mitochondrial membrane

25.25 ATP is convertyed to AMP and $2P_i$.

25.27 shuttles acyl groups across the inner mitochondrial membrane

25.29 a. alkane to alkene b. alkene to 2° alcohol c. 2° alcohol to ketone

25.31 *trans* isomer

25.33 a. Step 3, turn 1 b. Step 2, turn 2 c. Step 4, turn 2 d. Step 1, turn 1

25.35 compounds a and d

25.37 a. 7 turns b. 5 turns

25.39 A *cis-trans* isomerase converts a *cis*-(3,4) double bond to a *trans*-(2, 3) double bond.

25.41 a. glucose b. fatty acids

25.43 a. 4 turns b. 5 acetyl CoA c. 4 NADH d. 4 $FADH_2$ e. 2 high-energy bonds

25.45 64 ATP

25.47 They yield the same amount of $FADH_2$; the additional enzymes needed to process unsaturated fatty acids do not require FADH.

25.49 4 kcal versus 9 kcal

25.51 (1) dietary intakes high in fat and low in carbohydrates, (2) inadequate processing of glucose present, and (3) prolonged fasting

25.53 Ketone body formation occurs when oxaloacetate concentrations are low.

25.55 $CH_3-\overset{O}{\overset{\|}{C}}-CH_2-\overset{O}{\overset{\|}{C}}-O^-$, $CH_3-\overset{OH}{\overset{|}{CH}}-CH_2-\overset{O}{\overset{\|}{C}}-O^-$, $CH_3-\overset{O}{\overset{\|}{C}}-CH_3$

25.57 liver mitochondria

25.59 formation of acetoacetyl CoA

25.61 accumulation of ketone bodies in blood and urine

25.63 cytosol versus mitochondrial matrix

25.65 acyl carrier protein; polypeptide chain replaces phosphorylated ADP

25.67 liver, adipose tissue, mammary glands

25.69 a. Oxaloacetate reacts with acetyl CoA to produce citrate and CoA.
b. Citrate crosses the mitochondrial membranes, functioning as an acetyl group carrier.

25.71 carrier of C_2 units for a growing fatty acid chain

25.73 (1) condensation, (2) hydrogenation, (3) dehydration, and (4) hydrogenation

25.75 a. Step 1, cycle 1 b. Step 2, cycle 2 c. Step 3, cycle 2 d. Step 4, cycle 1

25.77 compounds b and d

25.79 C_{16} fatty acid

25.81 It is needed to convert saturated fatty acids to unsaturated fatty acids.

25.83 a. 6 rounds b. 6 malonyl ACP c. 6 ATP bonds d. 12 NADPH

25.85 a. 13%–17% b. 83%–87%

25.87 a. mevalonate b. isopentenyl pyrophosphate c. squalene

25.89 a. fewer than b. fewer than c. same as

25.91 a. fatty acid spiral b. fatty acid spiral c. lipogenesis
d. glycerol catabolism e. ketogenesis f. ketogenesis

25.93 a. Step 1 b. Step 1 c. Step 3 d. Step 4

25.95 a. true b. false c. false d. false

25.97 a. incorrect b. correct c. correct d. correct

Protein Metabolism — Chapter 26

Solutions to Selected Problems

26.1 occurs in the stomach with gastric juice as the denaturant

26.3 Pepsinogen is the inactive precursor of pepsin.

26.5 Gastric juice is acidic (1.5–2.0 pH), and pancreatic juice is basic (7–8 pH).

26.7 Shuttle molecules facilitate the passage of amino acids through the intestinal wall.

26.9 total supply of free amino acids available for use

26.11 cyclic process of protein degradation and resynthesis

26.13 Nitrogen intake exceeding nitrogen output produces a positive nitrogen balance; nitrogen output exceeding nitrogen intake produces a negative nitrogen balance.

26.15 It becomes negative; proteins are degraded to get the needed amino acid.

26.17 protein synthesis, synthesis of nonprotein nitrogen-containing compounds, nonessential amino acid synthesis, and energy production

26.19 a. essential b. nonessential c. nonessential d. essential

26.21 b and c

26.23 an amino acid and an α-keto acid

26.25 a.

$$\mathrm{OH{-}CH(CH_3){-}CH(\overset{+}{N}H_3){-}COO^-} + \mathrm{H_3C{-}C({=}O){-}COO^-} \longrightarrow \mathrm{OH{-}CH(CH_3){-}C({=}O){-}COO^-} + \mathrm{H_3C{-}CH(\overset{+}{N}H_3){-}COO^-}$$

b.

$$\mathrm{H_3C{-}CH(\overset{+}{N}H_3){-}COO^-} + \mathrm{{}^-OOC{-}CH_2{-}C({=}O){-}COO^-} \longrightarrow \mathrm{H_3C{-}C({=}O){-}COO^-} + \mathrm{{}^-OOC{-}CH_2{-}CH(\overset{+}{N}H_3){-}COO^-}$$

c.

$$\mathrm{H{-}CH(\overset{+}{N}H_3){-}COO^-} + \mathrm{{}^-OOC{-}CH_2{-}CH_2{-}C({=}O){-}COO^-} \longrightarrow \mathrm{H{-}C({=}O){-}COO^-} + \mathrm{{}^-OOC{-}CH_2{-}CH_2{-}CH(\overset{+}{N}H_3){-}COO^-}$$

d.

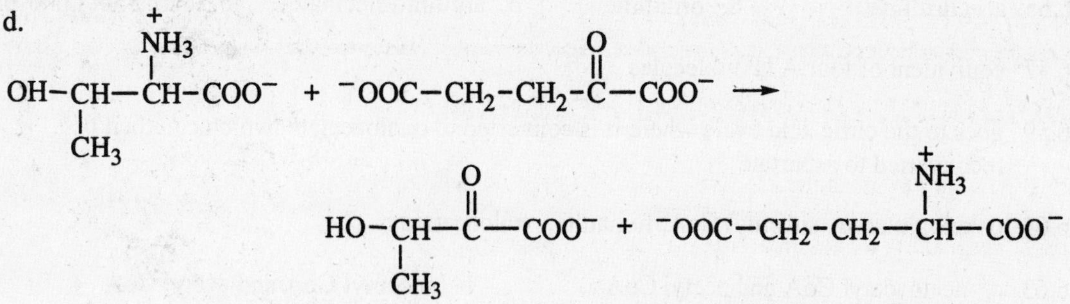

$$\mathrm{OH{-}\underset{\underset{CH_3}{|}}{CH}{-}\overset{\overset{+}{NH_3}}{\overset{|}{CH}}{-}COO^- + {}^-OOC{-}CH_2{-}CH_2{-}\overset{\overset{O}{\|}}{C}{-}COO^- \longrightarrow}$$

$$\mathrm{HO{-}\underset{\underset{CH_3}{|}}{CH}{-}\overset{\overset{O}{\|}}{C}{-}COO^- + {}^-OOC{-}CH_2{-}CH_2{-}\overset{\overset{+}{NH_3}}{\overset{|}{CH}}{-}COO^-}$$

26.27 pyruvate, α-ketoglutarate, oxaloacetate

26.29 coenzyme that participates in the amino group transfer

26.31 conversion of an amino acid into a keto acid with the release of ammonium ion

26.33 Oxidative deamination produces ammonium ion, and transamination produces an amino acid.

26.35 a. 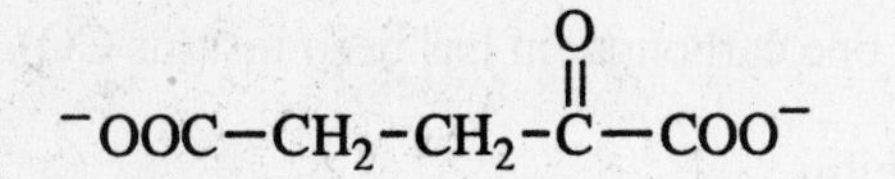 $\mathrm{{}^-OOC{-}CH_2{-}CH_2{-}\overset{\overset{O}{\|}}{C}{-}COO^-}$ b. $\mathrm{HS{-}CH_2{-}\overset{\overset{O}{\|}}{C}{-}COO^-}$

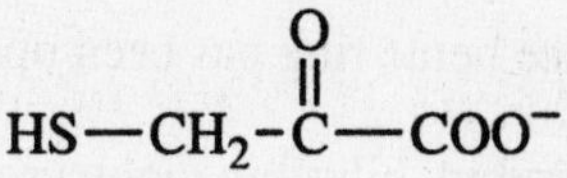

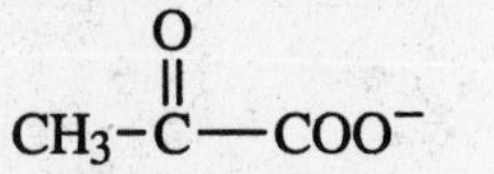

c. $\mathrm{CH_3{-}\overset{\overset{O}{\|}}{C}{-}COO^-}$ d. 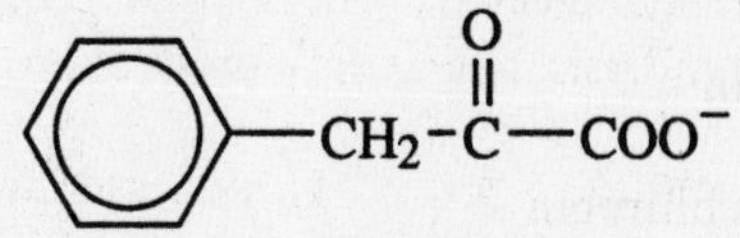 $\mathrm{C_6H_5{-}CH_2{-}\overset{\overset{O}{\|}}{C}{-}COO^-}$

26.37 Transamination of the α-keto acid produces the amino acid (leucine).

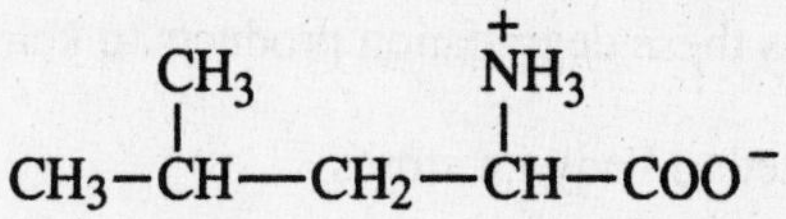

$$\mathrm{CH_3{-}\overset{\overset{CH_3}{|}}{CH}{-}CH_2{-}\overset{\overset{+}{NH_3}}{\overset{|}{CH}}{-}COO^-}$$

26.39 a. aspartate b. glutamate c. pyruvate d. α-ketoglutarate

26.41

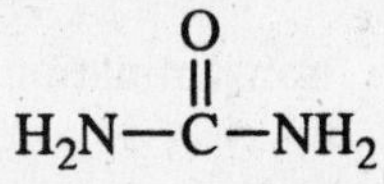

$$\mathrm{H_2N{-}\overset{\overset{O}{\|}}{C}{-}NH_2}$$

26.43 carbamoyl phosphate

26.45 an amide group

26.47

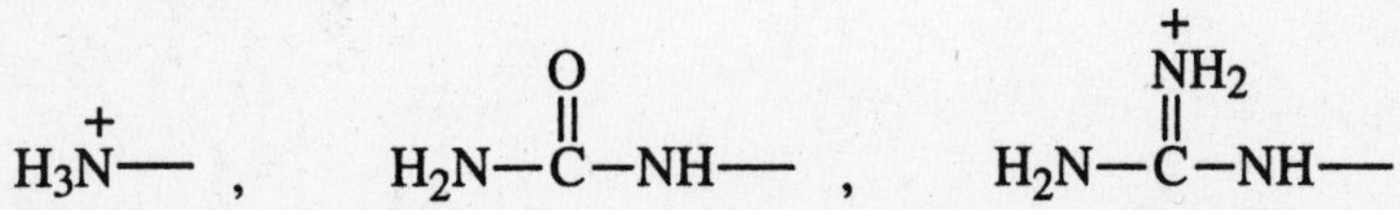

$$\mathrm{H_3\overset{+}{N}{-}\ ,\quad H_2N{-}\overset{\overset{O}{\|}}{C}{-}NH{-}\ ,\quad H_2N{-}\overset{\overset{+}{NH_2}}{\overset{\|}{C}}{-}NH{-}}$$

26.49 a. carbamoyl phosphate b. aspartate

26.51 carbamoyl phosphate

26.53 a. N_2 b. N_3 c. N_1 d. N_4

26.55 a. citrulline b. ornithine c. argininosuccinate d. carbamoyl phosphate

26.57 equivalent of four ATP molecules

26.59 goes to the citric acid cycle where it is converted to oxaloacetate, which can then be reconverted to aspartate

26.61 α-ketoglutarate, succinyl CoA, fumarate, oxaloacetate

26.63 a. acetoacetyl CoA and acetyl CoA b. succinyl CoA and acetyl CoA
c. fumarate and oxaloacetate d. α-ketoglutarate

26.65 Degradation products can be used to make glucose.

26.67 glutamate

26.69 pyruvate, α-ketoglutarate, 3-phosphoglycerate, oxaloacetate, and phenylalanine

26.71 hydrolyzed to amino acids

26.73 In biliverdin the heme ring has been opened, and one carbon atom has been lost (as CO).

26.75 biliverdin, bilirubin, bilirubin diglucuronide, urobilin

26.77 urobilin

26.79 excess bilirubin

26.81 Amino acid carbon skeletons are degraded to acetyl CoA or acetoacetyl CoA; ketogenesis converts these degradation products to ketone bodies.

26.83 converted to body fat stores

26.85 a. (1) b. (2) c. (3) d. (2) e. (1) and (3) f. (3)

26.87 (1), (3), (2), and (4)

26.89 a. transamination b. deamination d. deamination d. transamination

26.91 a. true b. false c. true d. true